Marketing Technology as a Service

Marketing Technology as a Service

*Proven Techniques that
Create Value*

Laurie Young and Bev Burgess

A John Wiley and Sons, Ltd., Publication

This edition first published in 2010
Copyright © 2010 John Wiley & Sons

Registered Office
John Wiley & Sons Ltd, The Atrium, Southern Gate, Chichester, West Sussex, PO19 8SQ, United Kingdom

For details of our global editorial offices, for customer services and for information about how to apply for permission to reuse the copyright material in this book please see our website at www.wiley.com

The right of the author to be identified as the author of this work has been asserted in accordance with the Copyright, Designs and Patents Act 1988.

Wiley also publishes its books in a variety of electronic formats. Some content that appears in print may not be available in electronic books.

Designations used by companies to distinguish their products are often claimed as trademarks. All brand names and product names used in this book are trade names, service marks, trademarks or registered trademarks of their respective owners. The publisher is not associated with any product or vendor mentioned in this book. This publication is designed to provide accurate and authoritative information in regard to the subject matter covered. It is sold on the understanding that the publisher is not engaged in rendering professional services. If professional advice or other expert assistance is required, the services of a competent professional should be sought.

A catalogue record for this book is available from the British Library.

ISBN 978-0-470-74840-4 (H/B)

Typeset in 10/12pt Times by Aptara Inc., New Delhi, India
Printed in Great Britain by CPI Antony Rowe, Chippenham, Wiltshire

Bev dedicates this book to her daughters, Katherine and Lauren, with love.

Contents

Foreword

This book is for a growing group of service businesses with common characteristics and challenges. They provide a service to buyers based on a network of technology. The most aggressive members of this category are computer service companies like IBM, HP or Fujitsu, that have made great noise about their move into more service-based business over the past two decades. It also includes the utilities: telephone, electricity, gas, oil distribution, water and cable TV companies. It embraces consultancies, like Accenture, whose advice depends on advanced knowledge of technology. Companies that use technology to provide transport services (like airlines, rail, road management, freight distribution and courier services) also tuck into this sector, as do companies such as GE that provide services based on medical technology.

These companies all exist to provide service to customers by exploiting an 'installed base' or infrastructure of technology. The service may be conceptual (like consultancy) or may provide specific support to a set of physical products (like maintenance). In most, a 'core service' (which risks becoming a commodity) runs on the network platform, and 'added-value' services are built on that core service. The perceived value of all three is interrelated and interdependent.

The technology industries have, to date, routinely failed to produce real value propositions which entice people to pay more for these core and added-value services than they probably should. They are the complete opposite to the luxury goods industry, which uses heritage, sex, design, distribution and celebrity to create aching desire for bits of leather and fragrant water, imbued with a mystique by names like Gucci and Chanel. The IT industry, for example, invests huge sums in science and research. Many have their own state-of-the-art laboratories and sponsor doctoral programmes in universities across the world. They will announce new breakthroughs and excite the world with dreams of new, 'life-transforming', technologies like, each in their day, broadband, cellular telephony and digital TV. They then market and sell their offer through worldwide distribution systems founded on the belief that everyone is changing very fast and wants everything cheap. They discount and throw away the value of their precious scientific advances because of the logical and systematic approaches that got them there in the first place.

Similar difficulties occur when technology firms turn to services. When they create a service, it is normally crafted with precision and care. It is likely to reflect latest thinking, to comprise well-considered processes and to involve modern tools or technology. It is just as likely, though, to look exactly the same as competitors' offers and to be priced using a 'cost plus' approach. As a result, the maintenance service of IBM has been substantially the same as that of HP; and the 'managed service' of BT has been very similar to that of Orange; while the electricity, gas or water services of different utilities have been exactly the same as those of their peers.

Creating a service inside a technology firm has been a little like high-school biology. Students are shown a frog, which jumps and croaks. They are taught about its physiology and evolution. They are then shown how to slap it down and dissect it. They can see how the tendons interconnect and where all the internal organs are. Yet, once they are finished, they are left with a dead frog. Time and again, technicians inside some of the most famous technology firms in the world do exactly the same thing with technology-based services. They examine the offer of others and use detailed processes to specify the features of the proposed service. They know exactly how it will work and what customers will receive. Yet, when it is launched, it is a dead frog, lifeless and valueless, over which customers haggle about price and moan about tiny errors.

Yet, throughout history, there have been outstanding individuals who have applied art, intuition and insight to science and technology. Leonardo da Vinci, Thomas Edison, Isambard Kingdom Brunel and Albert Einstein are just a few famous examples of outstanding individuals who have applied creative and adventurous thinking to scientific problems in order to create profit. Steve Jobs of Apple is probably the most famous modern example of the imaginative application of design and insight to create real value from technology. Users of the Apple Mac computer tend to be loyal enthusiasts who laud its ability to process information 'as if it's designed for a human being'. This desire to create innovative and enticing technology-based products gave Apple a difficult and chequered history until it launched the iPhone. The industry had been talking for many years about one device to integrate computing, mobile telephony and email. There had been several PDAs designed and launched with mixed success. But it was the (at the time) radical design and innovative approach that made this product so sexy and appealing. Shops were mobbed for it, and Apple became a byword for innovation.

Apart from consumer electronics, though, too much technology marketing remains depressingly tactical and erratic. Far too many technology marketers spend hours creating PowerPoints of the latest whizzo ideas to present to themselves, or running around administering depressingly tactical activities. There is often little sense of organisational learning, brand equity or any substantial investment in customer knowledge. As a result, many of the services of these firms are not valued by too many of the people who use them. Consumers who flock to new shopping centres or are delighted to pay inflated prices for luxury goods will grumble about water bills, broadband costs, rail tickets and flight charges. While business-to-business executives who pay high prices for McKinsey's advice and select the big four accountancy firms to be their auditors will pick at every detail of outsourcing or maintenance contracts.

By and large, the vast businesses of the technology industries demonstrate that it is possible to have all the functions of marketing, backed by large budgets and supportive leadership, but still be unable to create genuine perceived value. As they move towards more sophisticated marketing and need to create value propositions out of service skills, their adoption of the sort of systematic, progressive marketing competence seen in other sectors must increase. In short, they need to become market-led and leave behind tactical, erratic marketing.

There are now many individuals who 'engineer' (in the true sense of that word) a sexy and appealing service experience. Some are driven individuals and entrepreneurs like Steve Jobs (iPhone Apps), Richard Branson (Virgin Atlantic) and Hans Snook (Orange mobile phones). Others are organisations that steadily apply a skill to a service industry; as the design company

Nokia did to mobile phones. While others are marketers (like Chris Gent who put Vodafone on the world stage), applying marketing knowledge and principles to services that rely on technology or networks for their success.

It is the thesis of this book that the learning, insight and approaches of these unusual technologists will help the marketing of all those trying to position a service based on a technology platform; that they not only need to recognise that they are part of a wide and valuable services category, with common characteristics, but that they also need to move beyond 'dead-frog' approaches to marketing.

Unfortunately, there is little help for management teams facing this need to change. There is no serious book, little research and few academic scholars who specialise in this specific area. So, the issues they are facing must be tackled with little guidance from effective management theory or academics. However, the experience of companies in a range of different sectors shows there to be some clear steps involved. This book sets out to examine them and to find out what, from their experience, can be drawn into more generic lessons.

We have a number of people to thank for their help in the completion of this project.

Some were generous with their experience and provided case studies: BT (Ninder Takhar and Dinah McLeod), EDF (Colin Warne), Fujitsu (Vincent Rousselet and Philip Oliver), Interoute (Tony Rogers), Microsoft (Barbara Jenkins), Orange Business Services (Chris Ellis), Virgin Media (Ginnie Leatham and Ashley Stockwell) and Xerox (Julie Meyers, Tom Dolan and Robert Corbishley). We are also grateful to: Peter Snelling of Michelin; operations specialist Graham Clark of Cranfield University; advertising specialist Adrian Hosford; communications planner Leigh Stops; corporate communications specialist Carolyn Esser; and services marketing specialist Lynda Chambers for reading some of the work before publication. David Munn, president and CEO of the Information Technology Services Marketing Association in Boston, gave invaluable help and support on several case studies and Chapter 8, and provided detailed statistics of service marketing in the IT services sector. PR and communications specialist Alison Lambden was, as always, a key professional resource and friend for us in the development of the text. We are also grateful to the Wiley team, Claire Plimmer, Michaela Fey and Jo Golesworthy. In a challenging and changing industry, they have given us support, patience and encouragement.

Whenever we have been in line jobs, running part of a business or advising clients, we have both looked for access to insights, tools or techniques that would help round out decisions to reduce the risk of failure. In this work, we have trawled published academic papers and worthy publications like the *McKinsey Quarterly* or *Harvard Business Review*. We have particularly tried to understand the experience of leading firms because, actually, business is very simple and there are very few who have not faced similar issues before. We have tried to tease out and present the best of all those sources in this work and hope that their insight, freely acknowledged, will help the busy manager or business leader to create wealth by marketing sexy, desirable services to a curious public.

1

Technology Companies, Services and Networks

INTRODUCTION

Most of the developed economies have strong and vibrant service sectors. In fact, it is surprising how much they are dominated by service businesses, as opposed to manufacturing or agriculture. In the USA and much of Western Europe, around 75% of economic activity is thought to be service; and emerging economies like India and China have announced their intentions to increase the size of their service sectors too. This category includes: professional services, financial services, consumer services and facilities services. Each has been tracked, analysed and researched. Academics have produced tomes on buying behaviour in consumer services, the marketing of financial services and the management of professional services. Yet one vast, neglected category is those services provided from, or embedded in, technology companies. These are offered on the basis of a common technology platform like a communication network, a set of airline slots, a pumped utility or a global broadband network. They are distinct from other services with unique issues and opportunities. This chapter explores the characteristics of those services and their unique dynamics.

THE GROWTH OF THE SERVICE SECTOR

For the past 50 years the service sectors of developed countries have been an increasingly important part of their economies. For instance, American manufacturing is, at the time of writing, 13% of GDP (a decline from 26% in 1970), whereas the service sector represents around 75% of economic activity, depending on definition. According to the British government's statistics (*source*: UK Office for National Statistics), 76% of the UK economy is services compared with around 40% in 1948; and the service sector has been increasing in importance for many decades. In all the years from 1992 to 2004, for example, the 'finance and business services sector' provided the largest contribution to the UK's economy, rising to £344.5 billion out of a total £1044.2 billion. The manufacturing sector, by contrast, contributed more than 20% until 1998, but has fallen, as a proportion of the British economy, every year since then to around 4%. There have been several reasons for this explosion in service activities.

The first is growing wealth, changing lifestyles and the increasing consumer aspirations that come with it. As people become richer they want more services that enhance their lives, like education, health and leisure. A symbiotic relationship develops between consumers and entrepreneurs, who see new demand patterns in society. In common with much of Europe, Britain was still devastated by war in 1948 when its government reported service to be less than 50% of its economy and manufacturing relatively dominant. As it emerged from austerity, though, its population's expectations and standards grew. In some cases, services that had been used by the wealthy for many decades became more economic and accessible to broader groups of people. Private education, foreign travel, sports, leisure and beauty services, for example, are used by a much wider proportion of society today. In other cases, changes in lifestyle have given companies opportunity to create entirely new services. Self-service supermarkets and fast food chains are now very familiar but were created in the latter half of the 20th century as both disposable income and the pace of life increased. This pattern of rising standards, related to economic success and the expansion of service businesses, has been followed by other developed economies and is being repeated by developing nations today; prompting an explosion in service businesses around the world.

Not only has there been an increase in the number and variety of services, but a re-categorisation of large swathes of modern economies has caused competitive services to evolve. For instance, the service sector has been increased by the growth in 'outsourcing' (the tendering to external suppliers of services that were previously part of a company but are no longer considered to be the prime skill or a source of sufficient economic return for that business). These might include: security, reception, IT, secretarial services or staff catering. Activities that were previously part of a manufacturing company have been passed to companies that are expert in that field.

As a result, the service sector is thought to have grown and manufacturing declined; and, as this 'facilities market' is now thought to be worth over £100 billion in Britain alone, this is not an insignificant change. (Note, though, that the British Institute of Facilities Management is at pains to point out that estimates of this new market's size vary widely because definitions of it are also varied.) Yet, although the total value of economic activity has been relatively unchanged by these initiatives, the orientation of the service functions themselves has been transformed. Whereas they were once internal departments of large corporations, they became competitive service businesses in their own right and separately recorded as revenue-earning entities.

Privatisation has been another significant change in categorisation, which has increased service sectors throughout the world. This policy was pioneered by the 1980s British government (under Prime Minister Margaret Thatcher), and initiated a major turnaround of their economy using privatisation. Although several cultures are suspicious of it, the policy has been copied and adopted by many governments since. It is now, for instance, often a condition of lending from the World Bank and the International Monetary Fund that developing nations create programmes to open their economies to competitive forces.

The market introduces into privatised organisations the relentless drive for process improvement, cost reduction and innovation natural to competitive businesses. So, over the long term, privatised corporations have sought productivity improvement in response to pressure from shareholders, customers, competitors and public commentators. They have served customers better, raised needed capital to invest in infrastructure, created employment and paid handsomely in corporate taxation. One source (HM Treasury, 1992) estimates that the 11 major organisations privatised between 1979 and 1987 netted the British government a total of £11.8 billion; and, in the 1986/7 financial year, receipts from privatisation paid for 2.6% of government expenditure. In several cases the tax contribution after privatisation far outweighed the profit earned when these businesses were publicly owned.

It is the sheer size and diversity of the organisations moved into the private sector which is most striking. One academic (Parker, 2009) has pointed out that during the pioneering period alone, the British government sold, *inter alia*, organisations in: telecommunications (BT, Cable and Wireless); Energy (British Gas, Britoil, BP); Travel (BA); and manufacturing (Rolls Royce, Rover subsidiaries). The privatisation of these massive organisations has meant that a large amount of economic activity has transferred to the private sector and that entirely new enterprises have been created. Throughout the world, participants in the telecommunications, water, gas, electricity and airline industries have all become competitive service companies who need to take their propositions to market in new ways. Vast areas of economic activity became subject to market forces for the very first time and the service companies in them, often engineering-dominated monopolies, needed crash courses in sales and marketing. They needed to learn to go to market effectively.

Another important influence on the boom in services has been the increasing international interdependence between countries and changes in the conditions of international trade. Huge and growing international service markets have tempted many firms to set up service businesses. According to the United Nations Conference on Trade and Development (UN, 2008), world trade in services grew 18% in 2007. This is comparable to the previous two years, in which growth rates jumped after a spell of annual growth at around 6% per annum (in every year, except one, between 1990 and 2003). The United States was the lead service exporter during that period and commanded just under 15% of world service trade. The results also show that services were one of the most important sectors for the European Union, contributing to around two-thirds of GDP and employment. As a result, 25 European countries together took around 50% of world service trade. However, international trade was also important to developing countries, where several were gearing up to tackle international service opportunities. In 2006, they accounted for around 25% of world export in services; up from 22.6% in 2004.

International success in service exports varies according to the development of an economy, government policy and the education of the population. India is a recent example of international service export success. In 2006 it had raised its share of world service trade to 2.6% from 0.5% in 1990, despite the massive growth in overall international service business during that period. For instance, the country used its developing international competitiveness in IT skills to plan an incursion into the business process outsourcing market. One estimate put the number of Indian firms providing international, IT-enabled, services, at over 200 in 2004 (Javalgi *et al.*, 2004). As a result, its 'computer and information services' exports were 40% of its total service exports in 2006.

China too, at the time of writing, is starting to expand its service sector and to rival India as an outsourcing hub for routine tasks where conversational English is less of a requirement (like preparing tax returns and filing patents). In 2006, approximately 42% of its economy was officially recorded as service sector and it held 2.7% of world export in services. So, that year, it announced that it was planning to increase incentives and financial aid to boost service companies in a bid to increase the sector's contribution to 50% by 2020. The State Council said incentives (such as land approvals and capital bailouts) would be given to service companies engaged in logistics, information technology, software, electronic commerce, industrial design, law and accounting. This was driven by Beijing's eagerness to lessen its reliance on the manufacturing sector, which had led to problems such as energy shortages and pollution.

Services are and have been, then, a healthy and viable means of earning export revenues. Now, though, a combination of forces is making the international conduct of service business even easier and more attractive, stimulating further growth. As Professor Javalgi and his colleagues point out (Javalgi *et al.*, 2004), they include:

- **New technology like the Internet and e-commerce**. Cross-border trade in a wide variety of services, such as professional advice and travel, can be delivered cost-effectively through modern electronic highways.
- **Increasing sophistication and the growth of middle classes in a number of developing nations**. This creates demand for services in different parts of the world and socialises new concepts to a wide international community, paving the way for suppliers.

- **The opening of trade through negotiations in forums like the GATT rounds**. These encourage focus on the competitiveness of specific service markets.
- **Regional trading blocs like the European Union and North American Free Trade Area**. These seek to bring down trade barriers between their members, stimulating international trade in services like air traffic or financial services and affecting demand by creating larger markets.
- **Government legislation and support**. Governments in countries like China, India, Singapore, Indonesia, Brazil and Mexico are, at the time of writing, actively promoting initiatives to encourage their service sector.
- **Easier transport links, efficient international postal services and cheap flights**. All make the cost of international services cheap and viable.

As a result of these dynamics, a wide range of businesses are planning international market penetration, hoping to repeat in service markets the success of, say, the Japanese in consumer products during the 1980s and the tiger economies in electrical components during the 1990s.

The effect of the service economy is most clearly seen in employment patterns. In the autumn of 2005 the *Economist* magazine published an analysis of the number of people employed in manufacturing as a percentage of the total workforce. It estimated that 10% of American workers were employed in manufacturing as opposed to 25% in 1970 (employment in services was 80%). The estimates for Britain (14% compared to 35% in 1970), France (15%) and Canada (14%) were similar; with other big economies, like Japan, at 18%. They found that the only big economy where more than a fifth of workers were in manufacturing was Germany (23%), which has a lot of innovative companies and a high content of capital goods that are not as easy to copy. Since a number of workers within manufacturing companies still occupy service roles (like marketing, design and facilities management), the actual employment in manufacturing roles among the developed economies could be much less.

So, over the past 50 years, the balance in many economically developed countries has shifted to the point where services account for almost three-quarters of their gross domestic product. This comprises a wide array of services, including financial, utility, professional and consumer services. As large manufacturing processes have become more automated, the developed economies have turned to human expertise to create wealth. Some bemoan the decline in manufacturing as a percentage of their country's economy and ridicule the service sector, as if services do not increase GDP or do not add value. They do. The growth in this area of economic activity is a valid contribution to wealth creation and to society as a whole. Whereas a manufacturing process uses physical resources to create wealth, a service process uses human skill, effort or knowledge to do so.

THE DIFFERENT TYPES OF SERVICE

As the service sectors have boomed, government economists and business thinkers have sought to categorise them and understand their contribution to the wealth of nations. There are several broad categories. Financial services, for instance, are based on money and, in the 20th century, have been subject to massive growth and changes in regulation. Institutions which had been working to similar procedures for many decades have suddenly been subjected to new forces. Although they are yet to get to grips with a number of important issues facing their industry (e.g. the bonus culture and its effect on risk, the fallout from combining investment and traditional retail banks, post credit-crunch regulation and the challenge of costly retail premises), the marketing of their services has been researched and explored, if erratically applied. Thoughtful managers can easily find a body of work on the creation, positioning, selling and marketing of financial services. Science exists in this field if executives want to access it.

Professional services are, by contrast, based on the expertise of the service provider and participate in a market predicated on 'asymmetry of information' (the fact that the supplier knows more than the buyer). They include some of the most profitable, enduring and influential businesses that the world has seen. Vast international partnerships like Deloitte in accountancy and Clifford Chance in law are, for instance, in their second century of business. Throughout that time they have (in war, recession, depression and boom) returned to their owners very high net margins and routinely walk away with massive projects, resulting from high-level access to leaders of the world's most famous businesses.

This industry includes: consultancies, law firms, accountancy practices and architectural partnerships. Over time, they have raised a barrier to their market, and the value of their skill, by making their expertise complex. This has been done by creating standards, benchmarks and controls which aspirant suppliers have had to attain (and maintain) in order to participate in the market. These standards were set and controlled by representative bodies comprising members of the profession. Buyers tolerated the resultant higher price because they believed that it guaranteed quality. Experience of competition in this industry has, though, been patchy. Some professionals, like architects and consultants, have been competitive for a number of years and have had to market their services. Others, like the elite accountancy and law firms, have had more protected markets but, at the start of the 21st century, began to experience new competitive forces and have had to consider how to market themselves more systematically. Again, there is published and well-researched accumulated knowledge to help thoughtful practice leaders address issues of strategy, management and marketing.

Consumer services, by contrast, are a range of offers to individuals. They are normally based on high-volume distribution systems, some of which are amongst the most famous brands in the world. They include: fast food, travel, entertainment and leisure services. All have been studied and researched extensively by academics as part of the transition of modern economies from manufacturing to service dominance. Most of the work described in Chapter 2, which investigates the difference between product and service marketing, has been grounded in consumer services.

But there are some newer service categories where less is known. The relatively new facilities management market is one and the range of 'embedded services' is another. The latter are a highly specialised category connected with manufacturing or the processes of companies. They include maintenance, installation and various advisory services. In 2005, McKinsey estimated this sector to be worth $500 billion (Auguste et al., 2005). Along with them, newer services offered on a network by technology companies, such as the emerging cloud computing services, have experienced little analysis of their unique needs and perspectives.

SERVICES IN TECHNOLOGY OR ENGINEERING SECTORS: A NEGLECTED CATEGORY?

There are a group of service businesses that have common characteristics and challenges. They can be defined as:

A service provided to buyers, based on a network of common technology.

The most aggressive members of this category are computer service companies like IBM, HP or Fujitsu, that have made great noise about their move into more service-based business over the past two decades. Also included are: telephone companies, electricity companies, gas companies, oil distribution, water companies and cable TV companies. It embraces a number of consultancies like Accenture and Booz Allen whose advice is based on deep technology knowledge. Companies that use technology to provide transport services (like airlines, rail, road management, freight distribution and courier services) should also tuck into this category; as should those that provide services based on medical technology, such as GE, and the distribution arm of the oil companies.

They exist to provide service to customers by exploiting an installed base or infrastructure of technology, usually composed of dispersed geographical units. The service may be conceptual (like consultancy) or may provide specific support to a set of physical products (like maintenance). There is usually a means of combining these technical resources into a network. Because of the underlying infrastructure, the provision of service access (or the remedial work of resumption in service) may include the supply of spare parts or replacement units to geographical locations within tight deadlines.

As the development and application of the core technology can involve the deployment of massive capital sums, many of these services were started as public utilities. The supply of water or electricity, for instance, was considered to be of sufficient social significance that, in most countries, governments invested in new organisations to provide these infrastructures. In some cases, though, governments left it to the private sector to develop networks. They framed helpful legislation and sponsored regulatory regimes but invested little. Both the early rail networks and the recent Internet infrastructure were created in this way.

ENGINEERING AND SERVICE

In 1850, the famously successful banker Lord Rothschild said:

There are three ways of losing your money: women, gambling and engineers. The first two are pleasanter, but the last is more certain. (Friedel, 2007)

This is not only because, in his age, a number of ventures failed due to lack of acquired learning; nor was it because science was seen as new and adventurous as well as exploratory and risky. As an experienced and shrewd financier, he also knew, as others learned to their cost over the centuries, that engineering and technology progression tends to be incremental, carefully building on the back of other work. When, for instance, Robert Friedel (History Professor at the University of Maryland) set out to write his remarkable book on a millennium of technology development (Friedel, 2007), he called it *A Culture of Improvement*. He demonstrated that science, technology and progression tend to be incremental, building on the innovation of thousands of nameless technicians. This leads to an attitude of precision, care, repetition and calculation, reflected in large technology businesses. Yet it can also lead to the destruction of value.

Technology industries routinely fail to produce real value propositions that entice people to pay more for exciting products or services than they probably should. As a result, they are the complete opposite to the luxury goods industry which uses heritage, sex, design, distribution and celebrity to create aching desire for bits of leather and fragrant water imbued with a mystique by names like Gucci and Channel. The IT industry, for example, invests huge sums in science and research. Many companies have their own state-of-the-art laboratories and sponsor doctoral programmes across the world. They will announce new breakthroughs and excite the world with dreams of new, life-transforming technologies like, each in their day, broadband, cellular telephony and digital TV.

They then market and sell their offer through business organisations founded on the belief that everyone is changing very fast and wants to buy everything as cheaply as possible. They discount and throw away the value of their precious scientific advances because of the logical and systematic approaches that got them there in the first place. 'Moore's law', for example, is routinely quoted at IT conferences but there is no logical reason why the power of chips should continue to expand at such a rapid pace. It is probably not so much a scientific principle of physics, comparable to, say, Boyle's law. It is more a description of a worldwide social system with common beliefs and mind sets; an industry, behaving idiotically.

Yet, throughout history, there have been outstanding individuals who have applied art, intuition and insight to science and technology. Leonardo da Vinci, Thomas Edison, Isambard Kingdom Brunel and Albert Einstein are just a few famous examples of outstanding individuals who have applied creative and adventurous thinking to scientific problems in order to create profit. And there are many who are less famous. In Victorian Britain, for example, the city of Manchester had a problem in that it was overcharged by its nearby rival, Liverpool, for items shipped through its docks. It was a now relatively unknown engineer (Sir John Rennie) who originally proposed, in 1830, the outrageous idea of a canal capable of taking ships to Manchester. (It was dug by hand 50 years later.)

This phenomenon has led to some cultures of the world (Germany for instance) distinguishing between technical work and true engineering. The first is systematic, precise, process-driven and clearly defined whereas the latter is professional, insightful and science-based but also intuitive and creative. Careful, well-crafted technical work is important and essential. The basics have to work well. Yet a growing number of leading technology firms are now showing how important it is to engineer real magic into value propositions.

Steve Jobs of Apple is probably the most famous modern example of the application of design and insight to create real value from technology. Users of the Apple Mac computer tend to be loyal enthusiasts who laud its ability to process information 'as if it's designed for a human being'. This desire to create innovative and enticing technology-based products gave Apple a difficult and chequered history until it launched the iPod and iPhone. The industry had been talking for many years about one device to integrate computing, mobile telephony and email. There had been several PDAs designed and launched with mixed success. But it was the (at the time) radical design and innovative approach that made this product so sexy and appealing. Shops were mobbed for it and Apple became a byword for innovation.

Similar difficulties occur when technology firms turn to service. When they create a service it is normally crafted with precision and care. It is likely to reflect the latest thinking, to comprise well-considered, internally focused processes and to involve modern tools or technology. It is just as likely, though, to look exactly the same as competitors' offers and to be priced using a 'cost plus' approach. As a result, the maintenance service of IBM will be substantially the same as that of HP; or the 'managed service support' of BT will be very similar to that of

Orange; while the electricity, gas or water services of different utilities will be exactly the same as those of their peers.

Creating a service inside a technology firm is a little like high school biology. Students are given a lively green frog that jumps and croaks. They are taught about its physiology and evolution. They are then shown how to slap it down and dissect it. They can see how the tendons interconnect and where all the internal organs are. Yet they are left with a dead frog. Time and again, technicians inside some of the most famous technology firms in the world do exactly the same thing with technology-based services. They examine the offer of others and use different processes (like blueprinting and service mapping) to specify the detail of the proposed service. They know exactly how it will work and what customers will receive. Yet, when it is launched, it is a dead frog, lifeless and valueless, over which customers haggle about price and moan about tiny errors.

Yet, there are individuals who 'engineer' (in the true sense of that word) a sexy and appealing service experience. Some are driven individuals and entrepreneurs like Steve Jobs (the 'iPhone Apps' services), Richard Branson (Virgin Atlantic) and Hans Snook (Orange mobile phones). Others are organisations that steadily apply a skill to a service industry; as the design company Nokia did to mobile phones. While others are marketers (like Chris Gent who put Vodaphone on the world stage), applying marketing knowledge and principles to services which rely on technology or networks for their success. It is the thesis of this book that the learning, insight and approaches of these unusual technologists will help the marketing of all those trying to position a service based on a technology platform; that they not only need to recognise that they are part of a wide and valuable category of businesses but that they also need to move beyond dead-frog approaches to marketing.

COMMON CHARACTERISTICS AND ISSUES AMONG SERVICE BUSINESSES IN TECHNOLOGY SECTORS

The service businesses of technology firms have a number of characteristics in common which affect their approach to market. They influence the agenda and decision-making processes of top management and, as a result, create a group of businesses with a similar culture and similar set of challenges.

Common Characteristics

These businesses share a number of common characteristics that drive business priorities, culture and behaviours. The most important are the following.

A Technology Infrastructure

They have an infrastructure, a 'platform', which is constructed from some form of technology (like a computer network or a piped utility). This heritage was once a major innovation allowing a new industry to develop and to distribute a basic, 'core' service. In many cases, though, this basic service provision eventually became restrictive. For instance, the invention of the telephone was an innovation that allowed simple voice communication. For decades the industry was limited to this very basic use of the technology because of public sector controls, lack of investment and lack of competitive innovation. In Europe it was not until the 1980s that the industry could really begin to acknowledge that there were different service needs between different customer groups. Their public sector ethos meant that customers as different as low-user residential customers and international financial traders were treated to the same standards of service. At the same time, new technologies were not fully exploited. Three decades later, there is now a dynamic industry seeking profitable return by meeting the varied needs of many customer groups. Other technology services have been through a similar evolution. In some (like rail and utilities), government policy has separated the operation of the technology infrastructure from the core service to customers in order to force management companies to improve the value and responsiveness to customers.

A Network

This is a similar point to the first, but subtly different because networks exist in other industries which are not based on technologies or on an engineering culture. The telecommunications, cable TV and electricity companies have a cable-based network. The gas and water utilities have a piped network. The airlines, freight, courier and road transport services have a physical distribution network. As such they are subject to similar dynamics which demand the attention of senior management, unique to this group of services, fostering similar activities and similar cultures. An example is traffic flow.

The networks of these service companies have items that flow through them, which are fundamental to their customers' service experiences. If there is congestion caused by, say, a drop of pressure in the wrong distribution point (or a bunching of messages due to inadequate switch capacity, or a tail-back due to a narrow road), quality of service will be affected. So, senior management needs to know where to invest in the basic infrastructure of the network in order to meet peak demands, and there is an emphasis on the forecasting of demand for access to the service, as a basis for resource allocation.

For many of the newly privatised utilities this process has been dramatically affected by the change from public to private sector. In the past their monopolies and universal service requirement had meant that they approached this by understanding the basic growth patterns in the population and projected the demand for service based on historical usage patterns. The move into the private sector caused, however, a re-evaluation of this very basic planning process. Management now had to take note of varying usage patterns between different customer groups, changing usage patterns as customers learnt new ways to apply the basic service, and the effect of competition. The methods of demand forecasting had to be radically changed and aligned with new ways of allocating capital expenditure, such as discounted cash flow analysis.

In addition, many of these industries have also changed in the past few decades to a point where they can, proactively, manage traffic flow through their network. By setting up a capability where they can intervene in real time to improve movement through their network, they can improve quality of service, overcoming congestion. Initiatives by many governments, at the time of writing, to introduce proactive road management to their transport network demonstrate how important this can be. By introducing technology, which monitors the flow of traffic, congestion can be eased and less investment in new roads becomes necessary. It does, though, require that the customers are educated in new habits and behaviours, such as variable speed limits and better lane discipline.

An Engineering Culture

Organisations tend to take top management from the most important or dominant function, usually reflecting the core competence of their business. In these service organisations the dominant discipline has, historically, been technical. They tend to be run by engineers and this technical heritage creates common attitudes throughout their organisations. For instance, there is likely to be a hierarchical or complex matrix structure, with an emphasis on job descriptions, roles, responsibilities and internal stakeholders. They will often have a compartmentalised attitude to organisation design, breaking functions into departments (e.g. sales, human resources and operations). They are likely to emphasise detailed operational processes, be systematic in project and programme management, and be excited by the potential of new, or changing, technologies.

This engineering heritage means that there is a preponderance of people who prefer to have precise, measurable options to problems and opportunities; and such attitudes mean that these organisations tend to be risk adverse and slow to change. They also tend to be slow to accept new ideas into the infrastructure of the organisation and suspicious of the value of creativity, originality and individuality; particularly where this stems from intuition and not obvious logic. New business ideas often have to go through a period of secretive 'skunk works' for example, before they reach critical mass and are accepted into the organisation as a whole.

Social Significance

Many services that are based on a technical network have a significance to society that is greater than other forms of service. Water supply is self-evidently important, but only comes to the front of its customers' minds if it is interrupted or contaminated. Telecommunications is vital to business operations and essential in time of war. Gas and electricity are, similarly, important parts of modern society. Even computer service support is increasingly considered to be an important part of the social fabric of society, as computers play an increasing role in running key processes in the community.

Governments therefore tend to legislate carefully to ensure that these services fulfil their social obligations. Legislators require service providers to meet social needs such as the provision of service to remote areas, weaker members of society and emergency arrangements. Although these obligations might be funded by a government, they still demand the attention of management and affect the culture and strategy of the whole organisation.

As a result of their ubiquity, history and criticality, there is also a set of emotional expectations in the general public as to what these services should provide. These expectations are often unarticulated but, if unfulfilled, there comes a point where society begins to express, through public media, that a line has been reached. This happened in 2008 during a period of steadily increasing prices for utility services and fuel, which were met with demonstrations and strikes in several countries. So, senior management of these socially significant service companies cannot ignore the emotional expectations that their service creates in the general public.

The Importance of Logistics and Spare Parts

Unlike professional or financial services, services with a technical infrastructure comprise physical components. The provision and maintenance of service therefore depends on an efficient method of managing and distributing physical components. This means that physical distribution skills, with all their associated functions (buying, warehousing, replenishment, forecasting and geographical stock holding) are integral parts of the service. If firms are to provide service which meets their customers' expectations they must establish physical distribution systems and supply chains that rival the major retailers for their speed, cost and efficiency. Many have, although as it is an adjunct to their business, this achievement is not generally recognised by their industry or, sometimes, by the companies themselves. Yet this ability is an important competence in the successful management of a competitive, network-based service.

Fundamental Changes in the Core Technology

These service companies periodically face a fundamental change in the core technology upon which their service is provided. There are many examples of this. For instance, at the time of writing, television companies are upgrading from analogue transmission to digital and telecommunications companies are upgrading their 'local loop' to broadband based on optical fibre. This change is not normally a complete surprise, because major breakthroughs in technology are normally brought into commercial use over a period of years. Each firm then faces a significant policy decision as to how to phase in the change. It involves decisions about capital investment, human resources, project management and customer education.

This phenomenon is becoming more complex as the development cycle of new technology speeds up and as the opening up of services to global competition increases. In protected markets, the development of a new technology occurs at a pace that allows the industry to assess it and apply it at leisure. Yet, although government policy might seem a benign way of protecting customer interests, it leads to delays in the deployment of enhancements. With the arrival of competition, however, management can no longer be assured that their competitor is not deploying the new infrastructure and giving enhancements to customers.

Safety

It is a simple fact that, in many of these businesses, it is possible for either employees or customers to be injured, or even die, as a result of neglect. Some have life-threatening substances as part of their core service. Others, such as airlines, cannot operate without rigorous adherence to safety standards in order to avoid major accidents. All have people engaged in potentially hazardous or life-threatening tasks. So, management and staff at all levels of the organisation must give careful attention to operating practices and procedures that are carried out on a regular basis. Failure to lay down a written safety procedure can involve management in criminal prosecution, and failure to adhere to it can affect all who work in the organisation. This makes attention to processes and procedures unavoidable, affecting the culture of the whole organisation. In extreme cases it can make people rule-bound and secretive.

Intimacy with Defence Organisations

A large number of these businesses have involvement with their government's defence and security services to some extent; from data access to the supply of specific services. This makes unique demands upon them. There are likely to be, for instance, specially vetted employees reporting through a separate management line to dedicated senior executives. In fact, some companies are almost completely reliant on defence revenues for their survival. It may be that only a very small percentage of revenue comes from commercial, private customers. The marketer's job in these circumstances is often to window-dress, to disguise the reality of defence dependence by creating high-profile campaigns alluding to commercial issues and communicating those private contracts it does enjoy. This intimacy with defence or security does not necessarily imply anything sinister or corrupt. It does, though, affect the culture and style of the company. As there are very few technology services (including telecommunication networks, utilities and computer services) that do not have this dynamic, it is a characteristic of this category of businesses.

Common Issues

A number of issues have to be managed by these firms, mostly as a result of their common characteristics. The most significant are the following.

Managing Capital Investment

As services in the technology sectors are based upon a core network infrastructure, the management of capital investment is a major consideration and focus of business leaders. The fundamental access to the whole service relies upon the pace at which the core infrastructure can be grown, developed and renewed. So, senior management must set aside a proportion of earnings to invest in the development of the network. As a result, the judicious use of capital to keep pace with the needs of infrastructure development and the cost of capital are critical success factors for management teams competing in these markets.

In a government-owned utility, the requests for capital funding are made to the treasury where they are compared with other funding requirements. They queue for attention and the priority they receive depends on the policies of politicians. The funding decisions of these businesses are therefore based on the ability of the representative government minister or the ideology of the incumbent government (and the reality is that state industry is always behind either defence and law or social security and health in priority). When moving into the private sector, however, businesses have to raise their own capital, finding funds either from cash, equity or loans. So, the decisions to fund projects are much more closely related to the ability to earn a return on capital and that, of course, is dependent upon how closely they provide the benefits that customers are willing to pay for. As a result, the cost of capital, the lifecycle of capital assets and return on investment are important financial measures that affect the thinking and approach of top executives in these businesses. Many have massive investment programmes on which they have to report to investors and analysts. This affects the culture of these organisations, distinguishing them from other categories.

Facing a New Commercial Environment

Many of these organisations have experienced a radical change to their business and market during the past few decades, often prompted by government legislation. The most obvious example is the change from public to private sector ownership of many in Europe. This caused a fundamental change in the philosophy of senior management and in the way companies were run. The market began to dictate the priorities of the business, a more efficient mechanism for determining the profitable creation and distribution of product and service benefits.

Another major change for some has been industry maturity. When an industry reaches the maturity phase of its lifecycle it has to find new ways to present benefits to customers (see the Tools and Techniques appendix). The computer industry, for example, reached maturity for large mainframe systems in many of the developed economies at the end of the 1980s and the beginning of the 1990s. As a result it had to look for new ways to sell the benefits of its core product (processing power). It found that customers had sufficient installations of computers to process the information in their organisations (in some cases organisations had more processing power than they needed; some more than twice as much). But they did need the ability to process information held in different databases, across different technology platforms. In short, they needed to apply their processing power to different, changing business needs. They needed to buy human skills. So, as the industry matured, the supply side turned to services to create profit. People who had previously been technical support specialists to products needed to sell their skills in new ways.

A further change is the creation of new markets or new forms of competition. Staying with the early computer industry; in the early 1980s, a small group of entrepreneurs bought spare parts and recruited engineers in order to take some of the lucrative maintenance contracts that computer companies had with their customers. As a result, the billion dollar 'third-party maintenance' market was created. It became possible for customers to buy maintenance for their computers from several different organisations, often at less cost. This simple but fundamental change of market conditions meant that very large revenue streams, which had previously been protected, were now subject to vigorous competition and many managers (who had built their careers as engineering support specialists within large computer companies) found themselves running competitive service divisions of considerable worth. This required new skills, new management processes and the application of a new business philosophy.

Another example of fundamental change affecting service firms in technology sectors has been the advent of serious global competition. In the world telecommunications market this has caused a considerable change in the way global organisations function. In the 1970s most of the world's telecommunications supply was dominated by nationally owned utilities and monopolies. Even in America, telecommunications was dominated by the mighty monopoly: AT&T. However, since then, the American telecommunications market has been the subject of deregulation legislation and many European telecoms suppliers have been denationalised. As a result, new network competitors have appeared, and a race is on to gain significant world market share. Similar international competitive changes have been affecting the gas and electricity markets. The management of these newly competitive organisations had to build up a worldwide distribution and network infrastructure, by either organic growth or acquisition. They had to recruit senior management with international experience and began to develop worldwide services.

Managing the Corporate Brand

As discussed later, the corporate brand of a service is an important component in its success. The association of emotional benefits with the company's name and corporate identity is absolutely critical to the successful marketing of services. However, the name and image of the service organisation has associations which may not suit its aspirations in the market. One of the very first actions must therefore be to change the corporate image into a viable and attractive brand.

This is a challenge that most services with a technical heritage have had to tackle. BA, for example, had to change from being the rather staid 'British Airways' when it decided to compete on the world stage with service quality. One of its first acts was to create a new corporate image and name to suit its aspirations. Similarly, large computer manufacturers, like HP and IBM, found that they had to develop a reputation for service which was different to their previous image, in order to compete effectively in the new computer market. Historically, they had manufactured their own proprietary machines and were perceived to be solely interested in selling them. Now they had to take a stance that was genuinely in the interests of their customers, supplying skills that crossed over the technologies of other suppliers. IBM in particular invested heavily in initiatives to reposition itself around services, running a global branding campaign that featured its own employees, and then a series of initiatives to be seen as a leading technology consultancy and services supplier after its acquisition of PricewaterhouseCoopers' consulting business.

Maintaining Network Access

Access to the technology or network upon which these services are based is a fundamental requirement of the service. If access is not correctly priced, planned and managed, neither the core service, different versions of the service to different customers, nor added value services can be provided. Service providers must therefore place considerable emphasis on the planning of access to the technology and the handling of access interruptions. An example is broadband communications, where customers' tolerance to speed of access delays is, at the time of writing, changing to become more demanding. Similar efforts have had to be made with computer service companies giving remote diagnostic access to customers. In doing so, they build up a network of service support, which is of enormous benefit to both the supplier and its customers, but pioneering this concept takes investment, careful communication, time and costs.

Service providers must also place considerable emphasis on maintaining access to their network. They must dedicate staff, systems and procedures to restoring customers' access in the case of an unforeseen interruption of service. They can even use interruption of service as the ultimate sanction when customers are unable or unwilling to pay. Moreover, for many network providers, continuous access is now becoming a technical feasibility. It is possible to create, cost-effectively, network structures whereby, if a component fails, the service can be re-routed with minimal interruption. It is also possible to deploy technology that predicts network failures and isolates them before they become a problem to the customer. This capability means that users are increasingly unlikely to know, understand or care about the causes of failure in access. They are also unlikely to be knowledgeable about the processes and procedures used by the service providers to recover service. They will value and concentrate upon the use of the service and will be intolerant of interruptions.

So, a major cost and logistical problem to the service provider becomes an insignificant commodity to the customer. Some companies, such as Avaya in the telecommunications industry, have tackled this perception with 'value reports', which show customers the financial impact of downtime avoided through remote monitoring and preventative maintenance. Elsewhere though, this issue has given many of these businesses a major strategic problem: how to create and maintain the value of their 'core', basic service.

Making the Core Service Relevant and Valuable

Successful marketing is about identifying a group of customers with common needs and presenting them with a proposition in a way that creates profit. However, the heritage of many technology companies, like utilities or maintenance companies, is grounded in the supply of a "commodity" service based upon a general technology. The whole organisation assumes that their core service has low perceived value and it is therefore difficult for them to find ways to make that service more relevant or valuable to different groups of customers. Yet, businesses can only make substantial money by meeting a set of customer needs, and these may be latent needs that customers are unaware of until the product or service is presented to them. Service firms must find out the latent needs they might meet in different customer groups and either tailor the core service to meet them, or create added value services upon their technology platform, customised to each groups' needs (as Interoute has done, see the case study in Chapter 6). Due to an ingrained belief that the core service is a commodity, it is often left to new entrants or challengers in a market to do this as they evolve their competitive strategy; as Sir Richard Branson's Virgin group has done numerous times. Yet the ability to innovate around the core service remains a significant source of profit and competitive advantage in many technology sectors.

25 YEARS WITH VIRGIN ATLANTIC

Over the past 25 years, Virgin Atlantic has lived up to our expectations of a Virgin company: the small newcomer taking on the giant and complacent establishment; the people's champion introducing better service and lower costs for passengers with a reputation for quality and innovative product development. It was developed as an offshoot of Sir Richard Branson's Virgin Group, better known at the time in the world of pop and rock music.

Virgin Music, famed for its megastores, grew by being the first to spot new trends and offering their customers an exciting and fresh environment in which to buy their records. They only had one rule, 'the Andy Williams rule', which stated that they never stocked an Andy Williams record because they just weren't in that market. Virgin Atlantic set out to bring the same innovative and fun atmosphere into what was at the time a very tired and dull, engineering-led airline sector.

Founded in June 1984 with just one leased Boeing 747, Virgin Atlantic has grown to be the UK's second largest long-haul airline with a fleet of 38 planes. In 2008 Virgin Atlantic flew nearly 6 million passengers to 30 destinations worldwide and over 63 million passengers have flown Virgin in the 25 years since it opened for business.

But how did it manage to thrive in such a cut-throat marketplace that has seen the demise of so many airlines? Laker Airways, Dan Air, British Caledonian, Zoom, XL, Skybus to name but a few. The answer lies in the way it has built on the Virgin principles of excellent customer service, high quality and value for money, while being adept at handling the perilous cash flow problems inherent in running an airline.

Bringing a Touch of Magic to Air Travel

On Virgin's first ever flight, Maiden Voyager, the celebrity-packed passengers were famously treated to a view of the cockpit during takeoff. It showed the two pilots and flight engineer sharing a cigarette and chatting nonchalantly whilst paying no attention to the controls or runway as the plane took off. As the nose of the plane lifted into the air, the pilots turned around to face the camera: they were Ian Botham and Viv Richards. The flight engineer was one Richard Branson. This publicity stunt, as well as other gimmicks like handing out choc ices in the middle of movies, earned Virgin the reputation as a youthful, fun airline, clearly different from its main competitor, the rather dowdy British Airways.

Before the maiden flight, five small planes wrote 'wait for the English Virgin' across the skies of Manhattan to raise awareness of the new service. Following the maiden flight, numerous branded cross-Atlantic challenges took place, in speedboats and hot air balloons to continue to publicise the service. There was even a round-the-world balloon trip.

In his 2005 autobiography *Losing my Virginity*, Richard Branson explains the rationale for this high-profile approach. 'I realised that I would have to use myself to raise the profile of Virgin Atlantic and build the value of brand. Most companies don't acknowledge the press and have a tiny press office tucked away out of sight. . .' (Branson, 2005).

Then, in 1990, Iraq invaded Kuwait. The price of aviation fuel doubled and people were more reluctant to fly, fearing terrorist attacks. Yet even in adversity Virgin kept its profile high by removing all the seats from one of its 747s and loading the plane with blankets, rice and medical supplies to help the refugees who had fled to Jordan. The return trip carried a number of British nationals who had been stranded by the conflict. British Airways followed suit the following week. A few weeks later into the conflict, Saddam Hussein was keeping British nationals as hostages (as a human shield) around vital Iraqi installations. Through a personal relationship with King Hussein of Jordan, Branson was able to broker a deal whereby a Virgin Atlantic plane would fly into Baghdad, a war zone at this time, with medical supplies, in exchange for some of the hostages. Another PR coup for the airline resulted.

In early 1993, more column inches were garnered when Virgin Atlantic won their court case against BA in the 'dirty tricks' legal action. The UK's *Sun* newspaper headline read 'Virgin screws BA'. Editor Kelvin McKenzie was disappointed, saying he would have preferred it if BA had won, since it would have resulted in a better headline! Branson divided the £500,000 personal payout between all his staff, so that each employee received £166.

Competing on Value and Service Innovation

From the outset, Virgin Atlantic decided that it wouldn't be an exclusively no-frills economy service, as this would leave it vulnerable to a simple cost-cutting attack by its more established competitors. So while a proportion of every plane would carry economy fares, Virgin also set out to capture the business traveller by offering a first-class service at business-class fares.

Upper class is the equivalent of other airlines' business class. At Virgin, it includes extras such as complimentary ground transfers, state-of-the-art clubhouses, an exclusive on-board bar and one of the longest fully flat beds in the air. It also boasts in-seat laptop power and power leads for IPods, as well as offering customers a limousine pick-up service. On arrival at Heathrow, Gatwick or Johannesburg by chauffeur-driven car or LimoBike, the chauffeur will check customers in at the unique 'Drive Thru Check In', so that customers can bypass the terminal and head straight for the clubhouse.

Virgin realised the value of innovation to both the core and added value services, and launched a series of firsts for the industry. It was the first to offer seat-back TVs for every seat. It was also the first airline to offer a premium economy fare with added leg room, meals served on china and priority disembarkation for only a little more on the price of an economy ticket.

In February 2008, it became the first airline in the world to operate a commercial aircraft on a sustainable biofuel blend. This built on its 2007 environmental policy, which featured a carbon offsetting scheme as well as an order for 15 B787–9 Dreamliners that will burn 27% less fuel than the older planes they will replace.

Nearly all of these firsts are achieved through excellent investment in staff and training, which stems from Virgin's belief that if you look after your employees, they will look after customers and everyone will benefit. In 2005 VA became the first airline to be accredited by the Chartered Management Institute to provide its own Diploma and Certificate qualifications.

Outside of the core service, Virgin Atlantic has developed an industry-renowned 'Flying without Fear' programme. It has helped thousands of people overcome their fears, ranging from anxiety at takeoff, to a complete inability to board an aircraft, and customer feedback shows a 98% success rate. The programme is now supported by a new book, called *Flying Without Fear, 101 Fear of Flying Questions Answered*, for which Sir Richard Branson has written the foreword.

In true Virgin style, the programme's website announces that by special request, the Flying Without Fear team travelled to America to help actress Whoopi Goldberg overcome her fear of flying. She hadn't flown for a decade, but the opening of her new musical production *Sister Act* in London meant that she had to get on a plane again, which she was dreading. Whoopi received a lot of media attention both in the USA and UK about the help she received from Virgin Atlantic's Flying Without Fear programme, and this is what she said about it on British television's GMTV in April 2009: 'Virgin does this amazing program here in the UK and I'm begging them to bring it to the States where they can get people over their fear of flying... There are too many fail safes... Just knowing that was enough to sort of get me humming in the car on the way to the airport – something which used to get me clawing at people and scratching.'

Celebrating 25 Years in the Air

In June 2009, Virgin Atlantic unveiled a £6 million advertising campaign designed to beat off the credit-crunch blues and highlight why it's still red hot after 25 years of flying. The high-profile campaign, including press and TV ads, featured cheeky slogans such as '*More experience than the name suggests*', '*Extra inches where it counts*', '*Fly a younger fleet*' and the simple '*Hello Gorgeous*'. The campaign also featured Austin Powers with '*There's only one Virgin on this T-shirt baby*'.

In the TV ad, the cast is seen walking through the airport before boarding the airline's inaugural flight to New York. The advert features iconic images of the 1980s, such as the Rubik's Cube, brick-sized cell phones and the Asteroids video game.

Steve Ridgway, chief executive of Virgin Atlantic, said: 'When our competitors are feeling down in the dumps, and we enter into a year of economic uncertainty, you can always trust Virgin Atlantic to raise spirits and stare into the future with as much optimism as we did back in 1984.'

And there's much to be optimistic about. The airline was voted 'Best Transatlantic Business Class' by *Conde Nast* readers in 2009, on a wide variety of criteria including its in-flight service and efficiency of service. It beat off stiff competition from Singapore Airlines and Emirates to the top spot, with British Airways back in fourth place. It also won 'Best Long-Haul Airline' at the *Sunday Times Travel Magazine* Readers' Awards 2009.

In the midst of all this success the airline keeps its feet on the ground with a simple mission statement that keeps it focused: to grow a profitable airline where people love to fly and where people love to work.

Creating Added Value

When, in the 1970s, Gene Lodenberry launched his morality tale, *Star Trek*, he foresaw Captain Kirk talking into his wrist communicator throughout his 23rd century adventures. Yet, even he, a creative force, could not imagine loading 'Apps', like light-sabre battles, into IPhones, or many of the other applications being developed on mobiles at the time of writing. The fact that this has happened only 40 years later shows the power of numerous entrepreneurs, engineers and buyers interacting to create unimagined opportunities.

People have similar difficulty seeing the added value applications to be gained from the services which could be built on other technologies and networks. It seems that human beings have to first become accustomed to an innovative form of technology before they can really start to apply it to their life. Moreover, customers are not often able to imagine added value opportunities before they are presented to them, which means that traditional market research will not uncover their need. For instance, many adults, at the time of writing, have not spent the hours that their children have playing computer games. So, they find it difficult to understand the value of these programmes to future applications such as education and training.

In an environment that is dominated by engineers, it can be difficult to champion a new creative concept based on intuition and not hard data. As customer research often does not substantiate an innovative idea, innovators are often ignored or undermined. Yet, once concepts become familiar to an industry, they are then exploited by all. This tends to make these technology-based companies slow to reach for radical ways of bringing customer innovation into their organisation and to force them into creative partnerships; like Ericsson's with Sony in mobile phones.

Measuring Service Performance

There is a similarity of operational service measures amongst technology companies. These are produced at varying frequencies but have a fundamental purpose that is the same in each firm: to allow senior management to ensure that the operations of the company are in line with what it considers to be the key factors of success. Strangely, though, these measures might not be checked against the priorities of customers. Even more strangely, they might be changed at the whim of senior leaders without any real attempt to research customer needs or to calculate the consequences to massive complex organisations.

Typical measures include:

1. Provision of service access compared with demand.
2. Frequency of interruptions to service (mean time between failures).
3. Remedial actions taken within contact time or customer's expectation (mean time to repair).

Handling Catastrophe

Most services which rely on a technical infrastructure experience catastrophe. So, as these services are dependent upon physical networks and are so critical to society, they must have systems in place to handle unusual and devastating events. Experience in different disasters shows that the public will tolerate an interruption to service resulting from catastrophe if it is informed of what is happening, directed to emergency procedures and restored to service access in what it considers to be a reasonable time. However, as with the massive failure of emergency support in New Orleans after Hurricane Katrina, if the service provider fails in any of these areas at the time of catastrophe, customers will be unforgiving. It may be that the organisation responds to the operational problems quickly but does not direct customers to emergency procedures during the catastrophe or to alternative service provision afterwards. In this case, customers will resent and remember this for some time, affecting reputation, revenues and costs. So, managers must give real attention to planning and practicing disaster recovery.

Each of these characteristics affects the style, structure, culture and commercial priorities of service firms built on a network or technology platform. Each produces an environment that sets them apart from consumer product, manufacturing, professional service or financial service companies. The characteristics, and the issues they generate, give the marketers within these businesses some unique considerations. Moreover, the difficulties and unique challenges faced by this category of service companies have not been thoroughly explored by academic writers or specialists, so there is little useable science or substantiated accumulated knowledge on how to market them.

IS THIS REALLY A DISTINCT MARKET OR CATEGORY?

Service companies in technology sectors have many significant dynamics and characteristics in common. People, who move from one to another, find that there are real similarities in approach to work, even if the underlying technology is very different. Executives within them have very similar outlooks, beliefs and attitudes to work.

A market is not just about impersonal economic forces. It comprises numerous people taking decisions in competitive companies and in buying groups that will eventually develop common reference points. These service companies tend to believe, for example, that customers care about the technological infrastructure as much as they do, and that longstanding customers remain with them out of loyalty, rather than simple inertia (often the effort to look for a new supplier can outweigh the benefits of finding one).

The participants in these service businesses tend to see their own industry as uniquely complex; a narrow view which comes from their technological heritage. For instance, the telecommunications companies see the provision of competitive modern communications, particularly through the 'local loop', as uniquely complex and a balance of safety, technological excellence and return on capital invested. However, the safety requirements in balancing the different pressures, say, in the gas distribution network against the need to invest in modern plant are very similar. Also, the development of the electrical distribution infrastructure has similarities to the network of remote diagnostic access common in modern computer service contracts. Participants in these service industries are therefore grappling with similar issues, going through similar development, and finding similar solutions. There are lessons and insights that participants in these companies can learn from each other.

VICTORIAN JIM REFORMS THE MIDLAND

It is very hard for modern people to really understand (especially any who have been stuck on an Amtrak train during a snow storm) just how revolutionary, scientific and advanced the railways were when they first appeared. This was the first time in human history that mankind could travel faster than a galloping horse, and it is no exaggeration to say that they transformed society.

The first real railway line went operational in Britain (Darlington to Stockton) in 1825 and proved to be a fabulous investment for its Quaker owners (returns of 15% between 1839 and 1841). It prompted a railway mania and investment boom. By 1840, 200,000 people were involved in railway construction in the UK alone. British iron output doubled as a result of it and, by 1850, £240 million had been invested. By 1869, the first trans-continental railway had been completed in the USA and, by 1890, the massive trans-Siberian railway was finished in Russia.

The railways created new towns, new concepts and new jobs. In London, for instance, a young insolvency specialist called William Deloitte created a new system of accounting for these industrialised service businesses and, through such advanced thinking, created the major accounting firm that still bears his name today. The railways introduced consistent time, holidays, commuting and new concepts like the word 'class'. Historian Eric Hobsbawm says of them (Hobsbawm, 1999):

> By 1850 the railways had reached a standard of performance not seriously improved upon until the abandonment of steam in the mid twentieth century, their organisation and methods were on a scale unparalleled in any other industry, their use of novel and science-based technology (such as the electric telegraph) unprecedented. They appeared to be several generations ahead of the rest of the economy, and indeed 'railway' became a sort of synonym for ultra-modernity in the 1840s, as 'atomic' was to be after the second world war. Their sheer size and scale staggered the imagination and dwarfed the most gigantic public works of the past.

Despite the work of novelists like Thomas Hardy and Charles Dickens, it is also hard for modern audiences to understand the attitude of educated and wealthy people to the poor during that period. Many resisted, for example, any educational initiatives because they feared that it would cause unrest. There were, though, a number of enlightened souls pushing for reform. Victoria's Prince Albert caused outrage and concern, for instance, when he insisted on there being days when the poor and uneducated could visit his Great Exhibition. Another reformer was James Allport, who ran the Midland Railway in the mid century.

His first significant act caused as much outrage and concern as Prince Albert's. At the time, 'Third class' was for the poor and working people. It normally consisted of simple open carriages with wooden benches, which were given low priority. There are reported instances of Third class trains being shunted into sidings to let even cattle or freight pass them by. Allport abolished this. He had covered carriages, all of which had upholstered seats, partitioning and more leg room. His peers in the industry hated him for it because (as BA did with flat beds in business class a century later) he set a new standard for the basic service offered on this new and exciting network of technology; and they had to keep up. For him, it was not enough just to offer carriage any more. He wanted to serve people.

He contracted with a successful catering company, Spiers & Ponds, to ensure that wealthier travellers could enjoy a food service. They had begun their business with sporting events like Wimbledon and opened the Criterion restaurant which still operates in London's Piccadilly. Passengers could buy one of their hampers at one station and drop it at another after eating it on the train. Their service became a social occasion, famous in Victorian

England. In his extensive history of advertising (Sampson, 1875), Victorian author Henry Samson could have been describing the answer to airplane food:

Ten years ago no man in his senses would have dreamt of applying for food or drink at a railway buffet while he could go elsewhere; now Spiers & Pond daily serve thousands who desert the old familiar taverns and crowd the bars at the various city stations. . . the old regime of mouldy pork-pies and stale Banbury cakes has made us feel very well disposed to a firm whose name has already passed into a proverb.

The following personal advert from the *Daily Telegraph* of 1874 shows how much this service had become part of social life:

The lady, who travelled from Bedford to London by Midland train on the night of the 4th inst, can now meet the gentleman who shared with her the contents of his railway luncheon basket. She enjoys the recollection of that pleasant meal, and would like to know if he is going on another journey. Will keep any appointment made at the Criterion in Piccadilly.

Allport's other remarkable innovation was to create a premium service on his railway. He constructed an outsourcing contract (yes, an outsourcing contract in the 1870s!) with America's famous Pullman trains. They provided a 'hotel standard service' for an extra fee, using their own carriages and attendants. In fact, Allport had been so effective at increasing the return on the basic rail service that he was eventually asked to run another innovation, the railway clearing house, which handled ticketing and pricing across the whole national network. He repositioned the value of the core service based on a novel network infrastructure and created attractive added value services. An outstanding services marketer.

SUMMARY

In the past few decades, services and service businesses have become more and more important to developed economies. Economists have recognised several sectors of the service economy but those built on technology or networks, despite being a recognisable market, are a neglected category. This is a unique market with common attitudes, benefits and behaviours. Marketers in this field need to understand and allow for the unique characteristics of this market whilst deploying state-of-the-art marketing techniques. In particular, they need to create attractive, enticing services.

2
Marketing Services

INTRODUCTION

Although marketing is taught in universities and business schools across the world, it must be one of the least defined and most misunderstood of functions. With its roots firmly in the consumer product industries, its components cover a variety of activities such as advertising, sales and public relations. When well crafted, it can influence the thinking of a group of human beings, prompting them to buy. In fact, the cumulative knowledge and progressive organisational competence in marketing of many businesses has generated enormous wealth and, en route, created iconic images and memorable campaigns. So marketing positively affects both revenues and the cost of sales; and some technology companies are becoming increasingly competent at it, systematically building up a track record of meaningful contribution to profit. This chapter explores marketing and the reasons why the marketing of services needs to be so different from the marketing of products; and how these differences should be applied by technology marketers. It emphasises, in particular, the importance of growing organisational competence in marketing.

ONE MORE TIME: WHAT IS MARKETING?

Marketing has been variously called an art (because it is creative, requiring insight, judgement and experience), a science (because market-orientated companies use data and analysis to inform decision-making), a management discipline (one of the functions of many firms), an academic field of study (most universities have dedicated faculty) and a profession (because it requires deep knowledge of techniques with the experience to know how and when to apply them). In reality, it has elements of all of these but, above all else, it focuses on how a business can generate future revenue. One of the earliest marketing academics, R.S. Butler (quoted in Bartels, 1988), captured something when he said that marketing is:

> . . .everything that the promoter of a product has to do prior to his actual use of salesmen and advertising.

In other words, it is the thought and effort that goes into making a sale as easy and cost-effective as possible; and it is different in each market. Other opinion formers have defined it as:

- . . .the process of planning and executing the conception, pricing, promotion and distribution of ideas, goods and services to create exchanges that satisfy individual and organisational goals. American Marketing Association
- . . .the management process that identifies, anticipates and satisfies customer requirements profitably. UK Chartered Institute of Marketing
- . . .the social process by which individuals and groups obtain what they need and want through creating and exchanging products and value with others. Kotler and Armstrong (2003)

Presenting credentials to potential buyers, managing press relations, hosting seminars, creating new products and giving presentations at public conferences are all part of the marketing mix. It is a management process by which leaders of a firm draw in revenue and grow the business. Yet, it is not limited to the specialist function within a firm or to highly trained marketing managers. When an executive of any kind is engaging in activities or plans which affect the revenue line and streamline the moment of sale, they are marketing. So, whether they call it marketing or not, whether they have specialist marketing managers or not, all firms undertake a range of marketing activities to grow their business. As in all aspects of business life, these activities can be undertaken using common sense by anyone in the firm. However, they are likely to be more effective, and generate more income in a more cost-effective way, if appropriate, honed techniques are used and if experienced specialists are involved. In fact, experience shows that decades of increasing organisational competence in marketing, with successive generations of marketers passing on effective marketing processes, generates substantial wealth for shareholders.

DIFFERENT MANIFESTATIONS OF MARKETING

In publicly owned companies, marketing is likely to be the responsibility of a relatively small specialist unit comprising people who are qualified and experienced. They will be responsible to the management team for creating strategy, plans, budgets and programmes to grow the business. They will also be expected to balance the skills, resources and processes of the marketing department to optimum effect and for the benefit of the business in its market. They will have delegated responsibility to manage the function effectively, and will need to ensure that it has good knowledge of relevant concepts, develops appropriate competencies, uses reliable techniques and installs robust processes or systems. On top of this, they will need to engage with the whole firm to ensure that the customers' experiences are appropriate to meet the firm's objectives. In short, the marketing function needs to keep up-to-date and act as a catalyst for the business in its approach to market.

Figure 2.1 depicts typical functions within the unit but these can vary enormously. For instance, brand management might exist within market management (because it focuses on market 'categories', a retail term) or in marketing communications. Sales and sales support might be combined with marketing under one 'sales and marketing' director or vice president.

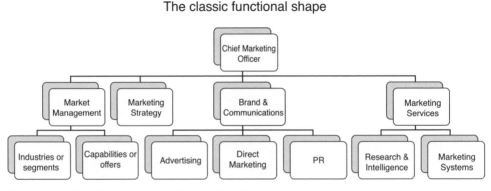

Figure 2.1   The classic functions of a marketing department

Also, there may be elements of the marketing mix in other parts of the organisation. For example, some firms place new product development and pricing under a separate director. Others have directors dedicated to corporate relations; responsible for reputation, brand, corporate social responsibility (CSR) and public relations.

The 'chief marketing officer' (CMO), as leader of the function, has several roles. The first is to provide marketing experience, judgement and advice to the firm's leaders about policies; to round out and inform decision-making and strategy development. The second is to be the voice of the market within the firm, challenging it to embrace opportunities and understand customer needs. The third is to lead the function, ensuring that it is properly resourced and contributes appropriately.

Yet, not all marketing is undertaken in the carefully organised departments of large publicly owned companies. In small firms and professional partnerships, the situation tends to be more fluid and less clearly defined. Different people will initiate marketing activities, which are frequently handled by executives who have no specialist marketing knowledge and may not be aware of other marketing initiatives in different parts of even their own company. If a marketing manager or marketing department exists, which is by no means certain, they may not have exclusive responsibility for all revenue generation tasks. The business world is not uniform. Companies succeed with a huge diversity of products and services with very different cultures and operating approaches in a wide range of markets. Marketing has to be adjusted to suit these very different environments if it is to be successful.

One of the main business types is, for example, the conglomerate or decentralised business. Figure 2.2 depicts a decentralised organisation with devolved business units, like IBM, Ericsson or Philips. These might be businesses specialising in different disciplines or in different geographies. The effectiveness and behaviour of the marketers depend on the degree of autonomy of the different units. For example, this structure is often adopted by large technology companies. In these, a 'strategic business unit' (such as a geographic region or a business line), will be expected to work within tight corporate guidelines. The marketer is likely to have a

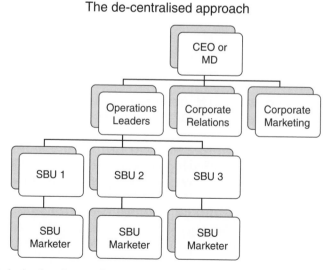

Figure 2.2 Marketing in a decentralised business

The single marketer or SBU structure

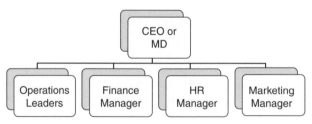

Figure 2.3 Marketing in the small firm

clear job description with defined competences and objectives. They are also likely to report regularly to the corporate function and participate in clear, firm-wide processes. At the other extreme, though, are large conglomerates that own separate firms with different profit pools (like Ingersoll Rand, Virgin or ABB). The marketers in these firms are likely to have greater autonomy within very broad corporate guidelines.

On the other hand, the majority of businesses range from tiny start-ups to medium-sized companies. In these, any specialist marketing units are limited by resource constraints; often to a single, or isolated, marketer, as depicted in Figure 2.3. One person is all that the business can afford, and that person is likely to report to the firm's leader, gaining authority to create programmes, strategy and change from the closeness of that relationship. Constrained by lack of resources, the single marketer needs influence to get the wider organisation to take on work and the budget to rely on external contractors. They need the capability to handle a wide variety of tasks and the humility to do much of the work themselves. In some firms, they report to a functional leader who manages all specialists (HR, finance, IT, etc.). This, however, tends to be less effective because it weakens the assumed authority behind the speciality, reducing the influence of marketing.

One difficulty in this environment is the marketing skills that small firms can afford. They frequently hire young professionals or marketers from middle management in other firms who have never really progressed in their career. Consequently, they tend to focus on promotional work, and are frequently unaware of service marketing techniques that might be more profitable and effective for their firm. They are often weak at giving credible strategic insight to influence the direction of the firm.

A wide range of companies, both large and small, have yet another form of governance: the intuitive business leader. These creative entrepreneurs tend to use the principles of marketing, based on their business experience and intuition. Individuals (from Richard Branson, Simon Cowell, Alan Sugar and Phillip Green in Britain to Donald Trump, Bill Gates, Larry Ellison, Michael Dell and Steve Jobs in the USA) become millionaires, or even billionaires, through an intuitive approach to markets. They are pre-eminent practitioners of marketing, creating enormous wealth for themselves, their investors, their employees and society as a whole.

They frequently succeed by breaking rules (Harvard would not have advised young pop entrepreneur Branson to go into the airline business, for instance) and shake up or create markets. Their emphasis is on customers and markets, so they push all the elements of their business to focus on external opportunities rather than internal activities. They also tolerate both risk and failure. As their businesses grow, the most successful learn to use specialists (like accountants, lawyers and merchant bankers) to round out or refine their vision and instinct; and most use marketing specialists (either as employees or through agencies) in a similar way.

The cultures of their highly successful, market-orientated businesses are focused on understanding and executing the leader's will. Much time is taken to tease out, develop and communicate their vision and implement it throughout the organisation. Their marketing is therefore most often about connecting the leader's vision with a market opportunity. In some cases, the marketing structure is a small, fluid team, in another a trusted agency and, in yet another, the more recognisable structures of big, publicly owned firms. They use novel marketing techniques, frequently exploiting publicity opportunities and, in some cases, breaking away from the rules of consumer product marketing to pioneer service marketing techniques.

Finally, much marketing occurs in the marketing supply industry, working on behalf of their clients. This is a very diverse range of businesses. It includes advertising, direct marketing and new product development agencies, as well as a variety of consultancies. They range from the huge conglomerates like WPP to single-person businesses. Despite the fact that most of the writing and theory of marketing is developed from and for marketing specialists in large companies, much of the real added value comes from agencies.

THE EVOLUTION OF MARKETING IN AN ORGANISATION

Much of the academic study of marketing focuses on issues of theory, concept or practice and makes assumptions about the way the function works in a firm. It is assumed that there is a well-developed organisation led by a marketing director or CMO, able to call on financial and human resources to undertake research, manage advertising or adjust product features in the light of rational, justified arguments. There is little talk of the need to convince organisations of marketing's importance, of competing for resources, or of organisational politics.

The situation is rarely so clear-cut. The marketing function is often under-developed. It has to argue for its role in the organisation and has to invest in processes, as well as running projects to generate revenue. Some companies do not understand their own need for certain marketing skills and restrict the contribution that the function makes, limiting it to, say, a minor promotional role. In many technology firms, marketers spend their time running around making PowerPoint presentations to each other or organising the mind-numbing detail of different events. This is not effective marketing and does not necessarily create wealth.

Piercy (2001) showed that there was rarely consistency about what activities marketing leaders have responsibility for. Nor was there consistency in the shape of organisations. He found four types of marketing departments.

- **Integrated/full service:** closest to the theoretical models and with a wide range of responsibilities and power in the organisation.
- **Strategic/services:** smaller units with less power and integration. Their influence is in the area of marketing support services or specific policies/strategies.
- **Selling overhead:** often large numbers and dispersed but primarily engaged in sales support activities.
- **Limited staff role:** small numbers with few responsibilities and engaged in specific staff support such as market research or media relations.

Professor Piercy's research also demonstrates that marketing departments evolve (from a limited staff role to fully integrated) as firms develop and marketing grows in importance within them. The organisation's competence in marketing and sales increases over time. When a company is initially formed, much of the marketing role is undertaken by the founders or specialist subcontractors (like PR agencies). Some time after that a marketing specialist will be hired to manage specific activities such as brochure production, new product launch and perhaps some advertising.

As the firm grows, marketing specialists are recruited into different places in the organisation, such as public relations and sales support. Later, this develops into the fully integrated marketing function seen in many corporate firms. A large corporate entity might have several hundred marketing specialists running campaigns, integrated by both a hierarchical senior management team (culminating in a CMO) and by the use of common processes and technology. The development of marketing in many leading service companies that are based on a technological infrastructure has been through this evolution as their market situation has changed.

Marketing's leading thinker and educator, Phillip Kotler, has also noticed, through his many years of engaging with different businesses, this evolution of marketing competence. He said:

> ...the nature of the marketing function varies significantly from company to company.
>
> Most small businesses (and most businesses are small) don't establish a formal marketing group at all. Their marketing ideas come from managers, the sales force, or an advertising agency. Eventually, successful small businesses add a marketing person (or persons) to help relieve the sales force of some chores... Both sales and marketing see the marketing group as an adjunct to the sales force at this stage.
>
> As companies become larger and more successful, executives recognise that there is more to marketing than setting the four p's... They determine that effective marketing calls for people skilled in segmentation, targeting and positioning. Once companies hire marketers with those skills, marketing becomes an independent player...
>
> Once the marketing group tackles higher-level tasks like segmentation, it starts to work more closely with other departments, particularly strategic planning, product development, finance and manufacturing. The company starts to think in terms of developing brands rather than products, brand managers become powerful players in the organisation. The marketing group is no longer a humble ancillary to the sales department. (Reprinted with permission from *Harvard Business Review*, Kotler, Rackham and Krishnaswamy, July/Aug, 2006)

So, the ability of the marketing function to contribute effectively to income generation and the health of the business depends on the role and strategy it has in the organisation. In many long-standing consumer goods businesses (Procter & Gamble, Unilever) or confectioners (Mars), it is a lead function of the business. The firm recruits the cream of the graduate crop into a professional apprenticeship scheme that provides some of the best marketing training in the world. Alumni of the programme progress through the firm from junior brand managers to marketing directors. Their role is to generate income from branded value propositions, using wide-ranging and proven marketing techniques. Over many decades, these companies have institutionalised marketing into a leading philosophy of the entire corporation.

In other industries, however, the contribution of specialist marketing people is limited to minor functional roles like sales support or brochure creation. The formal marketing function is not as valued because the income of the business is generated by other means; such as a strong and healthy natural demand for an innovative product or systematic account management. But, there is evidence that this changes as markets mature. For instance, most of the firms that have evolved first-rate and leading marketing units are consumer goods companies, because they experienced market maturity in the developed economies during the 1950s and 1960s. Similarly, marketing has become more important to the car industry, the computer industry and the telecommunications industry as they, in turn, have faced market maturity. As IBM has adapted to its changing market conditions, for instance, the marketing function within the company has developed its own competence to keep pace. As the case study shows, it is progressively becoming one of the most sophisticated in the IT sector; combining knowledge of marketing basics with impressive marketing people and real customer insights through growing organisational competence.

THE EVOLUTION OF SERVICES AND MARKETING AT IBM

Twenty years ago the proportion of IBM's business categorised as 'services' as opposed to products was an estimated 5%. Ten years later, it rose to roughly 15%. Now services account for over 57% of IBM's revenues (which were $104 billion in 2008) and IBM Global Services is the world's largest technology services business.

Becoming a Services-led Company

The beginnings of IBM's IT services development lie with the 1989 deal with Eastman Kodak Company, whereby IBM designed, built and managed a data centre for Kodak in Rochester. Then in 1991, IBM's Board approved a new worldwide services strategy, 'To make IBM a world-class services company.'

Lou Gerstner's subsequent arrival at IBM in April 1993 as chairman and chief executive officer (CEO) has been well documented – not least by himself, in his best seller *Who Says Elephants Can't Dance?* (Gerstner, 2002). Yet, although it is just one chapter in his book, it was his strategic decision not to break up the company but develop an 'integrator' model for IBM, putting IBM Global Services at the focal point of this model, that lies at the heart of the multinational's renaissance.

Perhaps the bravest, and riskiest, decision he made was, in his words, 'to place a big bet on services'. After all, IBM was the archetypal product-based IT company, with a culture rooted in successfully selling what it manufactured. But by the time Gerstner took over, he and a few of his lieutenants could sense that the market was changing fundamentally.

First, IT was becoming a commodity business, which meant that products alone would not offer the kind of growth demanded by shareholders. Moreover, customers were getting fed up with not seeing the promised financial benefits from their expensive IT installations. Even more importantly, as a former customer himself, he could see that demand was shifting towards IT companies supplying the answers, not the products; to integration, not the technology by itself. He called it 'the customer's overpowering desire for someone to provide integration'.

As he continued, 'At first that was just the integration of technologies. But as the networked computing model took hold, it created whole new dimensions around integration, forcing customers to integrate technologies with core business processes, and then to integrate processes – like pricing, fulfilment or logistics – with one another.'

IBM Global Services (IBM GS) was formed in 1995, instantly rationalising its offers from 2500 down to about 100 'solution categories'. These were mapped to customer needs and concerns; and turned into consistent global propositions.

In the late 1990s and early part of the new millennium, IBM GS expanded its portfolio to offer more e-business services and then increased its industry focus to create solutions for 16 industries grouped across five sectors.

Then, in October 2002, the company completed a remarkable and startling step in its move towards a more service-orientated business: its $3.9 billion acquisition of PwC Consulting, the biggest in IBM's history. It was combined with the existing Business Innovation Services division to form Business Consulting Services (BCS). This was a powerful blend of the specialised knowledge of the former PwC Consulting professionals with the existing services and technology expertise of IBM itself. It also, at a stroke, gave the company the manpower it needed to accelerate its penetration of the developing service market.

Today the company's operations consist of Global Services (incorporating Global Technology Services (GTS), and Global Business Services (GBS)), Software, Systems and Technology, and Global Financing. Global Services is still a vital component of the company's strategy – providing IT infrastructure, business insight and solutions to clients. GTS delivers IT infrastructure services and business process services, while GBS primarily provides professional services and application outsourcing services.

The journey into a service-led proposition has been a challenging one for IBM, for a number of reasons:

- It has meant that, on occasion, IBM recommended other companies' products if they were the best answer to a customer's needs. And that, along with maintaining and servicing those products, met fierce resistance in the sales force.
- The company had to get to grips with the economics of a services business, which are very different from those of a product-based business. For instance, a major services contract can last anywhere between six months to 12 years, having big implications for sales compensation and financial management in the company.
- A services business is much more difficult to manage, because it is human-intensive, with the selling of a capability, of knowledge, and creating solutions at the point of delivery.

As the company transformed, the marketing function within it had to up its game and learn new skills, moving from classic product-push techniques to sophisticated approaches, to corporate branding and "demand-pull" campaigns for services and solutions.

Moving to Services Marketing

When Lou Gerstner arrived at IBM in 1993, he said, perhaps surprisingly, that marketing as such 'did not exist'. One of IBM's core strengths had long been its account management system. Marketing was mainly sales support.

Gerstner, however, believed that a successful company had to have a customer/marketplace orientation and a strong marketing organisation. So he addressed this gap almost immediately by hiring Abby Kohnstamm, who had worked with him for years at American Express, as head of corporate marketing. She, in turn, swiftly began to define what the marketing discipline should mean to IBM and brought in the required expertise in areas such as branding, database marketing, communications and the development of new sales channels.

One of the early 'wins' the team enjoyed, and which earned it a lot of credibility from some of the more sceptical employees, was the development of a new channel, the telephone, to the mid market. This moving of account management from a physical to a telephone relationship, using database marketing to align the right employees to the right accounts with the right levels of contacts, proved very successful. It grew to a multi-billion dollar business in a relatively short space of time and was backed by a memorable advertising campaign, 'Solutions for a small planet.'

It also helped lay the groundwork for the shift to services since mid-market customers were demanding solutions, or the integration of products, rather than the products alone. They also wanted those solutions set into the context of their industries, which helped the company begin to think in terms of marketing along industry rather than product lines. So, well before the purchase of PwC Consulting propelled IBM into a new competitive arena, it was already gaining experience in working out how to bundle products together into solutions aimed at the right segments.

IBM's marketing now reflects the journey the company has made into a portfolio that balances software, hardware and services. Their integration of these components within a consulting and services perspective gives the company far more impact than the individual elements would do on their own.

The marketing structure, both within and across the main business streams, is complex. There are several hundred employees, covering all aspects of marketing from market intelligence, analysis, marketing communications and support. Even though the advertising is centralised, with almost all major campaigns originating in the USA, units in other countries can tailor them to suit their market. Also, because increasingly the advertisements are based on case studies, different markets can submit case studies for use in the international adverts to underline IBM's global reach.

These adverts also emphasise the company's strategic shift. For example, to communicate the range and depth of its business expertise following its acquisition of PwC Consulting, it launched a campaign for the combined entity, Business Consulting Services (BCS), called 'the other IBM'. It was very successful, raising awareness in, for example, the UK market by a remarkable 11%. IBM's marketing has resulted in its brand being ranked second in Interbrand's 2008 Best Global Brands index, ahead of Microsoft, and valued at $59 billion.

Creating Value into the Future

IBM is now setting out its agenda for a "smarter planet", supporting public and private sector organisations as they seek to reduce costs, drive innovation and transform their infrastructure. The company's 2008 annual report explains, 'The economic downturn has intensified this trend, as leaders seek not simply to repair what is broken, but to prepare for a 21st Century economy.'

Bit by bit, they claim, "our planet is getting smarter, driven by the proliferation of technology instruments we use in everyday life, the interconnections between these instruments and the information they generate, and the increasing intelligence that can be gleaned from this information."

To drive the agenda forward and aggressively pursue this global, transformational opportunity, IBM's marketing function will need to raise its game once more. At the time of writing work is underway to redefine what roles will be needed and what competencies must be developed over the coming years. Whatever shape this next marketing evolution within IBM takes, it will certainly be smarter than anything we've seen before.

MARKETING SERVICES

It is only in recent decades that companies in service sectors have begun to establish substantial marketing departments similar to those in consumer goods industries. Banks, utilities, travel firms and retailers have learnt to use these structures to market services as their industries have experienced greater competition or, as in the case of newly privatised utilities, market forces, for the first time. Some of these firms appear to be adopting the processes of good marketing, seen in consumer goods companies, and embarking on decades-long organisational learning in marketing effectiveness. BT, for example, had no marketing resource at all when, in 1984, it was a government-owned engineering organisation. It now has several hundred marketers working on brand, communication and strategy programmes; and has a number of highly successful, award-winning campaigns to its name. More importantly, though, it has sophisticated campaign management processes which are progressively improved alongside other operational capabilities.

In parallel with all this, though, the marketing profession as a whole has been exploring how different the marketing of services is from the marketing of consumer products. Until relatively recently, the majority of marketing theory, language and practice was based on the experience of firms offering physical goods (particularly consumer products like washing up liquid) or drawn from economic theory. Many marketing techniques are built on product examples and most marketing teachers at leading institutes had their early career in product manufacturing. However, the word 'service' implies a difference from manufacturing. It suggests personal support for others. As Professor Levitt pointed out:

> The concept of 'service' evokes, from the opaque recesses of the mind, time worn images of personal ministration and attendance. It refers generally to deeds one individual performs personally for another. It carries historical connotations of charity, gallantry, and selflessness, or of obedience, subordination and subjugation. (Levitt, 1972)

As a result, most people who have tried to run both take the view that service businesses are very different from product manufacturing companies. Typical, for instance, is Lou Gerstner:

> ...I have worked in service companies (McKinsey and American Express) and product companies (RJR Nabisco and IBM). I will state unequivocally that service businesses are much more difficult to manage. In services you don't make a product and then sell it. You sell capability. You sell knowledge. You create it at the same time you deliver it. The business is different. The economics are entirely different. (Gerstner, 2002)

It is only in the past three decades that marketers have begun to move from traditional product disciplines to service environments in any significant numbers. So, coupled with the emergence of competitive service sectors in the developed economies, there has been a growing interest in this field. Marketers are adapting practices to a wider range of service businesses (whether in business-to-business or consumer markets) and to an increasing number of activities within different service businesses. There is also a wide body of academic research, much of which is conducted with service businesses in a variety of countries and sectors; giving science and substance to the marketing of services. So, thoughtful business people wanting to market services can generally rely on the applicability and relevance of much that is published. The work suggests that the methods to create, run, communicate, sell and grow a service business are different from those used to handle a product business. But exactly how different and how can these differences be integrated into the marketing of service in technology businesses?

The major differences reported by managers working in the field and by researchers who have studied it are as follows.

Intangibility

Pure services are intangible. They have no physical presence and cannot be experienced or detected by the five senses. So it is not possible to taste, feel, see, hear or smell them. As a result, they are communicated, sold and bought in the customer's imagination.

G. Lynn Shostack, a senior banking marketer, was the first to emphasise this. She said:

> It is wrong to imply that services are just like products except for intangibility. By such logic, apples are just like oranges except for their 'appleness'. Intangibility is not a modifier; it is a state. Intangibles may come with tangible wrappings but no amount of money can buy physical owner-ship of ... experience ... time ... or process. Tangible means palpable and material. Intangible is ... impalpable and not corporeal. This distinction has profound implications. (Reprinted with permission from *The Journal of Marketing*, published by the American Marketing Association, G.L. Shostack, April, 1977)

This lack of physical components affects marketing in several ways, the first being pre-purchase assessment. Human beings seem to need physical clues to help them assess benefits and compare value propositions. People find it hard to accept or engage with a concept until it is reality, presented to them in physical form. They need to see it and touch it to understand it. In fact, the comparison of physical elements (in a new car for example) often clinches the sale. So, potential customers of a service business need help if they are to understand and grasp the offer. When buying an intangible service, they will often seek the opinion of people they respect before purchase and will buy again only if their initial experience of the service is good. So the reputation of the service, the quality of the experience and the appeal of the packaging are all very important. The supplier needs to make its intangible service seem both tangible and testable.

Some service marketers create sales promotion materials that represent the service. For example, a recent trend in the retail market is the sale of 'experiences' as gifts. People can buy days at spas and race tracks for others. These services are sold in attractive packages which can be displayed on promotion racks and easily taken, by the buyer, to tills. In business-to-business markets there have been similar packaging attempts. Over the past two decades, for instance, a number of the leading IT firms have offered their maintenance and project management services in attractive, packaged form. Even leading professional service providers, who generally detest the idea of their service being 'marketed', use collateral in the form of expensive brochures and case studies to convey the spirit and skill of their offer. Packaging and promotion are therefore one of the prime communication tactics used to market intangible services.

More important, though, is the need to understand and emphasise the buyers' experiences of an intangible service. Just as people want to understand how a product might work in practice, they need to understand what their experience of a service might be; and this should be an important focus of service marketing. When launching a new service, for instance, many suppliers create ways to encourage potential buyers to take part in free trials or test runs. They are an essential part of the launch. Some famous hairdressers, like Vidal Sassoon and Nicky Clark, started their remarkably successful franchises by giving free service to celebrities and other opinion formers.

Once the buyer has experienced the service, they are in a good position to assess it for their own future needs and, just as importantly, for the needs of friends or relatives. So, free trials overcome many of the difficulties caused by intangibility because they reduce the perceived risk of purchase. They are also a way of starting one of the most important service marketing tools: word-of-mouth. If customers have a good experience they will talk about it to others and the reputation of the new service will begin to spread. (A bad experience, by contrast, will create negative stories that undermine any effort to grow the business.)

After launch, the most important marketing strategy is to amplify the natural reputation of the service created by accumulated service experiences. This can be captured in testimonials or printed case studies but it is most effective when word-of-mouth (gossip) is amplified through a communications strategy called viral marketing. This technique spreads positive recommendations of the service, and different messages, through informal customer networks, and can even create excitement, respect and demand for a service. It is a deliberate attempt to amplify word-of-mouth and is, at the time of writing, one of the newest areas of emphasis among marketers. Although successful leaders of service businesses have been intuitively nurturing word-of-mouth for many years, marketing specialists are, only now, beginning to properly understand and codify techniques to exploit it.

Also, as there is no physical product, sales people cannot emphasise "benefits" (an approach routinely adopted by manufacturing companies) but must communicate the experience and outcome of the service process. The selling of an intangible service must therefore be very different from a product sale. In fact, if the routine closing techniques of product sales are used to clinch service deals, due to intangibility, the buyers tend to feel coerced or cheated and, as a result, might challenge the price or have second thoughts. In some cases, the deal can unravel altogether.

The intangibility of services exaggerates the effect of a phenomenon called 'post-purchase distress'. People experience this if the purchase of any item, product or service, is emotionally challenging or expensive. Yet, anxiety caused by a large or important product purchase can be allayed by the buyer admiring it or showing it to others. However, as services are intangible, there is nothing to offset this anxiety and nothing to show off. Business buyers may be concerned about the effect on their budgets, the effort to justify the item to others, damage to their credibility or risk to their political capital. If this anxiety is not managed, then problems occur. It can cause them to re-examine the deal and query value. Intangibility can cause difficulty in the time between a contract being signed and a service being delivered. So, suppliers need to develop techniques by which they mitigate post-purchase distress. Many put emphasis on good after-care and others produce printed examples of previous contented buyers, which are given to new buyers soon after purchase. Both provide valuable reassurance.

Finally, as it is often not possible to patent intangible services, suppliers have to find other mechanisms to protect their investment. Failure to do so will mean that services commoditise quickly and markets will be dominated by price wars. Successful service companies therefore create powerful brands, like Virgin or Wal-Mart, which cannot be copied. Companies that are not skilful at brand creation and management, like some in the IT industry, have damaged their business by moving into service with very little means of maintaining differentiation.

Yet, as with a number of the differences between products and services that are proposed by academics, there are 'shades of grey' which complicate life for service marketers. In a number of services, for instance, there are physical components which mitigate the effects of intangibility. It might be that the service itself exists to support a physical product (as in the case of a maintenance service) or has important physical components (like the seat, cabin and meal in an airline service). They are tangible manifestations of the intangible service and can be used in marketing to help underline its benefits and overcome buyers' fears or hesitations. As a result, airlines will advertise the seat or the meal and consultancies their people.

In fact, despite academic assertions that intangibility fundamentally changes marketing, there can be little difference in practice between the marketing of product-based services and the marketing of some products. Many products are a mix of physical, conceptual and service components. A washing machine, for example, is bought because of its functionality, the physical components, the design, the brand and the service support package. It is also bought through the service system of a retailer. So, the white goods industry has learnt to offer extended warranty as part of product purchase, maintaining total return from the product line through service margins whilst actual product prices have fallen. Although the warranty service is intangible, it is presented as any other retail concept tied to a major product purchase.

Moreover, the marketing industry already has great experience in communicating conceptual, intangible offers. Most brand managers in successful consumer goods companies, like Unilever or Procter & Gamble, would argue that they are in the business of making intangible offers; that a branded product is a concept offering valuable intangible benefits to buyers. A product such as Heinz baked beans is, for example, a combination of: haricot beans, sauce, packaging, easy distribution, image and consistency of familiar taste. It is tangible in the sense that it contains physical components that customers can eat but enormous value is created for Heinz by careful management of the intangible, brand concept. Over years the company has invested time, money and skill in building the brand in the minds of a group of customers. As a result, there is a heritage and equity that causes customers to buy repeatedly at higher prices.

A wide range of brands (from clothing like Hugo Boss and Nike, through perfume like Chanel and diamonds like Harry Winston, to magazines like the *Economist* or *Vogue*) appeal to the aspirations and emotions of different groups of people. Customers associate their lifestyle with these offers and are seeking intangible, unarticulated benefits (like 'belonging' or 'recognition') when they buy. The battle for profit is fought in customers' imagination and the offers are, therefore, in many ways, intangible. So, in order to tackle intangibility, service marketing specialists should imitate many of the successful techniques of branded goods, such as: excellent brand management, positioning, packaging and progressive investment in fame. In fact, one of the prime differences between the marketing of intangible services and intangible-dominated product brands is probably the relative organisational incompetence of many service industries in marketing processes or campaigns. They have simply not been mature enough to have gone through the progressive organisational learning which puts such sophisticated and valuable offers in the market; but the best are catching up.

Variability (or Heterogeneity)

Once a product has been designed and the manufacturing process set up, it will be produced time and again by a factory with little variation. Customers know what they are buying and it is consistently delivered. Any aberrations in product quality tend to be few and easily (or relatively easily) driven out by quality control and improvement processes.

Services, though, are rarely as consistently produced as manufactured products. Even those service businesses that try to 'industrialise' their offer through efficient and robust processes find it hard to deliver reliable, consistent service every time. For instance, one of the essential components of many services is the people who serve customers. But people tend to think for themselves, adjusting behaviours and outcomes to suit unique circumstances. Moreover, in certain cases, customers are involved in the process of service production. As people take initiative and change or customise the service, it is unusual for one service experience to be identical to another.

This has implications for the way services are taken to market. For instance, people need to be given clear guidelines on how much they can vary the service. Some businesses want little variation and give their employees little discretion. Others, 'mass service shops' like Burger King, for example, see competitive advantage in allowing a degree of customisation but are aware that too much will damage the economics of their business. So, marketers need to design processes which streamline as much as possible but allow people to vary delivery and anticipate failures.

At the same time, the firm's marketing team need to set expectations and communicate the degree of consistency that customers might expect. Their main tool in this is their brand because it sets expectations of a standard of quality. Consumer brands like easyJet and business brands like Accenture set different expectations and the brand positioning, attracts different buyers. A cut-price airline will not be expected to provide high-quality food or good landing access to the heart of key cities, whereas the service experience from Accenture's consultants must be as excellent as the business outcomes they offer.

One particularly powerful brand strategy, which sets expectations to suit service variability, is to introduce an element of humanity or responsiveness, which acknowledges that errors might occur. Many high-volume consumer services emphasise in their advertising that, if errors occur, buyers should complain, so they can be quickly put right. The Virgin organisation, for example, grew through a commitment to contemporary service, which was prepared to be *avant garde* while asserting that any human errors would be remedied well. Their humour and humanity, embodied by their brand icon Sir Richard Branson, ensured that buyers were unconcerned by minor quality errors.

Branded volume services use advertising and signage in their premises to signal their intent to provide as consistent and as reliable a service as possible; but also to redress any errors. Some show examples of the processes or systems they use to produce the service. Others use practical documents, like a railway timetable, to set expectations of delivery. As important, though, is the appearance of the service and any physical elements in it. The design and cleanliness of trains and the uniform of employees, for instance, must be carefully managed. A rundown, scruffy service creates entirely different expectations to one which looks organised and professional. A volume service which is clean, reliable and well packaged will be more readily forgiven for unusual and minor inconsistencies.

At the other end of the scale, the prime mechanism by which professional services, like consultancy, try to achieve consistency is through the recruitment and training of people who will deliver a certain quality and style of service. The delivery of the service might be highly individual and customised to a client's unique situation, but the style of service and manner of delivery must be consistent. Young professionals are schooled in both the technical skills of the firm and its approach to business. Many practices have rigorous control procedures and detailed ethical guidelines that attempt to deliver consistent standards. Yet the heart of their business is the recruitment, development and motivation of high-quality 'human capital' which, without close supervision, will deliver the style of service that reflects the firm's position in the market. As a result, the main device used by leading firms to obviate heterogeneity of service is their investment in sophisticated and consistent recruitment marketing aimed at graduates in leading universities.

Other, attempts to handle variability include the nature of the contract and guarantees about quality standards, offering financial compensation if errors occur. One of the world's leading business gurus, David Maister, for example, offers a clear money-back guarantee to clients. Excellent, responsive post-purchase customer care is also seen as essential to the effective marketing of variable services and forces service marketers to focus on quality of service, service experience and service recovery.

Simultaneous Consumption

A product can be manufactured and stored or passed through a distribution system until it is bought. Once bought, it can then be stored by the customer and used at a later date. By far the majority of services, however, have to be used as they are created. They cannot be stored by either the supplier or buyer. As Donald Cowell said as long ago as 1984:

> Goods are produced, sold and consumed whereas services are sold and then produced and consumed. (Cowell, 1984)

In order to deliver, resources must be prepared and deployed ready for erratic demand. A maintenance service, for instance, must have computer systems able to receive fault reports, trained technicians with appropriate tools able to tackle technical difficulties, and carefully calculated caches of spare parts. Moreover, much of this investment is unseen by the customer before, during or after the service experience. This means that service companies need to develop techniques that communicate the value of this stored investment.

Simultaneous consumption also means that service marketers must focus on demand forecasting and demand management so that resources are available when needed by buyers. They frequently use statistical modelling techniques to predict likely demand based on previous purchase patterns. Some, though, work with key buyers to share the risk of demand fluctuations while others try to influence demand by different pricing patterns. The airline industry, for example, uses different ticket types to get commitment to purchase weeks in advance. In many cases they over-book flights, using incentives to get some passengers to change their flight if necessary. Some service marketers also seek to communicate the extent of their stored infrastructure as an asset at their customer's disposal and others use that asset to underscore pricing and value messages. Many use downtime to create other value for their customers. Maintenance services, for example, will fill troughs in demand with routine or preventative maintenance.

As the service is produced while the buyer uses it, the customer is effectively 'in the factory', able to see the production process and unable to see the finished result. So the production process must be well prepared and tested, able to host customers and engage them professionally without causing anxiety. Simultaneous consumption also contributes to the need for service marketers to manage expectations. They must explain the key steps in the service process and any tasks that the customer must undertake. In fact, one mistake made by many service companies, particularly in technology sectors, is their failure to sufficiently plan or design the process through which their customers will move.

This has implications for quality management because recovery must happen in real time. Quality processes cannot be the same as manufacturers' post-production sampling and correction because any error will be immediately experienced by the customer. Any marketing messages will be undermined if the actual experience is poor or the firm is slow to remedy service errors. So simultaneous consumption concentrates service marketers' thoughts on ensuring a quality experience, it reinforces the need to understand and exploit quality of service issues in their work.

Inseparability from human behaviour

Credible academic researchers (see, for instance, Zeithaml and Bitner, 2003) suggest that, in the customer's mind, most services cannot be separated from the person they encounter when they buy and use them. So people are not only important to the design and management of a service, they are a critical part of the service itself. Their motivation, behaviour and appearance are part of the benefit package offered to customers. In many ways they are an essential element in the value that customers seek, part of the bought service, and their behaviour affects customers' perceptions of price and quality. They are so intimately involved in delivering the service experience that their body language and appearance will communicate messages to customers about the service as much as their words or the firm's marketing claims.

So, at the front line of interaction with customers, a firm's employees must embody its intentions and, as a result, the appropriate treatment of employees is very important to service businesses. If employees of a production company are treated as mechanised 'units of production', their boredom or dissatisfaction will not be passed on to the buyers. But in a service business it will. If their leaders treat them badly then it will be communicated, even if they try to be professional and disguise their unease. Conversely, if they are treated like human beings there is a strong chance that the customers will be too.

Managing employee behaviour in line with changing customer expectations is a major focus and challenge for service businesses. Good service managers understand that front-line people can cause customers to turn to or away from a service. Many put real effort, investment and resources into programmes designed to improve the impressions caused by their people. This is one of the basic tenets of the Virgin group of companies, for instance. In fact, it is probably no exaggeration to say that service companies invest in this 'intangible asset' the way manufacturers invest in tangible assets.

So, it is not surprising that, with the rise of the service economy, marketers have created internal marketing functions. Used in some of the leading service businesses with increasing sophistication, these aim to communicate with the internal audience as professionally as the firm communicates with its external audiences. They set up internal communication media, which can range from simple management briefings to sophisticated TV channels or intranet broadcasts. Many create internal communication campaigns linked to external campaigns. This involves all the usual best practices of marketing communication, including: a specific communications strategy, the creation of clear messages, the segmentation of audiences and the management of responses from them; all directed at the company's own employees.

On the other hand, an increasing number of services are based on self-service technology, with no people involved in day-to-day delivery. From web services to airline check-in, the range of self-administered services is steadily growing. Clearly, in this rapidly increasing category of services, the 'people' element of the service marketing mix is less prevalent. For these, the personality of the company brand is used to replace the people who normally deliver service. Service marketers must invest heavily in branding this type of service at all points of visibility. The emotional reassurance of major, high-quality brands, like IBM, Ericsson or Virgin, implies a reliable response should self-service technology fail.

However, there is evidence that self-service technology only penetrates a market if the first buyers are shown how to use it by personable people or their proxy. They need help with the initial socialisation process, and in overcoming any technology fear, from specifically designed marketing initiatives. At the time of writing, for instance, most airlines are introducing self-service check-in technology that is very simple to use. They generally place the machines near the 'hand luggage only' queue. Employees are trained to approach people in the queue, often frequent flyers, and show them how to use the fast, easy machines. Although many refuse the original offer to use the technology, they are happy to comply once a demonstration overcomes their technology fear and potential embarrassment. Once sufficient number of people have leant the process, they communicate it to others and the technology penetrates the market. As leading suppliers in the IT industry gear up their different "cloud computing" offers this method of overcoming the technology adoptions fear of both business users and consumers is likely to be an important determinant of which suppliers really succeed.

Perishability

Generally, services cannot be stored or saved for other occasions. Once an empty airline seat has flown over the Atlantic, the opportunity is passed; the moment cannot be recaptured. So, services are time-bound and the management of time is an important task of a service business. In fact, one researcher (see Ruskin Brown, 2005) suggests that there are several 'flavours of time' that service businesses need to manage: punctuality, duration, availability, speed of response and speed of innovation. Of these, punctuality or reliability, in the sense of delivering the service when promised, has been shown in many studies to be one of the most important influences on the way customers judge quality of a service.

Perishable services must be marketed in the same way as perishable products like fresh food or flowers. Quality difficulties must be dealt with through an immediate, responsive recovery process and inventory must be carefully managed through forecasting and capacity management. This characteristic reinforces the need for good demand forecasting in a service business. It probably has to be more accurate than in a manufacturing company, which can store excess items in a warehouse until they are used. Maximum capacity must be carefully calculated, balancing the highest anticipated demand against the cost of investment. Marketers have to cope with variations in demand and develop techniques to sell spare capacity, such as empty hotel rooms for off-season or weekend capacity. Emphasis should also be put on easy and efficient distribution and the value of a new, premium service.

However, the real implication of this difference is that the supplier must become expert in differential pricing. They need clear communication methods to potential buyers, which demonstrate why the immediate, 'fresh', service needs to cost a different amount to those accessing it at a later time. They need to produce special price offers, reductions to use up spare capacity and understand marginal pricing. Moreover, they need to communicate the rationale and, particularly, the fairness of these pricing practices to their customers. This is particularly important for companies with repeat customers who become increasingly sophisticated, holding off purchase to get better offers.

Ownership

When people buy a product, ownership of it passes to them. They can store it, use it or give it away. This simply doesn't happen with a service though. The people who are part of a training or consultancy service cannot be owned by the buyer. Nor are airline or train seats bought; they are rented for a moment in time as part of the service. Payment is for use, access or hire of items. People will talk as if there is a sense of ownership (my flight, my broadband service or our training) yet there is not the same sense of possession as with a product purchase.

This lack of ownership will cause buyers to question value and price. They will look for greater added value or ask for price reductions. As a result, the service marketer must dramatise the moment of contract and find mechanisms to emphasise value as the service progresses or they will struggle continually with price pressure. This is one area which many technology firms find hard to manage. In a variety of businesses, numerous valuable services have been allowed to evolve into commodities because the suppliers have been incompetent at communicating value. It will be a particular issue for those businesses moving into cloud computing. The use of the service rather than the ownership of IT capability will have to be handled very carefully if they are to create a service which customers value.

Process

One of the major differences between the purchase of a product and the experience of a service is the process through which the buyer moves. Products are entities, which are bought through a process but are independent of it. Customers can use it, break it, give it away and ignore the instructions on how to use it. Services, however, have a process inherent in their design through which the customer must pass. So when people use a service they must submit themselves to the service provider's process.

One of the world's leading researchers in this field, Finland's Christian Grönroos, has put great emphasis on the importance of the service process as a result of his many years of study. He says:

> Services are processes consisting of activities or a series of activities rather than things. In order to understand service management and the marketing of services it is critical that one realises that the consumption of a service is process consumption rather than outcome consumption. The consumer perceives the service process (or production process) as part of the service consumption, not simply the outcome of that process, as in traditional marketing of physical goods ... the consumption process leads to an outcome for the customer, which is the result of the service process. Thus, the consumption of the service process is a critical part of the service experience. (Grönroos, 2003)

So, service marketers must plan the service process in detail and educate their customers in the parts of the service they will experience. Potential customers must know how to access the supplier's service system, which has to be designed to encourage use. The firm needs to use clear signage and communicate its meaning to all potential buyers. Educating them to access and use the supplier's process is, therefore, an important aspect of service marketing. This ranges from the deployment of branded consumer signage, like McDonald's, to the use of customised websites for client extranets in business-to-business services.

Once 'in the premises', though, the buyer needs direction (even in a virtual environment). If it is not immediately apparent how to use the system (the equivalent of standing help-lessly in a foreign shop), the buyer becomes embarrassed and gives up. So, clear signage into the delivery process and a step-by-step guide through it are equally important. On land-ing, for example, many flights show a video guide through the airport. This ensures that the stress levels of new passengers stay low and any difficulties are seen as aberrations, which can be corrected by excellent service recovery. Finally, marketing communication must concentrate on the 'outcomes' of the process rather than the benefits emphasised by product marketers.

Control

When a customer buys a product, they normally have complete control as to when and how to use it. They might keep it, use it immediately or give it to someone else. When they buy services, though, they do not have such freedom. They must surrender themselves to the service delivery process, which the supplier has designed for them. In doing this they cede control of themselves to the service provider and human beings detest being out of control. The purchase and use of a service therefore invades their personal space and raises emotional issues that service suppliers must learn to tackle (see Bateson, 1985; Bateson and Hoffman, 1999).

This lack of control means that service marketers must take steps to allay the anxiety of new users. They can design, for example, a simple, clear process and choose competent employees to perform the service. Both will help to alleviate anxiety. In the early decades of computing, for example, IBM inspired confidence through two important methods: leasing and quality of service. Customers knew that, if there were a problem with a purchased item, it would be resolved without any question or difficulty. This gave IBM significant competitive advantage for a number of years, turning it into global market leader. Many of the well-known customer service and after-care techniques must be used to allay fears and criticisms arising from this underlying emotion. Yet, although the character, reputation and professionalism of service employees can inspire confidence in the quality of the service to be received, the corporate brand also allays stress.

Unfortunately, customers who use a service repeatedly assert an unconscious need to regain control. They try to cut corners, look to serve themselves and get irritated if the process is inflexible. In fact, they can become annoyed with the simple, clear steps that they first found so enticing and necessary. So, service for the experienced user is very different to that for the new customer. It must be inclusive and respectful; a streamlined club, similar to frequent-flyer programmes, which allows as much self-service as possible. This has profound implications. In fact, in some markets, the arrival of a supplier offering, for the first time, a service designed for the experienced user has had major strategic impact.

For experienced buyers, service marketers must ensure that mechanisms are designed whereby, as the customer becomes familiar with the service process, he or she can do more themselves or cut out steps. The design of progressive self-service into the service system will reduce costs because some of the effort of performing the service will come from the customer. Yet, it will also improve perceived quality because the service performance is within the control of the customer and thus closer to perceptions of timely delivery. So, it is possible for service marketers to provide repeat customers with better service at lower cost.

The growing use of technology as a means of delivering services, and especially self-service, is visible across most industries. For example, in rail and air travel, customers can increasingly take control of their purchase and key interactions in their journeys through self-service technologies like those detailed in the case study on Portuguese rail company, Carminhos de Ferros Portuguese.

CAMINHOS DE FERRO PORTUGUESES STREAMLINES SERVICE

CP was set up as a state owned enterprise in 1975 to manage the railways in Portugal. Changes in domestic and community legislation led to the transport services and the infrastructure management being separated in 1997. While CP continues to run the transport operations, it underwent a profound change in order to adapt itself to the market, by organising itself into business units capable of satisfying the needs of different types of customer – Freight Transport (CP Carga), Urban services (CP Lisboa and CP Porto), Long Distance services (CP Longo Curso and CP Regional) and High Speed services.

Since restructuring, CP has invested in adding value to its services and strengthening its position in the transport sector. It has significantly improved the quality of services, invested heavily in rolling stock, started the process of working with other transport operators, and become a much more customer-oriented company.

In its drive to become more customer oriented, CP wanted to implement a range of new IT systems that would streamline and enhance its interaction with customers.

Using IT in the service experience

CP worked closely with its IT partner, Fujitsu, to use IT more effectively to enhance both sales and the service experience for its customers. First, CP developed an on-board selling system for use by its Long Distance and Regional units. Then, it developed other ticketing projects, including an on-board selling system for CP Porto, sales kiosks for CP Long Distance and CP Regional, and the reformulation of the basic system for long distance ticket selling and seat reservations. The company also created the systems needed to enable travel agents to access CP's travel information, so that they can now sell tickets directly by accessing the necessary information using a web-based system.

Eng° Jose Gaspar, IT Director at CP explains, *"All our ticketing systems were reformulated and new sales channels opened. In addition to traditional ticket offices and automatic vending machine sales, new selling systems were established in all bank ATMs in Portugal – I think this is unique at an international level – plus we've created internet sales and web service interfaces for travel agencies and have adapted some of these channels to provide other services, like car rentals, for our customers."*

Another notable success was the system created to manage the growth in demand generated by the 2004 European Football Championship. CP was under mounting pressure from many football teams to help them arrange for a group of supporters to travel to Portugal, and then to get to the venues that their team were playing at. Jose Gaspar continues, *"In record time, we created a dedicated web portal for selling train tickets that allowed football fans to make all of their travel arrangements for the Euro 2004 event."*

CP has also launched MyCP, a customisable browsing tool for the CP website that constantly picks up customers' preferences and habits so that they can get the information they need more quickly. Customers have to register with MyCP to gain access to information, but once registered, can make on-line purchases, subscribe to CP's news service, and receive the company's newsletter.

MyCP, apart from giving clear advantages for customers, is also important for CP. By getting an accurate view of what people are looking for, it can adapt its services accordingly, which will have added benefits for everyone in the future.

Preparing for a privatised future

In 2006, CP carried 133 million passengers and 9.75 million tonnes of freight over the 2,830 kilometres of the country's rail network, with the Urban Units alone catering for 86.4% of all passengers. However, EU guidelines mean that all of Europe's railways must be privatised by 2015. So, CP has had once again to develop new strategies, structures and systems to face the onslaught of new competition.

Jose Gaspar says. *"Our aim is to deliver more efficient rail services, simpler fares and ticketing, a more integrated transport system, greater uptake of public transport and significantly more satisfied customers. Improving the quality of the customer journey is central to this effort and so we've started to implement ticketing systems based on contactless technology."*

The first project of this kind was undertaken by CP Lisbon, covering three components:

- Sales booths – equipped with PCs that are connected to a central reader, so that travel cards can be validated and travel information loaded;
- Onboard sales – portable equipment enables sales to be made on-board and also validates people's tickets;
- Central systems – to gather all of the relevant sales and commercial information and provide it to the central transport management organisation OPLIS.

Once completed, the new system will make CP's tickets compatible with other city operators like the bus, subway, private rail and ferry operators. As a result, customers will be able to simply buy a contactless ticket once and then reuse it again and again by loading it with the required travel credits.

Jose Caspar continues, *"Contactless technology is very convenient for passengers and creates better efficiency and control for CP, together with reductions in operational costs. It is also opening up new business opportunities. For example, in the future contactless cards or tickets could be used for hotel reservations and car rentals, there are all sorts of possibilities – so long as these projects are implemented with maximum quality and security levels."*

Delivering benefits to customers

The ticketing solutions are delivering a number of key benefits to customers. Customer service has improved as CP can now sell and validate tickets in a wide variety of locations, including on board trains, making it much simpler for customers to travel. The possible inclusion of value-adding services, such as car parking and access to leisure facilities, in the contactless ticket will further enhance the overall customer experience. In addition, the close integration of systems and use of mobile devices cuts out unnecessary manual processes, delays and administration for customers. It also reduces operating costs significantly for CP.

In fact, in the Lisbon and Porto Urban areas more than 80% of tickets are now bought by customers using automatic vending machines.

Once the ticketing system is standardised across Portugal's rail network, CP will have much more accurate information on daily rail usage, which will enable it to prioritise its investment in those stations carrying the most passengers and also reduce service on those areas where they are not required.

"Ongoing automation allows employees to be allocated to the sale of other value added services. To CP it allows other services to be available and to our employees it creates a more diversified, less monotonous and more motivating workplace. To our customers it means that they have more alternatives, greater convenience and an increase in the availability of tickets" Gaspar concludes.

Extracted from a published case study with kind permission from Fujitsu.

Environment

For many services the environment in which they are performed is an important part of the experience for customers. The design and ambience of a restaurant and the layout and style of an airline cabin set expectations of quality and value. Retailers, hoteliers and managers of holiday resorts are just a few of the service businesses who have to think carefully about the physical setting of their service. It affects the behaviour of employees, sets the expectations of buyers and can be a source of differentiation.

There is a wealth of research on the effect of the physical setting on the health of a service business. A complex and sophisticated process is involved in creating a new one because the supplier must take into account and balance a range of factors, including: 'sight appeals', 'size perception', shape, colour, sound, scent, spatial layout, flexibility, brand and signs or symbols. It must also calculate operational factors such as capacity, crowding and queuing. In fact, some businesses use complex statistical techniques like queuing theory (that are used to design the capacity of sophisticated communications and computer equipment) on the flow of people through their premises. Yet in international or ethnic markets, they also have to allow for subtle cultural preferences which influence a range of ambient and behavioural factors.

For those services where the physical setting is important, these complex considerations are unavoidable because they affect buying behaviour. Credible experiments have demonstrated the effect of these different factors on sales. People buy more, for instance, if different music is played, if different colours are used and even if different smells are deployed.

Some services, though, are experienced in more subtle and flexible environments: the virtual vagaries of the Internet or the customers' imaginations. Yet, just as in a real, physical environment, impressions need to be created which set expectations of value. This needs careful design and sophisticated marketing if it is going to influence sales and price perceptions.

After initial design and launch, the 'service-scape', the environment visible to customers, must be kept fresh and inviting by marketers. It must continue to reflect the ambitions of the service provider. Marketers must design sales promotion campaigns and point-of-sale materials to stimulate sales, to increase the margin of purchases and to maintain excitement in the service. They must routinely audit the facility, ensuring that it maintains its enticing nature to customers.

Performance

Services tend to occur at a moment in time and involve the attention of one person on another. Many are 'performed', not produced. From representation in negotiation, to service at the restaurant table, people perform a service for others and the style of performance affects the customers' views of both quality and value. Moreover, the energy, art and style of the performance influence the price that can be charged and the degree of customer satisfaction. As a result, the service industry puts great emphasis on the 'service encounter'; that 'moment of truth' when the supplier's employees interact with the buyers. It is a moment when the firm has a chance to impress and deepen a relationship with a buyer, in addition to delivering the promised benefit. It is this area, probably, more than any of the others, that creates exhilarating, enticing services. A service can be researched, blue-printed and engineered but those which go the extra step of creating a performance tend to entrance their customers and earn higher margins.

Marketing campaigns can be used to keep the atmosphere, motivation and experience of the service encounter fresh and exhilarating. For example, internal marketing campaigns are frequently designed in businesses with large, diversified workforces to create internal competition and achieve really good service. In consultancy firms, by contrast, leaders frequently initiate new concepts, strategies or ideas to stimulate the interest of restless brains, and these often involve marketing skills.

MARKETING IN TECHNOLOGY SECTORS

In the late 1950s a man with the unlikely name of Buck Rodgers joined the mighty IBM as a sales trainee. During his career the company moved heavily into computing, became global market leader and (during Buck's tenure as marketing head) grew from 'a $10billion to a $50billion company' (Rodgers, 1986). As Buck retired, the PC revolution was starting and the monopolistic hold of proprietary equipment was being undermined. He claims to have initiated IBM's first significant consumer advertising and pioneered the concept of business technology 'solutions'. Technology firms have used a range of marketing techniques since. 'Intel inside' is still, for instance, one of the sector's most effective brand programmes (brave and risky when first muted) and 'no one ever got fired for buying IBM' remains one of the clearest statements of emotional promise in business-to-business marketing. Dell pioneered mass-customisation while Apple's difficult journey to become the world's most famous innovation shop means that it is routinely quoted as a creative case study.

Amongst the large public networks, gas boards had their showrooms, electricity companies sold appliances and British Rail companies flaunted the posters from their 'golden age'. But it was the privatisations of the 1980s that introduced competition and really began their venture into marketing. Most were large engineering companies with vast workforces, an obsession with detailed processes and huge budgets. They applied that culture to marketing and became big-budget advertisers. BA became 'the world's favourite airline' and BT positioned itself in human communications with Maureen Lipman and Bob Hoskins.

Above all, technologists have deftly made up, sold and exploited ideas by using a remarkably powerful technique called "thought leadership". Their marketing of concepts like process re-engineering, data warehousing, the millennium bug, CRM and customer experience management has sold numerous computers and other gadgets. Their two major achievements have probably been to convince humanity that everyone needs a PC and to liberate human imagination in the Internet. For Buck and his peers, though, "marketing" was primarily about new technology and sales; and that ingrained attitude has been the reason that marketing's incursion into the technology sector has been so erratic.

Technologists adore new breakthroughs and routinely claim to be changing the world. Their approach is to invest huge sums in technological advance and to bring it to market through worldwide distribution chains founded on the assumption that fast-changing customers only want cheapness; a business model which is as daft and stubbornly unchanging as that of the Detroit car manufacturers. As a result, apart from consumer electronics, much technology marketing remains depressingly tactical and erratic. There is often little sense of organisational learning, brand equity or any substantial investment in customer knowledge.

In beginning to take service seriously, though, this sector has learnt the hard way that their customers' experiences determine success. Nearly all have had disasters in customer service (particularly those that were privatised) and were forced to make massive investment to catch up with expectations. Even now, most are not valued by too many of the people who use their services. Consumers who flock to new shopping centres or are delighted to pay inflated prices for luxury goods will grumble about water bills, broadband costs, rail tickets and flight charges. Whereas business leaders who pay high margin for McKinsey's advice and the audits of the big four accountancy firms will pick at the detail of outsourcing or maintenance contracts.

By and large, the vast businesses of the technology industries demonstrate that it is possible to have all the functions of marketing, backed by large budgets and supportive leadership but still be unable to create genuine perceived value. As they move towards more sophisticated marketing and need to create value propositions out of service skills, their adoption of the sort of systematic, progressive marketing competence seen in other sectors must increase. In short, they need to become market-led and increase their organisational competence in marketing, leaving behind tactical, erratic, 'promotions'.

SUMMARY

In the past few decades, service marketing has become a more precisely understood activity as the developed economies have become more service-dominant. There have been high-quality academic research projects and good cooperation between industry and academia to understand the differences between product and services marketing. Technology companies offering a service need to understand these differences and build them into their marketing organisation's approach.

3

Gaining Strategic Insight into Service Markets

INTRODUCTION

Fundamental to the success of any business are the conditions of the market in which it operates and the position it holds within it. Profit, cash flow and shareholder earnings can be damaged if the leadership takes wrong decisions due to a lack of understanding of market dynamics. While fast decisions based on "gut feel" can be successful in markets where business leaders have spent the majority of their career, they can be extremely risky in unfamiliar markets; and service markets can have singular characteristics which need to be understood if any marketing is to be successful. Moreover, as much modern marketing practice was developed in consumer goods markets and many service businesses in technology sectors are part of larger product companies, a number of the concepts, techniques and processes by which marketing works in these companies are likely to be inappropriate. So, this chapter outlines the main issues (like market definition, market maturity and positioning) to look for and a number of tools and techniques that can help firms gain an objective perspective of service markets (such as a market audit or scenario planning). It also looks at methods to turn that knowledge into a basis for making decisions about exploiting opportunities revealed by the analysis.

WHY MARKET UNDERSTANDING IS ESSENTIAL

Whether a firm is a large, sophisticated and international organisation or a small specialised unit, whether it is highly rational and procedural in its approach to business, or largely intuitive, its leaders need to understand the market it operates in if they are to safeguard the future health of their business. There is a vulnerability here that arises from what is generally the greatest strength of service firms: closeness to the customer.

Front-line service personnel are in close contact with their customers, responding to them every day. This experience dominates their job, influencing decisions and policy formulation. So it is natural for them to assume that they know the market in which they operate. However, one group of buyers (or one large customer) does not necessarily reflect the trends and forces at work across a market as a whole. It can create a dangerous bias in the minds of executives.

Also, if market conditions are changing, existing assumptions about what works are dangerous. Poor decision-making or erratic leadership, as people change priorities in the light of the latest encounter with a customer, can result from a myopic perspective based on direct experience of only a limited number of buyers. So it is sensible for the leaders step back and take stock of their market using as objective and analytical an approach as possible; and it is the marketer's job to make that happen.

It is particularly important to get an analytical perspective on any new market that the company intends to move into and to use a recognised process which collects a range of data in order to provide as clear a perspective as possible. There are straightforward methods to collect relevant information and garner market-based insights, which need not take long to complete. Some take an economic perspective of markets and some a behavioural approach; each yields valuable insights which marketers can exploit and use to reduce risks.

IMPORTANT MARKET DYNAMICS TO UNDERSTAND

Market Definition

The first and most fundamental issue that marketers need to think about is how they define the market they operate in. This may sound simple, and it is, but there are numerous examples of businesses being destroyed because their owners had not defined their market correctly. It is important because there is an assumption built into the fabric of a business that gives direction to its activities. If this assumption is not aligned to the market, the business will, ultimately, fail.

Marketing and sales people can be so overwhelmed by the demands of their day-to-day job that they have very little time to investigate other sectors or think about the relevance of different ideas to their market. As one of the classic marketing thinkers, Wroe Alderson, pointed out (Alderson, 1957), this results in group-think and similar behaviour within markets, some of which is decidedly odd. Different beliefs grow up within different industry sectors which become the basis of business policy, such as a fascination with technology rather than what it can deliver. Some of these are profoundly idiotic and out of touch with the realities of the world. As a result, marketers, competitors and buyers collude to conform to established ideas within markets, which can be restrictive.

A consumer might, for example, shop one morning for a branded, luxury good, perhaps as an important gift. In that market, long-established brand names like Gucci or Chanel will be sought out. The shopper will expect their purchase to look and feel expensive; reflecting partly its durability and heritage. Some will even spend large sums on second-hand branded pens or watches that are decades old. Yet in the same shopping trip they might also buy an over-the-counter drug (where markets are slow to change and investment is on a seven-year lifecycle) and a personal computer (where technologists believe the customers are fast changing and only want cheapness). The precepts of each of these markets contradict, but are genuinely believed by the suppliers and, as a result, accepted by shoppers while they engage in that market.

Zaltman (2003) has called this set of beliefs 'the mind of the market' and it inhibits business opportunities. Occasionally, though, new propositions enter markets that shake up these odd, ingrained attitudes. Richard Branson, for example, showed with his Virgin Atlantic service that it was possible to change the competitive landscape of the airline market by thinking differently and contradicting established group-think. In their turn, propositions like First Direct (banking) and Direct Line (insurance) have shown the wealth that marketers can make by challenging the status quo. So, technology marketers should take a moment to step out of the established beliefs in their market to see if new approaches can shake it up a little, open up different opportunities and create substantial wealth for their shareholders.

Many do not, and spend their time projecting trends in sales from recent history, jumping to bash out e-marketing campaigns in the current quarter and combing the Internet for reports from respected industry analysts which will confirm their preconceptions. As Henry Mintzberg has said (Mintzberg *et al.* 2005), they are like primitive tribes throwing bones or consulting oracles on which way to go hunting. Instinct, history and luck means that they are sometimes right, but they frequently inhibit the potential earnings for their business owners by not stepping back and thinking. The exasperated business professor, Theodore Levitt (Levitt, 1960), called this phenomenon 'marketing myopia' and used the history of American railway companies (the computer companies of their age and one of the very first technology and network-based services) to dramatise the damage that such thoughtlessness does.

When they first appeared, they were handling a stunning new technology that would revolutionise life in the 19th century as much as (perhaps even more than) computer power did in the 20th century. By the end of the 19th century, there were hundreds of thousands of miles of track in the USA alone, some crossing the entire country. The American railway companies then earned huge profits. So anyone approaching the chief executive of one of these companies in, say, 1910, and predicting that there were major new threats to the business that could see them virtually bankrupt by the 1930s/1940s, would be dismissed out of hand. And yet, thanks to the development of the car and the airplane, that is exactly what happened. According to Levitt's analysis, the American railway companies struggled because they defined their businesses as 'trains' rather than 'transportation'. Had they focused their businesses on the 'transportation market' they would have invested in these new technologies (perhaps backing a young Henry Ford when he needed funds) and moved their businesses in exciting, different directions.

So marketers must define for their company the market on which they will focus and this must be done in clear, customer-centric terms. A medical equipment service, for example, is in the business of ensuring that healthcare remains available, not 'repairing kit'. Experience shows that this can be enormously difficult to clarify but, once agreed, gives direction to innovation, leadership, investment and service quality. It can also give a newcomer real advantage if it is entering an established market, which is dominated by complacent suppliers who all have a common view of the market need. The new entrant can quickly gain share by defining its offer more closely to customer needs. So, market definition, or re-definition, is a route to profound market insight.

Insights from the Phases in the Growth of a Market

The phenomenon of market maturity (see the Tools and Techniques appendix) reflects changing patterns in demand and supply; and occurs when there are multiple suppliers and multiple buyers. It is normally represented by the diagram shown in Figure 3.1, an "S-curve". This is sometimes confused with the 'product life cycle', which argues that the sales volume of individual products or services normally follows a similar pattern. Yet, although that idea has a strong hold on management thinking and executives can often be heard to talk about their product as, for instance, 'mature', it remains unproven and controversial for individual products or services (see the Tools and Techniques appendix). The maturity of markets, by contrast, is based on a reasonably well-established sociological phenomenon called 'the diffusion of innovation' (see Tools and Techniques appendix). It tracks the course of new ideas and concepts, like water purification or environmental concern, as they grow in influence through a society. It is often used by public bodies to help with significant social initiatives like changes in health practice or agricultural techniques. It is directly relevant to marketers because the products and services their companies offer are themselves technological innovations which are diffusing across different societies in the world. Taking a moment to understand the evolution of the macro market in which they operate should help with the development of marketing strategy and campaigns.

This is a learning process between suppliers and buyers which develops over time. When a completely new idea or offer is first made (a true innovation), buyers need to be shown how

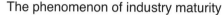

The phenomenon of industry maturity

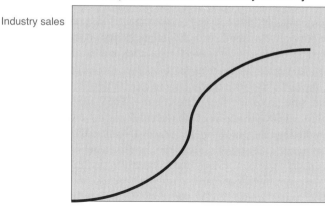

Industry sales

Time

Figure 3.1   A representation of industry maturity

to use it and how it is relevant to their lives (broadband services at the start of this century for instance). At that point, suppliers need to educate the market in the concept and grow their business by inducing customers to try it. In this phase of a market's life, costs are likely to be high and firms can be unprofitable. As a result, it may be wise for any businesses wanting to enter new service markets to wait for upstarts to burn investment and establish the concept. They can then enter by buying them out. Companies who have successfully marketed true innovations, like Apple and its iPhone, have used a range of concepts to induce trial and spread word-of-mouth like a contagion. They rely heavily, for instance, on PR to create demand and mavens (see Tools and Techniques appendix), like Apple's 'super users' to excite communities, as explored in Chapter 9.

As time goes by, customers become familiar with the concept and more buy it. In these early stages they frequently adapt it for different or unforeseen uses. When, for instance, the mobile phone markets began to expand in Western societies, suppliers tried to open new market niches, like teenagers, with different styles of products. At the time, it was strange for youngsters to be given what was, until then, a luxury item often used only for business. These younger markets, with dextrous fingers from constant use of computer games, did something unexpected. They started to use a little-known facility to communicate (texting) which, at the time, was free in many markets. This opened up a new way to communicate, new patterns of behaviour and even a new written language. The suppliers observed this behaviour and adjusted their products accordingly. They made keyboards friendlier towards texting, created new tools to help (predictive text) and changed tariffs to make it profitable.

These adjustments to innovations, in turn, cause other people to buy and adapt them to their lifestyle. This increases the volume and the offer enters the growth phase of its market. There is strong natural demand and, in many cases, sales just walk through the door. In this phase of the market's life the firm has to concentrate on servicing demand, so established suppliers will be focused upon obtaining and deploying skilled resource. They must also ensure that there are efficient processes to capture and meet orders. It is possible to concentrate on internal issues, ignoring competitor moves because there is sufficient demand for all. As a result, these markets are relatively easy to enter in a low-risk way. In fact, natural growth with a small venture is probably the best method of entry, particularly if a supplier has a well-known brand or a large number of existing customers who are likely to want their offer.

Yet, many services (such as outsourcing in some countries, and some aspects of consultancy) now operate in markets that are 'mature'. In other words, most buyers are familiar with the concept and have made their initial purchases. Growth comes largely from replacement or added value sales. However, although established businesses in this type of market may be struggling to gain substantial sales, this is often a good time for new entrants to gain a foothold by attracting the attention of buyers by offering an experience that is truly different. They can, for instance, command the attention of a distinct niche of customers, carving out a profitable and enduring position for their firm.

Normally, suppliers in a market are buffeted by the forces at play in each phase of its development. However, if marketers understand what stage of development their service market is in, they can set strategic direction for their firm in the light of that insight and perform more effectively.

Market Segmentation

The segmentation of markets into groups of buyers that can be easily reached by suppliers is a powerful concept which has improved the profit of many businesses. It suggests that buyers can be grouped around common needs. Then, by customising the firm's offer to meet these common needs, suppliers can both gain competitive advantage and save costs because they are only addressing a portion of the market. The generic concept and how to manage a segmentation process is described in the Tools and Techniques appendix. It is, though, of profound significance and one of the fundamental basics of marketing which, if properly conceived, gives insight and direction to different marketing tasks (like new service creation, brand development and marketing communication).

Unfortunately, it has often been neglected or poorly executed in technology businesses. Some segmentation in the category is, for example, rudimentary and limited. A number of companies simply group their buyers around the products and services they have bought. But a 'small-system' buyer might be a big buyer of other firms' products and may become a bigger buyer if approached in a more relevant way. On the other hand, many companies categorise their buyers according to size, as: global accounts, corporate business, small-to-medium enterprises (SME) and consumer. Yet it may be impossible to identify useful common needs from such a broad categorisation. For example, grouping all businesses under a certain revenue level as 'SME' fails to recognise the different buying motivations of these businesses. The growth rate, management talent and business strategy of small businesses are as different as the ideas they are built on. The needs of an IT start-up with venture capital backing are very different from those of, say, a local, independent pharmacy, even if their initial revenues are similar.

Industrial sectors are another frequently used segmentation type. Many companies, of all shapes and sizes, have 'line of business' or 'industry' specialists who focus on an industrial sector; trying to understand issues within them and customise the firm's offer to the companies in the sector. However, even these groupings can be limiting. For instance, some industry sectors are breaking down as new technologies change categorisation. As a result, it is increasingly difficult to tell which company is in which industry sector. Some retailers are moving into banking (e.g. Tesco), so are they now in the financial services sector? And, with the convergence of telecommunications and computing, exactly what sectors are Internet retailers or publishing companies in? Often, the definition of industrial groups to handle these changes is so broad as to make it meaningless. It cannot be used to identify common, useful issues or insights. In addition, not all businesses within the same industry sector will have the same requirements. For example, a service firm trying to sell training, or management consultancy, might be better placed to identify 'innovative' companies receptive to new ideas. Yet this promising segment could be in any industry sector.

Good market segmentation is a means of grouping buyers by their common needs, wants or aspirations, emphasising their humanity in a way which encourages them to respond to an offer. Whether the supplier is operating in the business-to-business or business-to-consumer environment, it is important to consider the characteristics of the people targeted. Properly done, it can be predictive, highlighting the future buying intent of the people in each group. Moreover, although human beings are unpredictable, difficult and irrational, they tend to group naturally. In the real world, markets segment themselves. Some suppliers, once they have identified a group, even create marketing strategies which make membership of it a badge of honour. In both business and consumer markets they turn the natural desire of some human beings to belong to a group into an aspiration to be part of a segment that suits their marketing strategies.

Interestingly, some of the differences between services and products, which were explored in Chapter 2, may be used as a basis for market segmentation, unique to service markets. These include:

- **The differences between new and experienced buyers.** New customers, using a service for the first time, become anxious and look for reassurance (for instance, when people first take a flight). Experienced customers, though, driven by their unarticulated dislike of being out of control, try to take shortcuts and try to improve on the service supplier's process (Bateson and Hoffman, 1999). Service for them is more likely to be about streamlined processes or even self-service. So, this difference between naïve and experienced customers allows different strategies and different services to be crafted.
- **High-tech, high-touch.** Some people prefer a service which is 'high-touch' because they like contact with people. They gravitate towards services that are highly customised and use human beings as part of the service offer. Often they are attracted by the fact that the people involved give them high status during the service. Other people, though, prefer a technology-based service. They like to use technology or tools to investigate and deliver their needs. They are self-reliant and prefer to meet their own needs. So, in many developed economies, while the majority of the population is willing to use ATM technology for their banking services, some niche banks thrive by providing personal service to retirees or high net worth individuals who tend to resist machines.
- **Mind sets.** Customers have different 'mind sets' when they approach a service. Their attitude will be different if they regard it as 'day-to-day use' or 'an emergency'. Some people, for instance, regard a taxi service as a normal part of their day-to-day life. Others use it only in dire need, when other modes of transportation have let them down.
- **Ambience.** People have different styles they prefer from services. Some people like an understated, respectful service while others prefer a noisy, fun experience.
- **Willingness/ability to cooperate in getting service.** Many people are pleased to share the service task and to participate in the service process. Others are not. Some might diagnose faults themselves, while others want a 'performed service' that saves their time.

Each of these differences means that segmentation can be based on issues unique to service markets. Moreover, as each firm has a distinct culture and its own distinct market position, it can use these insights to develop its own, unique segmentation. It is probably more important to use a process to create a unique segmentation than to steal a previously designed segmentation type. The process outlined in the Tools and Techniques appendix has been drawn from academia and tested in the reality of several projects conducted in service businesses over the years. It has produced, for the firms involved, a practical method of segmentation which has given them competitive advantage through a singular approach to their customers. Marketers can use it to understand potential buyer groups in the service markets they want to participate; the resultant segmentation should give them profound, valuable insights which become an important basis for much of their work.

Positioning the Business

The 'position' of businesses in a market is based on buyers' perceptions of the value of their offer and is often clearer after segmentation analysis has been completed. By gaining and holding a clear position in a market, a firm can maximise its margins. And, just as important to service firms is the fact that a clear position communicates to the recruitment market, attracting the right calibre of service employees; this, in turn, enhances quality and margins.

Figure 3.2 represents the scattering of different buyers as they seek different mixes of features and price in a market. The horizontal axis is 'perceived price' and the vertical 'perceived quality'; both of which are components of value. Some, for example, will want a no-frills, least-cost service, whereas others want a features-rich experience. A shopper at Harrods food hall in London will think the quality of product justifies the higher prices, whereas someone who shops at a local market, because they are more interested in lower prices, will be just as satisfied. Both think that they get value for money. This scattering occurs in any given market and allows suppliers to create different offers at different prices.

Figure 3.3 is a 'perceptual map' of a market, showing the positions that different suppliers take as they seek to command the attention of different buyer groups. Each company, either

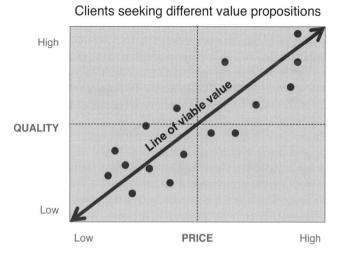

Figure 3.2 The different values sought by buyers

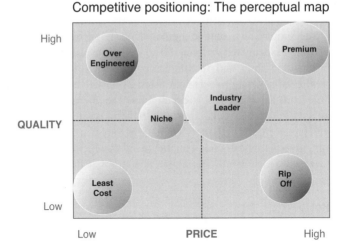

Figure 3.3   The perceptual/positioning map (Adaptation of diagram in Lambin 2000)

by design or default, eventually commands its own position in a market by serving one group of customers with a particular value proposition. It might be as a market leader (like IBM, BA and Centrica) with dominant share and the ability to influence the rules of engagement. These companies serve the majority of buyers and, as a result, set both price expectations and service standards for the whole market. Or it might be as a follower (like Fujitsu, Virgin Atlantic or Scottish Power), which is smaller than the market leader and able to earn a profitable existence by providing a healthy alternative either in terms of price or mix of features. Other sustainable competitive positions are: least cost (easyJet), premium (Emirates) or niche (bmi). The two unsustainable positions are in the top left and bottom right corners. Low price and high quality mean that the service is over-engineered (as many services in technology industries can be) and won't survive in the long term. On the other hand, the combination of high prices and low quality in the lower right quadrant is often caused by a distortion in the market, such as a monopoly, which means that buyers can't make clear comparisons.

The position taken represents the competitive position of the whole firm and ought to be the orientation of everyone in it. Also, the price and features of its service offers ought to suit that position and be different from others. For example, the service of a least-cost supplier ought to be very different from that of a features-rich premium supplier because each makes money in a different way from different groups of buyers.

The perceptual map can be used in several ways. Firstly, it can be a catalyst for debate about market positioning and the firm's strategy can then be based on it. The tool can be used to determine the number of buyers the firm should expect and what the value proposition should be. For instance, it may find that, through acquisition, it can become the volume market leader of an industry. However, to maintain that position, it must adopt the behaviour of a market leader, taking a stance on price, quality and leading industry issues. Alternatively, a firm might find that it can maximise margins by taking a position as a niche provider. In this case it needs to determine exactly how its service experience will be different from the market leader, communicating that to buyers.

These perceptual maps can also be used to understand competitive forces (the likely manoeuvres of other businesses in the market) and to work out competitive strategy. Markets are not static. It is dangerous, for example, for the market leader to assume its position is inviolable, since niche providers can progressively capture segments of the market and mount a challenge for leadership. Or a follower, which finds itself number two in a market with a vulnerable market leader, might decide to mount a challenge. So this tool can be used to anticipate the likely reaction of competitors when the firm makes any move. It is a powerful strategic insight, which can clarify strategic intent and give real focus to marketing programmes.

Understanding the Standard of After-care and Service Quality in a Market

It is surprising how many companies that consider the quality of their after-care to be an important part of their offer, do not think through the place of service in their marketing strategy. Service programmes are often imprecise and vague, losing any chance to contribute to competitive advantage. Often a business aims for the 'best quality' or the 'highest performance' whatever its market position. It is not uncommon for a profitable niche supplier to have a quality plan aimed at promoting the 'best' service in the industry or 'delighting' all customers. This is a stance which is likely to be as damaging as it is inappropriate. It is as costly as it is ridiculous to offer the same bland after-care service as all other suppliers. It is much more effective to develop a style of after-care which matches the general business strategy and reflects the intended market position. The style of support must match the competitive position and strategic intent of the firm. If the firm is to be a least-cost, premium or niche supplier, its competitive position is reinforced if its after-care is undertaken in a style similar to that positioning.

As important is to recognise the moments in the evolution of a market when there is an opportunity to gain ground by offering a new standard of service. Industries appear to develop through an evolution of thought and strategy with regard to service standards and some suppliers have reaped enormous rewards when taking advantage of those moments. For instance, service in some industries is below the standard expected by the general population, a national 'par' for quality (air travel in America and rail travel in Britain, for example). This is assumed, ill-defined and emotionally based, but is a common perception of the population nonetheless. Service quality which is below this national par may have been caused by general recessionary cut backs, a disregard or ignorance of changing consumer tastes, an historic monopoly or legislative distortion but it will eventually become the subject of public ridicule. It will become the target of comedians and journalists because they recognise that there is a common national experience which can be exploited. At the time of writing, for instance, many consumers in Western markets are dissatisfied with the service they receive from call centres, particularly those 'off-shored' to other countries.

In an industry where all suppliers are criticised, the first supplier to move to meet national par will gain market share. This has happened in numerous industries in both a national and international context. For example, in the 1980s, BA made radical improvements to its service which gained it market share and allowed it to grow for nearly two decades after, while other international carriers floundered. It moved to meet the service expectations of buyers when others in the industry did not. As a result it attracted new buyers, gaining competitive advantage. From Chicago's Marshall Field in the 1900s and Britain's Marks & Spencer in the 1950s to Amazon in the 1990s, many firms have demonstrated the economic power of a new style and standard of service.

Another example of competitive advantage derived from service quality is the difference between naïve and experienced segments of buyers. In an industry where all services are designed for naïve clients, the service standard comprises: good client care, clear process, smiling people and a reassuring brand. If a supplier enters a market dominated by offers to naïve buyers, it can gain strategic advantage by offering a new standard of service designed specifically for experienced customers. It will attract frustrated buyers from other providers, gaining share as it enters the market. So, marketers should look for insight from an analysis of service standards as much as from more traditionally recognised marketing perspectives like segmentation.

Critical Success Factors

Any market has established rules of engagement by which the participants in the market survive or prosper. These rules may be imposed by regulatory or market pressures, and usually by a combination of the two. For example, all the participants in technology industries must meet legal and technical standards in order to be able to participate. Meeting these criteria is a critical success factor as it is not possible to trade without them. Marketers should work through some sensible environmental analysis to understand whether new opportunities or new rules of the game are being developed. Are new technology standards proposed? Is economics changing patterns of demand? Are politicians changing long-standing rules? Marketers in European countries often miss, for example, opportunities to lobby for laws to be adjusted at EU level. It takes no real time to work through a recognised mnemonic, like 'PESTEL' (see the Tools and Techniques appendix) of the various external forces that can alter the critical success factors of a market. This analysis can reveal profound insights.

Beyond these basic requirements for being in the game, service providers also face certain criteria that enable them to succeed in the marketplace. However, these critical success factors are commercial imperatives resulting from the evolution of forces within the market. For example, a fast-moving market may call for the ability to bring new services to market quickly. Again, these should be carefully reviewed in order to exploit them as much as possible.

A Realistic Perspective on Market Distortions and False Markets

There is probably no such thing as a perfect market with perfect competition affecting supply and demand. Distortions occur in nearly every market, caused by things like dominant market share, strong reputations and long-standing brands. The heritage and positioning of premier consultancies like Accenture allows them, for example, to command higher than average prices. Distortion can also be caused by some form of monopolistic or oligopolistic grip on the market. The African mining company De Beers has, for example, had a grip on the international diamond market for two centuries and the 'big four' accountancy partnerships command nearly 95% of American and British financial audits of leading public companies.

Another distortion might arise from a misunderstanding of demand or preconceived ideas in the industry, distorting the services on offer. For example, a good proportion of the demand for service in the computer industry over recent decades has been caused by complexity of product and poor inter-operability between components in company technologies. Customers have had to pay extra to make their new purchases work in their premises. Something which, incidentally, would not be tolerated in other industries. Instead of paying extra for service skills, it would be possible to meet this demand by engineering reliable and simple products, rather than 'becoming a solutions supplier'. At the time of writing, some analysts think that Oracle's acquisition strategy shows that they intend to offer products that work easily in business-to-business markets and make profit by deflating this false service demand. In early 2009, the *Economist* said, for example:

> For some time, Mr Ellison's vision for Oracle has been to become the Apple of the enterprise, hiding complexity from customers, just as Apple does with its powerful but easy to use consumer products. Taking over Sun, he said, provides Oracle with all the pieces to put together systems that reach from application to disk. Oracle's engineers are already brainstorming about how to build industries in a box – complete computer systems that come fine tuned for, say, banking or retailing. (© The *Economist* Newspaper Limited, London, April 23 2009)

This strategy will ultimately affect the false service demand in that market for consultancy and technical assistance that has existed for more than two decades from poorly specified and poorly inter-operable products. It is remarkable that no technology firm has seen this opportunity sooner.

One of the biggest distortions of markets, though, is human behaviour. For many decades, economics has been taught on the basis of the rational behaviour of buyers in efficient, impersonal markets. Yet, even that discipline has realised that human beings, even when buying for businesses, are not necessarily logical or rational; emotion, beliefs, trusted relationships and prejudices play a large part in their buying processes. This has given rise to 'behavioural economics' and supplemented much of the work of 'behavioural marketing'. It recognises the methods by which entrepreneurs and marketers use programmes and techniques which distort markets to their advantage. Understanding these issues in depth can allow marketers to craft strategy which takes share of their market quickly by undercutting or altering the balance of market dynamics. On the other hand, misunderstanding human behaviour can cause a venture to fail altogether.

Asymmetry of Information

This is an important aspect of many service markets that is so obvious it is often taken for granted and neglected. Service markets, particularly professional service markets like consultancy, are structured around asymmetry of information: the fact that the supplier knows more than the buyer about the service. The market might, for example, exist because the supplier has technical knowledge that the buyer needs to pay for. The value of this knowledge, the price the supplier can charge, depends on the scarcity of the skill, perceived quality and how critical the service is to the customer. But superiority of knowledge also extends to the industry dynamics and the performance of suppliers, both of which affect the nature and structure of the market.

When buyers approach service markets for the first time, they lack knowledge of the various suppliers and, unlike product markets, are unable to test quality and value in advance. This causes the buyer two risks: 'adverse selection' (they may choose a poor or expensive supplier) and 'moral hazard' (poor behaviour by dubious suppliers, like post-contract opportunism). Service markets must therefore evolve mechanisms to induce trust and counter these problems. They include: regulation, personal reputation of prominent individuals, industry associations and the firms' reputations or brands. The buying process itself might also mitigate the risks of this lack of knowledge. For instance, many businesses use a selection process, which asks suppliers to present their credentials and proposed approach to a task. During this, they can explain their understanding of the problem and explore different methods to tackle it. This will enhance the knowledge and the decision-making of the buyer.

Asymmetry of information is at the heart of an unarticulated tussle between the buyer and the service provider, which is part of the marketer's job. At its most simple, this is reflected in the changing attitude of the general population in the developed economies to professionals over the past few decades. Since the 1950s, reverence for professionals, particularly medical practitioners, has been eroded as society in general has become better educated. Buyers are much more willing to question professionals, challenge prices and get second opinions. This, in turn, affects their willingness to pay different fee levels. If the buyer is unfamiliar with the industry and is confronted with a small number of elite suppliers, fees are likely to be high. If, however, the market contains a plethora of competing suppliers offering a familiar and easily understood process, prices are likely to drop and services commoditise.

As buyers become familiar with a sector of the service industry through repeat purchase, they become familiar with the individuals, processes and characteristics of competing firms. They then seek to get better value for money by cutting parts of the process, facilitating competition or doing some work themselves. The ultimate expression of this is the introduction of specialists into the buying process within business markets. They are more able to judge the nature of the technical skill offered and its value to the buyer's firm. When an HR director buys training or recruitment; when in-house counsel buys legal services; when a CIO buys servers or applications; when a marketing director buys research or advertising; or when a purchasing manager negotiates a formal contract – prices will tend to go down and quality up.

So, it is in the interests of service suppliers to maintain a degree of mystery and restraint about themselves and their approach. This is a balance. The suppliers need to give sufficient clues about their quality and their approach while retaining their intellectual property, their competitive advantage and a degree of mystery. As a result, certain industries (like executive search) and certain firms (like McKinsey) seek to maintain a mystique about themselves and their processes; and, as a result, seem to earn higher margins than many services offered by technology companies. They have shown that it can be a mistake to explain every step in a service process and to detail the entire firm's capabilities. Brand is a device which helps with this. A member of a large, branded consultancy business is able to allude to industry knowledge or technical skill that would be questioned more closely if operating as a sole practitioner for instance. By guarding their publicity and controlling communications, they exploit the buyers' lack of knowledge, within ethical boundaries, and maintain a price differential.

So, marketers should ensure that research is conducted into this specific issue. By segmenting buyers according to their insight into the industry, the basis of differential pricing can be understood. Opportunities to exploit asymmetry of information are likely to create higher perceived value and earn good margins.

The Market as Relationships or Personal Networks

Relationship marketing theory began to develop in the 1980s as the importance of human relationships in the buying process began to be more fully understood. As this thinking was applied to business-to-business markets, research focused upon the interconnection of business relationships in a network. Researchers and theorists were beginning to substantiate decades of experience amongst service suppliers. For instance, their work has covered the fact that, in a project-based industry, working relationships may only occur during the duration of a project whereas personal relationships have to continue over time. They distinguished between bonds between key participants, links over activities and ties over resources.

This is a behavioural view of markets. It sees a market as, primarily, a set of personal networks within which these mutually profitable business relationships occur, and suggests that firms can use existing customer relationships as a means of generating service revenues. They examine their market position by profiling relationships and use methods to build on them; for example, by encouraging referrals. For years, marketing planning and strategy has been dominated by people with backgrounds in economics, who have promoted a logical, systematic and economic approach to market analysis. Yet the experience of many service businesses, especially consultancies, is that they can thrive despite adverse economic conditions. Through the trust they have with their customers, they can stimulate and inspire them to try new ideas and to buy services, even though it has not occurred to them before. So, it is sensible to understand the depth and strength of customer relationships in order to create an effective marketing programme.

MEANS OF GAINING AN OBJECTIVE MARKET PERSPECTIVE

Figure 3.4 represents the perspective that a business needs to gain on a market; showing most of the relevant issues that ought to be understood. There are several different approaches to gaining these market insights, each of which has advantages and disadvantages. They are:

The Market Research Study

Research projects can give insights into changes in a market, buyer needs and business opportunities but, to be successful, they must be properly managed. An ideal process is outlined in the Tools and Techniques appendix but, whatever steps are taken, the leadership team should be engaged at an early stage. They should be consulted before the project is

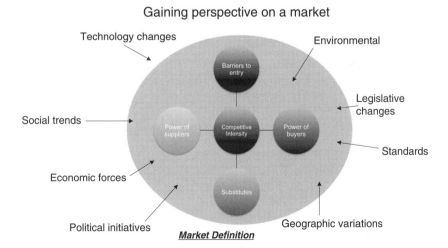

Figure 3.4 A pictorial representation of good market perspective

started and be involved in the selection of interviewees. They are then more likely to support the insights revealed by the research and consequent actions.

In many technology companies, operating in silos or with a distributed marketing approach, any number of research projects can be underway at any time. To prevent annoying the customers subjected to these studies, and to make the most of the information these different research projects reveal, it is good practice to organise them into one programme, led by someone close to the business's executive, as Orange Business Services has done.

CUSTOMER COLLABORATION SHAPES ORANGE BUSINESS SERVICES

Orange is one of the most important brands in the France Telecom Group, one of the world's leading telecommunications operators. Originally launched as a mobile communications brand in the UK, the Orange brand now covers: mobile and fixed communications, Internet and television services in the majority of countries where the group operates. It is the number three mobile operator and the number one provider of broadband Internet services in Europe.

At the end of 2008, France Telecom had consolidated sales of 53.5 billion euros (12.7 billion euros for the first quarter of 2009) and, at 31st March 2009, the group had a customer base of almost 184 million customers in 32 countries. These included 123 million mobile customers worldwide and 13 million broadband Internet (ADSL) customers in Europe.

The group's Orange Business Services division is a leader in providing telecommunication services to multinational companies. For the past three years, Chris Ellis has led their customer marketing group, developing programmes that not only capture customer feedback and ideas, but actively encourage customers to collaborate in improving the company's operations, portfolio of offers and strategy.

Customer Satisfaction is not Enough

Orange Business Services has always taken the need to listen to its customers seriously, but like most companies, much of this occurred on an ad hoc basis, mostly through one-off projects commissioned by different parts of the organisation whenever they needed input to a decision. The most systematic research was into customer satisfaction, sponsored by the CEO and stemming from her belief that giving customers a good experience is key to the success of any business; that 'happy customers' are the sign of a healthy business. As a result, the company's customer satisfaction research has always received sustained attention and resources.

But, while this approach to customer research was systematic, it dealt mainly with operational issues and relied on an analysis of average statistics to identify activities or functions needing improvement. It was an effective programme but it was just one element of a wider customer listening programme that could put the voice of the customer at the heart of everything Orange Business Services did.

This realisation dawned on Ellis at the same time as a significant shift in marketing was starting to take place: the growing importance of personal networking and peer referral in business development. Ellis had watched companies like Amazon make access to peer referral a key part of their service to customers. But while this was a growing trend among consumers, it was unclear how important it was to business buyers. So, the company commissioned more research into what their business customers' buying criteria were, and how much peers and other third parties influenced their decisions. The results showed that what was happening in the consumer market was mirrored in the business world too.

They reorganised the marketing activities to take this trend into account, and concentrated on three areas: direct marketing, influencer marketing and customer marketing. The first created direct communications programmes between Orange Business Services and its customers; the second worked on third-party influencers (analysts and consultants); and the third focused on the communication between customers and on capturing their collective insight.

Building Insight through Customer Collaboration

To really put the customer at the heart of Orange activities, and to stimulate advocacy in a world where peer referral was important, Ellis realised that the company's different research activities needed to be structured into a systematic listening programme. They needed to capture the voice of the customer and relay it directly to the company's executives. So, existing research projects were structured into a multi-level programme, with some additional components. It provided insight at three levels for Orange to act upon: at operational level, at the offer portfolio level and at a strategic level across the business. (See Figure 3.5)

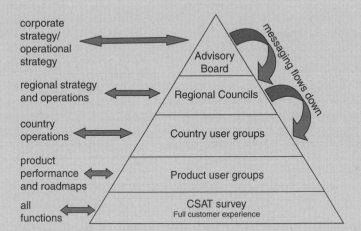

Figure 3.5 Orange's multi-level listening programmes

At an operational level, the customer satisfaction programme was improved by tightening up the sampling approach. Instead of a fairly random sample, 80% of the new sample came from the 5% of customers who delivered the majority of the company's profit. These were surveyed by a third party every six months. In order to be even more precise, the key person within each customer account approved the proposed interviewees within their organisation, to ensure that the right people were being contacted and listened to. This new sampling and 'validation' step made it very hard to dismiss the feedback as unimportant. It was coming from the people who matter in the most important customer organisations. And, while the feedback was still reviewed at a general functional or geographical level to identify what broad actions were needed, it was also fed into account planning workshops and used to develop appropriate actions in the account plans.

Regional and country meetings, bringing together the local executives of important customers, are now held quarterly to review local operational issues. This provides more qualitative and freeform feedback for the company, and begins to look beyond past performance into shaping future plans.

To capture insight at a portfolio level, Orange Business Services now runs user groups and panels that discuss specific offers in terms of their features and benefits, and the value of these offers to customers. This technique is used by most companies in the technology sector, and helps to shape the forward 'roadmap' of developments for any offer or set of offers.

But the jewel in the crown in the new listening programme is the global customer advisory board (CAB), set up to work across all three levels, but most importantly to provide input at a strategic level. Often, such boards are set up to reward valuable customers for their business and involve sumptuous dinners or hospitality with token presentations of already well-formed ideas for review. The focus of this group is different: it is a genuine advisory council, meeting with the company's senior executives to influence strategic decisions through a collaborative, joint working group approach.

Membership of the CAB is closely managed, with customers invited to participate both because of their importance to Orange's future business and their ability to personally make a valuable contribution. Up to 25 CIOs from around the world are members, with most physically attending the two-day board meetings once or twice a year. But just as importantly, nominated experts from their teams participate in collaborative working groups that tackle specific issues before reporting back via tele- or web-conferencing on a quarterly basis.

Genuine Interaction and Involvement Delivers Results

The chairman of the CAB is a customer, who helps to set the agenda for the meetings and invites the other customers along to them. Once there, customers lead at least half of the sessions, which creates genuine interaction and involvement.

The customers' commitment has built over time. As they saw the initial, more operationally focused issues actioned, and consequently the service to them improve, they realised that the company's executives took the CAB seriously and were prepared to act on its recommendations. This became a virtuous circle, increasing customer participation and the value of the CAB to Orange.

The CAB has tackled a number of issues over the years. At an operational level, it has worked on the complete 'quote to bill' cycle, simplifying and speeding up the process of getting complex international services up and running. A working group explored the process from the customer's point of view, identifying all activities and steps, and then made recommendations back to the CAB. As a result, lead times have been shortened and the punctuality of installations increased.

The CAB also looked at the direction in which the company's portfolio was moving, reviewing new offer areas, such as Telepresence. Following a collaborative approach to the design of the service, more flexibility has been built in and the value proposition for customers developed. Orange Business Services is now launching 'Open video presence' to the market as a result.

Today the CAB is working on more strategic issues. Ellis estimates that while initial discussions were mostly about operational issues, this is now down to less than half of the CAB's focus, and strategic issues take up most of the time available.

More broadly, the company has benefited hugely from this more structured and collaborative approach to customer insight. Its customer satisfaction scores are now at the top of industry benchmarks for service. Executives value the full set of customer insight programmes and are starting to experience the impact they can have on the direction of the company and business results. And while there's always scope to improve (by improving the sample of customer satisfaction surveys or by broadening the geographical and sector representation on the CAB for example), Ellis is happy that collaboration is now part of the company's DNA, and that the customer is truly at the heart of everything Orange Business Services does.

The Market Audit

Unfortunately, field research understands only one aspect of market dynamics. Other forces, such as regulation or technological change, also affect it. Limiting analysis to just research will risk missing potential insights that can be gained from a broader analysis. A more thorough and objective way to gain an economic perspective on a market is to carry out a full market audit (see the Tools and Techniques appendix); a concept pioneered largely by Malcolm McDonald (McDonald, 2002). Properly conducted, it is as objective and thorough as a financial audit, and provides a reasonable basis for planning. The process can take several weeks to complete but provides very detailed analysis and, more importantly, valuable insights. Data about the market is gathered under significant headings, using analytical techniques to tease out insights which can be the basis for strategic direction and competitive advantage. The information needed is surprisingly easy to obtain. In addition to generic Internet searches, within business libraries, professional institutes and government departments, there is a gold mine of valuable information, on almost any market in the world, which can be obtained at relatively low cost. In fact, with the advent of the Internet and search engines, the main cost is personal effort and time. The market audit can be conducted by a specialist engaged by the firm or, with assistance, by a sensible, experienced employee. If the culture of the firm is resistant to an analysis of the market, and it is not prepared to spend time gathering data, the process can be run as an interactive session with the leadership team. A half, or one-day session, working through the various subject areas, in data-assisted discussions, is likely to yield insights that can improve the quality of subsequent strategic decisions.

Scenario Planning

Scenario planning is another analytical tool (see the Tools and Techniques appendix). It helps firms to think about different potential futures in the light of change, complexity and uncertainty. Less linear than the market audit, it allows management teams to explore likely scenarios which might develop from current market forces. With its roots in the military and first developed successfully as a management tool by oil giant Shell, scenario planning creates a framework in which potential strategies can be developed and tested in the light of future uncertainties. The scenarios themselves are stories which help managers to develop different potential futures based on both knowledge and people's assumptions about the present. They are not forecasts, but help provide a common perspective and language (Ringland, 1997). Usually, a team comprising people from across the firm is formed to construct them. In a session led by someone experienced in the process, the participants brainstorm potential futures for their firm. Input to their debate might include evidence from futurologists, customer views and pertinent data. The team is normally encouraged to think widely before scenarios are grouped and ranked. The firm can create different scenarios of how its market might develop and how competitors might react to their plans; these different scenarios can then be worked into a market plan that anticipates risk.

Relationship Profiling

Those suppliers who take a behavioural, rather than an economic view of their market can audit the strength and depths of relationships between them and their buyers. They can understand the degree of trust, the intimacy of relationship and the effect on buyer behaviour. They can then use this information to strengthen relationships or plan viral marketing campaigns, which spread word-of-mouth messages through interconnected networks in the market. A tool that can be used to do this is the 'ARR' model (see the Tools and Techniques appendix). This was developed during the early 1980s, by researchers and theorists interested in both business-to-business marketing and network marketing. It can be used in practical business contexts as well as just pure theoretical research.

OPPORTUNITY ANALYSIS

The reason for undertaking an analysis of the market is to gain insights into new opportunities or to find justification for closing down declining areas of business. So, it is necessary to draw the analysis into an agreed view of the opportunities facing the firm to which resources need to be allocated. In this area, above all else, it is essential to combine analysis with market insights and the experiential judgement of the leadership team. While a powerful perspective can be gained from informal discussion, amongst leaders there are a number of tools that can be used to structure thinking and debate. Used properly, they lead to a consensus and healthy perspective. Neither they, nor well-crafted marketing plans, are ends in themselves though. It might be more important to identify, define and address one lucrative market opportunity than to write a full marketing plan, stuffed with opportunities, which would please an academic, but which the company is unable to digest.

The best known and most straightforward tool is the 'SWOT' analysis (see the Tools and Techniques appendix). This helps a firm identify internal strengths (S) and weaknesses (W) along with external opportunities (O) and threats (T). This information can be used to construct the SWOT matrix, plotting the opportunities and threats against the strengths and weaknesses. It can be done interactively, in discussion with the leadership team, and is most useful when used with the market analysis for reference. A surprising number of management teams start out with a 'SWOT' brainstorming session and no analysis. If they have a distorted or mistaken perspective on their market, this is counter-productive, since it builds that entrenched view of their competitive position into the firm's strategy.

Another tool, the 'Ansoff matrix' (see the Tools and Techniques appendix) was crafted by corporate strategy pioneer, H. Igor Ansoff (Ansoff, 1957). It aims to help marketers examine both current and potential offers in current and potential markets. Ansoff based it on very detailed, long-term research that he conducted into American company behaviours. Unfortunately, over the decades, this idea has collected academic dust and Ansoff's original energy has been lost. It was originally about creativity, innovation and entrepreneurship. He encouraged business leaders and marketers to think as widely as possible, then to fill identified opportunities with either product adaptations or acquisitions.

This tool collects opportunities into four different growth strategies, which are listed below in ascending order of risk:

- **Market penetration.** Increasing market share with existing propositions to current customers.
- **Market extension.** Taking existing propositions to new markets.
- **Product development.** Developing new propositions for existing groups of customers.
- **Diversification.** Growing new businesses with new propositions for new markets.

The matrix helps to clarify leaders' thinking and to illustrate the very different strategic approaches needed for each of the four strategies. Ideally, an operational marketing plan should be constructed for each strategic option that is approved.

The 'directional policy matrix' (see the Tools and Techniques appendix), by contrast, helps the marketer to balance a firm's strengths against the attractiveness of a market. Its power lies in the ability to create decision criteria unique to the firm, which can form the basis for the prioritisation of opportunities. For example, it could be that a particular market offers potential growth, or that it contains customers willing to pay high prices. It might also be attractive because of the ease with which the firm can access it.

These tools and techniques help create a framework to structure the leadership team's thinking about its markets by developing a common language, by testing assumptions, and by helping to reach consensus. Using them won't guarantee success or eliminate risk. But it will provide the basis for a more thoughtful and potentially lucrative approach to strategic marketing issues. They are also likely to reduce risk and improve the quality of decision-making.

SUMMARY

There is evidence that truly attractive service experiences are based on insight into a unique opportunity, a market insight. In pursuit of lucrative insights it is sensible for marketers to gain a thorough and objective view using techniques that will help uncover the unique conditions of service markets. They need to understand the competitive forces and any opportunities (or dangers) they offer. This chapter has discussed a number of tools and techniques available to help do this, which need not be too costly or take too long. Used properly, they provide a deeper understanding of the market and insights that will help craft effective programmes. More importantly, perhaps, they do so by giving leaders a common language with which to debate and discuss the market. The prime objective, though, is to tease out unique insights into juicy market opportunities to which service skills can be applied.

4

Internal Perspectives and their Strategic Impact

INTRODUCTION

Academics and idealistic marketers like to argue that marketing begins with the market and the customer. Although that is in a sense true, the market to concentrate on and the customers to aim at are framed by the organisation's products, services, history and reputation. In reality, every business, no matter how market-orientated, approaches its customers with a proposition, an idea of what it can offer. So, as marketers set out to work for service firms, they must also understand a range of issues about their own organisation's capabilities. Some of these will be the source of strategic insights that can lead to profound opportunities; others will be impediments. This chapter outlines the most significant strategic issues that marketers should seek to clarify as they address their market. It particularly emphasises the need to clarify the "strategic intent" or purpose of each service business.

STRATEGIC CONTEXT

It is very rare that a service business operates at will with no limitations, direction or inhibitions set by its owners. There is normally a context in which it works. The owners of the firm will set, for example, a direction for the whole organisation it is part of. Called "corporate strategy", this defines a group of businesses; setting their purpose, priorities and resources. The business strategy for the service business itself, by contrast, will focus on issues like: its resources, its revenues, its customers and the competition in its own market.

The way an organisation sets its corporate strategy and its changing priorities will affect the service businesses within it. A service manager at IBM, for example, will have had very different priorities and roles over the last 30 years as that massive organisation has changed its direction and offers, sometimes painfully. In the 1980s their job would have been about supporting the equipment sold by IBM; running, for instance, an efficient maintenance service so that customers would have a good image of the company's quality and willingly buy again. In the early 1990s they would have supported newer technologies such as networks and servers, much of which will have been manufactured by other suppliers. In the early part of this century, as the company became one of the world's biggest service organisations, it is just as likely that a service manager would be running a worldwide network of technology, available around the clock, often on behalf of outsourced customers. The priorities and capabilities of the service divisions within this huge company have chopped and changed as corporate strategy has been adjusted in the light of market opportunity.

Yet in organisations like IBM, corporate planning is generally well defined and undertaken by specialists in line with a rational planning timetable. In others, priorities can be less clear. The company may, for instance, have been bought by a group or individual that has other priorities. Or it may be part of an organisation owned by a national culture that fundamentally believes in local delegation and consensus, such as that in Swedish multinationals. In these, although the service business is a part of a broader organisation, it has much more delegated responsibility. In fact, in some cases, it may not even be told that it has delegated authority, its independence is assumed and it is left to chart its own course; sometimes in conflict with other units in the same organisation.

Whatever the position, size or attitude of the larger organisation, it is essential that the leaders of a service business clarify the over-arching priorities of the owners as they develop their own marketing strategy. In an organisation with detailed and rational planning procedures this is likely to involve dialogue at an appropriate point in a clear timetable. In a less elaborately controlled organisation, the leaders of the service business may need to prompt the conversation. In both, though, they will need to know the aims and objectives for the larger organisation and how they impact their own business. Are there, for example, changes to networks, technologies or products, which will affect service requirements? Are there financial constraints or investment opportunities, which will affect plans? What are the constraints on their unit and what resources might be available to help them? It may be that the leaders will have to create a scenario of their business or an outline plan to show to the owners. Some senior executives will say to these people: 'I didn't know that was what I wanted, but now I see it I buy in.'

The first step to developing effective marketing, then, is to understand the resources, constraints and aims that the owners of the business impose on the service business (whether they are venture capitalists, a family, an individual or shareholders who have delegated management to executives through different governance structures).

STRATEGIC INTENT: THE NEED FOR CLARITY OF PURPOSE

At some point the firm needs to clarify the purpose of its service business or businesses if it is to make good profits and compete successfully. This is much more than a few words in a bland statement. Strategic intent is the *raison d'être*, the over-riding purpose of a business. It gives direction to the management team and all operational decisions, avoiding extraneous activities and unnecessary costs. As the famous strategy writers Hamel and Prahalad said in their article, which coined the phrase:

> Strategic intent is more than unfettered ambition. The concept also encompasses an active management process that includes focusing the organisation's attention on the essence of winning; motivating people by communicating the value of the target, leaving room for individual and team contributions; sustaining enthusiasm by providing new operational definitions as circumstances change; and using intent consistently to guide resource allocations. Strategic intent captures the essence of winning. (Reprinted with permission from *Harvard Business Review*, Hamel and Prahalad, May/June 1989)

The benefits of clarifying the strategic intent of a service business were examined, and dramatised, in a report called 'Service companies: focus or falter' by two academics (Davidow and Uttal, 1989). They used several examples of service businesses (the most memorable being a Canadian hospital which specialised in hernia operations) to demonstrate the remarkable effect that strategic focus has on profit and service quality. They showed that it leads naturally to the co-production of service with customers, investment in technology, the industrialisation of erratic service components, productivity improvement and the setting of expectations with potential buyers. They said:

> ...fuzzy or conflicting strategies make good customer service impossible ... Without a strategy, you can't develop a concept of service to rally employees or catch conflicts between corporate strategy and customer service or come up with ways to measure service performance and perceived

quality. In short, without a strategy you can't get to first base. (Reprinted with permission from *Harvard Business Review*, Davidow and Uttal, July/Aug 1989)

A clear strategic focus has paid dividends in both customer satisfaction and business results at EDF Energy, providing a rallying cry for employees and a focus on the ambition to be number one for service quality in its business market.

A CLEAR STRATEGIC INTENT IS THE BUSINESS AT EDF ENERGY

EDF Energy is the largest supplier of electricity to British business and has over five million residential customers. It is present throughout the electricity value chain – from generation and trading, to network management and customer supply. It is also a sizeable gas supplier, particularly in the residential market. The company is a fully owned subsidiary of France's EDF SA, which had 38 million customers across Europe and revenues of €64.3 billion in 2008. The group's goal is to be the leading provider of low-carbon, high-performance energy solutions.

EDF Group entered the British market in 1999 by acquiring London Electricity (LE). The British electricity market had only recently become fully competitive, the market having opened up in three phases: in 1990, the 5000 biggest customers had been given the ability to choose their supplier; followed in 1994 by the next 50,000; with the remainder of the business market and all of the residential market opening in geographical phases in 1998.

At the time of the acquisition, LE was not seen as a serious competitor. For the previous two years, it had been owned by US energy company Entergy, who had quickly become disenchanted with the British market and decided to exit, putting the company into strategic limbo at an important stage of market development.

Adapting to Market Deregulation

Strategy rethinks were nothing new at the company. Back in 1990, LE was one of the newly privatised regional electricity companies, supplying businesses and households in its London franchise area. Its initial reaction to the opening of the market could be described as lukewarm, seeing little profit potential and underestimating the extent of the competitive threat. So it did not prepare seriously for the new market and, as a consequence, started to lose business customers.

This same attitude was reflected in preparations for the second phase of market opening. With competitors moving to sign up customers early, LE found that, by mid-1994, it had lost around 40% of its eligible customers and had won hardly any new ones.

The situation prompted a major reappraisal and the fight-back began. A new business unit was formed, marketing expertise was recruited externally and a thoroughly professional, marketing-led programme of pricing and product development, plus mould-breaking marketing communications, commenced. In the 1995–96 financial year, 50% volume growth was achieved, with a further 35% the following year.

The highly competitive nature of the market, however, made for thin margins and by the time the 1997–98 financial year had started, the company had changed strategy again. 'Volume is vanity, margin is magic' became the new mantra. And so the downward cycle of customer numbers and revenues began again.

Cue the arrival of EDF Group. At a time when the industry was still debating the most effective business model (did you need production assets to succeed?), EDF had a clear perspective that you *did* need assets, plus scale, and quickly moved not only to acquire LE, but to add power stations, the supply business of SWEB, the whole of Seeboard and the network business of Eastern Electricity. In 2005, following several stages of transition, the company adopted the EDF Energy name for its supply businesses.

In the business market, EDF ownership meant that LE immediately became seen as a serious player once again. Moreover, EDF's desire to be able to serve its large French industrial customers outside of its home territory opened up significant opportunities to enter new sectors (having stuck principally to its traditional commercial and public sector customer base). EDF also brought access to additional sources of renewable energy, a scarce commodity that many business customers were starting to demand. As a result, even with an emphasis still on profitability, the decline in customers and revenues was rapidly reversed and a period of sustained growth began, which would culminate in the company becoming the largest supplier to business.

So what lay behind this success story?

Getting the Basics Right

Electricity is a true commodity. The physical product is exactly the same whoever the customer buys it from (regulation forcing companies to operate their network and 'energy retailing' businesses as separate entities). So there is no escaping the fact that price plays a fundamental part in winning and retaining business. But customers also have other needs, including accurate billing, special contract and payment terms, the provision of management information and, as the market has become more sophisticated and volatile, help with their purchasing and risk management strategies.

Customers also demand a quick response to their request for a quotation. Large customers, in particular, may request a quote at any time, depending on the level and volatility of the wholesale market, and indeed may request numerous repeat quotes. The ability to quickly and accurately process the huge amount of data necessary to calculate an energy price is therefore a critical success factor. Creating and managing a highly efficient 'sausage machine' to process high volumes of quotes is a prerequisite for a large, successful supplier. Service excellence, based on process efficiency, is also an important success factor in the 'energy retailing' business. It was this concentration on critical success factors and on trying to get the basics right that formed the centrepiece of the company's strategy.

Re-energising the Core Service

The company's 2000 marketing plan for the business market set out a vision to be 'a profitable, top-five energy retailing company (at a time when there were still around a dozen major players), delivering slick, customer-friendly service'. The plan expressed the belief that 'the eventual winners in the business market will be those companies which stick to the core business and focus on getting the basics right. The winners will operate to a simple formula: they will offer competitive prices, a limited range of added value services closely linked to gas and electricity but, above all, will effectively manage the increasingly complex processes which underpin the gas and electricity industries'.

A number of initiatives began immediately to bring the strategy to life. The first stage, driven principally by the need to integrate the newly acquired SWEB supply business, focused on switching large numbers of customers to telephone account managers (rather than more expensive face-to-face sales staff). Underpinning this move was the recognition that many customers simply had no desire for regular sales contact. A dedicated team was also established to target the construction industry, providing a one-stop shop for major developers who were being hindered by the fragmented structure of the new market. A European sales team was also set up to serve those customers that EDF wanted to follow outside of France.

Attention then turned to the processes and IT systems, which had been developed to tight deadlines with just the basic functionality needed to participate in the new market. Full market opening had brought a growing pile of customer problems, which were initially tackled by forming a 'customer action group', where sales staff could bring their customer issues and see that they were being addressed. Now the time had come to find more permanent solutions and 'Project Slick' was launched, providing a focus for the numerous changes that were needed.

The marketing team had also noticed how the rapidly evolving world of IT was creating new customer demands. Research was undertaken to help prioritise the ideas to be developed, and in 2004 the business launched 'Energy Zone', its first online service. 2004 was also highly significant on the pricing front, with the company launching its ground-breaking 'Flexibility' product. Up until then, electricity customers had signed fixed-term, fixed-price contracts; now they could buy energy in tranches, spreading the price risk. The product was an immediate success, winning significant numbers of customers. Similar products have since become the norm for larger customers.

Customer service remained a constant theme, however. The stakes were raised when the company CEO, Vincent de Rivaz, set the business the target of becoming number one for customer service (as measured by the satisfaction survey carried out independently by Datamonitor). A raft of new initiatives was spawned, including 'Project Delta', where front-line staff went away in groups to tackle a chosen service problem. They would start by identifying the causes, then work to develop 10 solutions, which they presented in a 'dragons den'-style presentation to a panel of executives (one of whom had to volunteer to sponsor any idea that was accepted). Vast numbers of simple ideas emerged from the process, making a big difference to the operations of the business.

Helping Businesses Become More Energy Efficient

In 2005, escalating energy prices hit many businesses hard. This, combined with growing acceptance of the reality of global warming, suddenly made energy efficiency a serious boardroom topic for the first time since the 1970's oil crisis. EDF responded by creating a hugely successful DVD-based resource, that gave business customers a step-by-step guide to achieving efficiencies. In addition, technical resources were recruited, to provide on-site assistance to businesses.

With increasing regulation and legislation surrounding energy use (such as the compulsory fitting of smart meters and the introduction of the CRC Energy Efficiency Scheme), energy efficiency is becoming a topic that businesses cannot ignore. This area is set to be a fascinating new battleground for the energy companies, as they seek to extract value from the opportunity and strengthen their customer relationships.

Electrifying Results

In March 2009, EDF Energy was voted the best electricity supplier in the UK for its business service, achieving the objective they had pursued so doggedly over the years. Commercially, the mass-customisation strategy has paid off too, with the energy sold to business customers tripling from 13TWH in 1998 to 34TWH in 2006, mostly through organic growth.

There are a number of different types of business focus open to service businesses with a technical heritage. They include:

Tied Maintenance and After-care

It is rare that a modern company does not have an arrangement to repair any faults that might arise with the technical products it sells. No matter how high the quality of engineering, the selection of components or the control of manufacturing processes, a number of products are likely to develop faults. A tied maintenance unit exists to repair those faults. Whether directly employed by the company or subcontracted to dedicated dealers, it is tied to one manufacturer and its range of products.

Simple maintenance seems to be a straightforward, even boring, part of most product companies. Yet it is an important ingredient in customer satisfaction and has an impact on repurchase intent. The warranty and maintenance services provide buyers with an emotional reassurance that any difficulties in the first years of product use will be resolved without fuss; and this reassurance helps sales. So, this type of business exists to aid repurchase and its prime focus ought to be competitive defence.

In the early months of product use, support is normally provided through a guarantee or warranty. Any failures are quickly addressed by a repair or, in many instances, by product replacement. Usually this warranty work is undertaken by the firm's maintenance resource. However, at some stage, the buyer has to contract specifically for maintenance support as the product ages. Often a variety of contracts are available offering different standards of response depending on the needs and wallet of the buyers.

There are, though, several conflicting tensions within this seemingly simple business. Firstly, a number of suppliers try to sell extended warranty at point of purchase. This can undermine the future sale of a maintenance contract, extending the warranty period sometimes up to the life of the product. A second tension arises because investment in excellent maintenance can delay the moment when buyers think they need to replace an old product, affecting the revenue from new product sales. Good maintenance, routine equipment checks and modern approaches to preventative maintenance can extend product life and reduce new product sales. The maintenance business needs to communicate carefully with customers who own old equipment, progressively increasing maintenance costs as obsolescence approaches. They also need to signal carefully the time at which an obsolete technology will no longer be supported. Failure to manage communication with these customers is likely to affect their propensity to buy again. Another tension is the way in which the increasing stability of new technologies affects the buyers' willingness to invest in maintenance or extended warranty. People are less likely to buy cover as faults in the generic product group become more infrequent. Few, for example, buy extended warranty with modern televisions while the falling price of personal computers for business makes maintenance almost uneconomic.

Real conflict occurs, though, when the purpose of this business unit is muddled. Maintenance businesses can have high-margin annuity revenue, which props up the whole business during times of economic difficulty or seasonally poor sales. As a result, its leaders can be asked to pursue revenue from any source to help cash flow. Yet, by allowing the maintenance unit to offer new services, perhaps supporting equipment sold by different suppliers, conflict is caused with other business units while, at the same time, confusing its funding status. Previously its costs would have been covered by product margins. Whether recorded as a specific maintenance cost centre or washed in the generic business margin, its aim was to support business sales. Now, however, it is allowed to grow revenue and its confused status can lead it to take uneconomic business or to undermine product sales. So clarity of strategic intent is important in even this straightforward business.

Tied Technical Advice

Many businesses hire technical specialists to advise and assist their customers. Called 'professional services', they have been employed to work exclusively on the products supplied by their employers, helping sales and becoming part of the total offer. The computer industry has, for example, employed programmers to analyse their customers' needs and design systems for them for many decades. Some companies also employ specialists with more soft skills, like experienced project managers, to help buyers design, plan and implement complex technical projects. Others (car dealers, for instance) employ specialist finance people to offer a range of loan and leasing options to help purchase. All are part of the cost of business funded by product margins.

Like tied maintenance units, these specialists are primarily employed to support product sales and, in addition to ensuring that the product performs to specification, their value lies in the effect they have on customer satisfaction. If the customer is so pleased with specification, delivery and implementation that they become a repeat buyer, then the service has proved its worth. However, in a number of industries, these businesses have been encouraged to grow revenues by the sale of independent advice. Yet, as with tied maintenance divisions, conflict can occur if they are successful. Moreover, despite being intelligent and highly proficient people, they can lack the personality traits needed in competitive consultants. If they have had a career that is based on providing the correct technical advice to customers who have bought something, they can find it hard to challenge or inspire these customers to buy independent advice. It is difficult to move a large workforce of this kind into competitive, independent consulting; technical excellence does not necessarily mean excellent consulting.

'Third-party' Maintenance

These businesses exist to gain competitive advantage from technical repair skills by supporting a range of different products. They vary from the consumer maintenance companies which support domestic goods to technical companies repairing the complex computer and communications equipment owned by businesses. Many of the functions of the business (from field engineering management and logistic supplies to call reception and diagnostic processes) are similar to the tied maintenance arm of manufacturers. However, the focus is on a range of products owned by one customer. They therefore need to be particularly good at managing an asset register of equipment owned by customers and at the process of extending their business competencies to take on and service new equipment.

Their benefit, though, is that the focus of their business is on the customer and the customer's installation rather than any particular supplier's equipment. In situations where there are installations of different types of a manufacturer's equipment, they can get to know the customer well and anticipate their needs. Their other strategic advantage is in the experience they gain from their repair work. Their tools, systems and processes tend to give them cost advantage over time. As a result, many can undercut manufacturers' maintenance units, keeping prices low.

Suppliers of Scalable Independent Services

These businesses are unrelated to any specific product range, but are based on the supply of a service that handles volume transactions and have, at their core, a process or a technology. Some, like Accenture, are dedicated businesses and others, like those in GE or IBM, exist as business units within broader companies. The supplier earns revenue by replicating the service and increasing the number of customers but improves profit by deploying process and technology improvements to reduce costs.

In fact, any inability to replicate causes these businesses to earn lower margins or fail altogether. They must 'industrialise' their offer, reducing costs over time. This has been a growing phenomenon in the last two decades. Based on the principles of outsourcing, suppliers as varied as Michelin, Accenture, BT, Nokia, Capita and IBM have won contracts to run parts of their customers' business. Some call them 'managed services' and some 'business process management' – but all are based on the proposition that the supplier can, due to specialisation, knowledge or business experience, run the business process better than the firm itself. Apart from huge service contracts in the IT industry, this relatively new attitude to the management of big business created the multimillion-dollar facilities management industry. Business after business has chosen to contract specialists to run their security, estate management and catering services; in fact, most processes that for them are not 'core' business.

Outsourcing is itself based on another concept called the 'experience curve' (see the Tools and Techniques appendix). This deceptively simple idea suggests that any organisation will, through productivity improvements, reduce costs over time in the area of its prime competence, its business focus. As shown in Figure 4.1, a firm whose business is focused on a particular function or process is likely to be further down the experience curve than a customer's in-house team. A computer company will be better at managing computing operations than the in-house IT division of, say, a bank; whereas a security company will be better at controlling access than, say, an airline and a cleaning company better at cost-effective office refreshment than, say, an accountancy firm. If the in-house operations are passed to the supplier, the customer gains the advantage of the supplier's experience and the savings can be dramatic.

Unfortunately, after the initial deal, further dramatic cost gains are unlikely. The continuing success of the contract depends on a number of different factors. One is the ability of the supplier to win other contracts and, through volume and economies of scale, continue to maintain a cost advantage for its customers. Second is the ability of the supplier to gain continuous improvements and efficiencies in the process (such as Fujitsu's approach to 'lean' services). The third, and probably the most crucial, is the nature of the relationship between the two parties. The customer needs to have a senior manager responsible for managing the

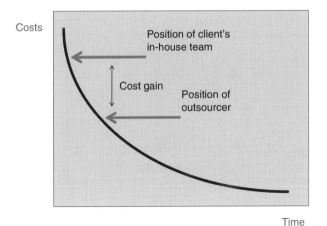

Figure 4.1   The experience curve as an explanation of outsourcing

ongoing relationship with their outsourcer. This person should not only monitor performance but also brief the contractor on the company's developing strategy so that plans can be adjusted. The failure to do this has caused many relationships of this kind to break down. Yet, if that happens, the customer loses the supplier's experience and begins to build cost back into its business. For their part, suppliers of managed services need contract managers who are trained in business strategy and understand their customers in depth if they are to maintain their business position.

The Knowledge-based Professional Service Supplier

The number of independent professional services is vast and varied. One thing they have in common is that their knowledge is the barrier to entry into their business, and they make vast sums from this asymmetry of information. Another is that their businesses are based on skill and people rather than scalable processes. It is an entirely different service business to process management, with very different margins and performance criteria; so they need a different strategic focus.

The professions include some of the world's most successful, enduring and profitable businesses. They have, through experience, common sense and brilliance, evolved a number of unique strategic approaches (like reputation management, demand pull and thought leadership). These approaches developed from the fact that the industry is dominated by a very different type of business ownership: mutual partnership. The partners of a firm share its ownership, frequently having to earn their way into the business through outstanding long-term performance and capital investment. Yet they are also the elite leaders of the service. Not only do they own the business, sharing profit and loss, they lead client engagements, actively participating in the work. They head up cells of business units (called practices) which, because they are run directly by an owner of the business, evolve and respond to market changes; even when the leadership of the firm makes fundamental errors of strategy or management. They are, perhaps, one of the business world's few self-righting organisational structures.

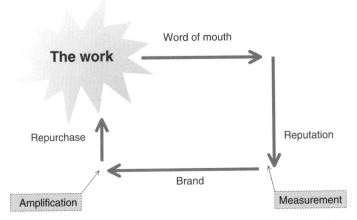

Figure 4.2   The role of reputation in creating demand pull for high-end professional services

Technology companies that want to engage in this type of business, perhaps as consultants, must have a clear idea of its dynamics. One dramatic example is the approach to revenue generation, which is based on enhancing the natural reputation of the practice. Most first-rate professionals will say that 'all marketing starts with the work'. For a number of reasons clients talk about a professional service after it is finished. This creates a strong reputation (which may eventually turn into a brand) and this, in turn, draws in more work, as illustrated in Figure 4.2.

This demand pull is the complete opposite of the product push used in, say, product sales and has two very powerful benefits. First, it keeps the cost of sales low (because the firm does not have to go out and get work) and second, it keeps prices high (because practitioners can focus on diagnosing need and pricing becomes a consequence, not a focus, of discussion). Within the professions this makes the difference between an elite profitable practice and a grubby ambulance chaser. As a result, all successful professionals focus their attention on it. They ensure that any strategy or business initiative is aimed at enhancing the firm's reputation or preserving it. Those that don't simply do not make good margins. So, a consultancy business set up by a technology firm ought to adopt this dynamic as its strategic focus or it will be condemned to be a low-margin, cheap alternative, constantly worrying about cash flow. It cannot be run the same way as a volume, process-based service.

The Hybrid Offer

A hybrid proposition is usually described as a combination of products and services. Systems integration projects, for example, meet the customer's need by examining the installation of existing equipment and using technical service skills to upgrade it, deploying any new equipment or software as part of the project, even if it belongs to other suppliers. 'Solutions' are also hybrids. They knit together products, services and skills to meet unique needs. They should create value for both the customer and the supplier, which is beyond the price of each of the components.

Another type of hybrid is the service hybrid. This is a combination of scalable service processes and professional skills. The financial adviser in a car dealership has to be skilled in financial management and act as a professional service adviser. They also rely on clear, scalable processes to design and deliver their different financial products. Similarly, some technical support services are provided by sophisticated, world-class technical advisers while being based on advanced streamlined processes.

The success of a hybrid relies on the division of front- and back-office processes, together with a clear idea of the 'perceived transaction time'. Profit for an airline, a software distributor and an international network support centre relies on the continual improvement of robust processes. The operations of the service must be as streamlined, effective and efficient as any manufacturing plant. Yet the 'front office' needs to cope with the erratic arrival of customers and the need to engage with them. Failure to spend appropriate time with them will damage future revenue growth.

METHODS OF DETERMINING STRATEGIC FOCUS

Although there might be variations in the service for different groups of customers or different levels of contract, the business needs to decide where its competence lies and which form of service maximises those skills to ensure healthy margins. It might focus on high-end, customised professional services or more process-based advisory work; or volume-driven process services; or pure after-care. Each of these business types has its own dynamics and success criteria.

Figure 4.3 shows the range of business types that exist in the service industry. It uses axes that represent the volume of buyers and the time spent with each one to categorise the different offers. Professional services, for instance, range from high-end customised services, at the top left of the diagram (such as some types of strategy consultancy) to the volume retail offerings of mobile phone services in the 'professional service shop'. Volume-based services, on the other hand, occupy the bottom right corner of the diagram. Many of these are trying to customise their offer a little through data warehousing and customer relationship management systems in order to give their buyers a more individual offer and to improve perceptions of service,

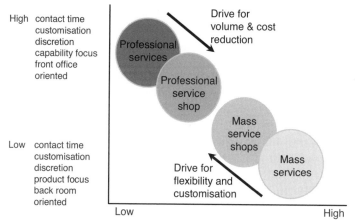

Figure 4.3 The service industry
Reproduced with permission from Pearson Education from *Service Operations Management*, Johnston and Clark.

such as in air travel. They are trying to become 'mass service shops'. The margins, approach to market, degree of engagement with buyers and personnel used vary enormously between these vastly different offers. Suppliers must be clear about their business model – they must have clear strategic focus – if they are to maximise profit and survive over the long term.

There are several concepts that have been developed by strategy specialists to help leaders select the right business focus, whether in charge of one service business or responsible for a large business with multiple service units. They include:

- **Business mission.** Creating one simple statement of what the business is about.
- **Business vision.** Used best in an environment where change is needed; this creates a scenario of where the business is headed in order to engage all in the change process.
- **Core competence.** This simple, but well-researched, idea suggests that businesses have one, or maybe two, functions at which they really excel and, if they are made an explicit focus of executives, can provide real competitive advantage. In many cases the core competence is not explicitly understood or managed by the people within the organisation. It is, however, the one unspoken priority and, over time, managers put time, money and expertise into improving it. There is though, it seems, real advantage in explicitly clarifying and articulating the business's core competence. The company can concentrate on it and subcontract all other areas, gaining enormous advantage in terms of cost control, clarity of talent development and priority of investment. This will make cost savings because the services which are contracted out will be given to companies whose core competence is in that area, so they will be able to perform them well at less cost. The core competence can even be reapplied to a different market opportunity and a range of new services built around it (Hamel and Prahalad, 1990). For example, in the late twentieth century the European service division of American computer giant Unisys settled, via brainstorming, on a core competence of 'the ability to manage the processes of computer service'. They were excellent at analysing a situation and putting in place a robust process. They called this 'Service Management'. By specialising in this area it was possible to offer to run diagnostic processes for their customers. They won a contract with a major brewer to replace the computer equipment of all of its three thousand bars throughout one country, a large and complex project. They designed the overall project, set up (through sophisticated electronic methodologies) the upgrade of each individual outlet and training courses for staff. They then went on to apply exactly the same skill to other industries, such as financial services, where they were able to offer clients more effective service at better costs. They had moved, very effectively and very quickly, from being a repairer of proprietary equipment to the manager of diagnostic processes in large distributed companies. Unfortunately, important though it is, there appears to be no recognised technique for identifying the core competence of an organisation. Once identified, however, it should pass three tests: creating value for customers; proving difficult for competitors to imitate; and providing access to multiple markets (like Honda's skill in creating small engines, or Canon's competence in imaging technology).
- **Experience curve.** Referred to earlier and detailed in the Tools and Techniques appendix, this concept demonstrates that a business gains cost improvements and efficiency gains in its prime area of focus. By identifying the overriding source of competitive advantage, the business can set targets to improve further. It can even estimate the experience curve of competitors. This tool is particularly appropriate to firms that offer high-volume managed services because the need to continue to drive down costs, which they can pass onto customers, is at the heart of their market proposition.

- **Service concept.** It is no coincidence that some of the very best service companies, which are routinely used as case studies in business schools and conferences, have a clear articulation to the world of what they are about. IKEA, Disney Resorts, Virgin Airlines and First Direct: all have a very clear idea of their service style, which interests employees, commentators, other businesses and customers. They create interest, excitement and wealth by setting out their business in a way that appeals to buyers. Remarkably few technology companies, by contrast, have such a clear and appealing vision of what their services are about. The service concept has been defined as:

> . . .a shared understanding of the nature of the service provided and received, which should encapsulate information about: the organising idea, the service experience, the service outcome, the service operation, the value of the service. . . It is something that is more emotional than a business model, deeper than a brand, more complex than a good idea and more solid than a vision. It is also something that can unite employees and customers; and create a business advantage. (Johnston and Clark, 2005)

There is a difference between a well-designed, high-quality service and an appealing service experience that entices and pleases customers; and the difference results in wealth. The service concept seems to be at the heart of companies that are able to create a unique experience. People seek out and spend money on service experiences like Virgin Atlantic and Disneyland because the idea of these services excites and interests them. Marketers should set out to create a shared understanding of their service concept. It should have depth, excitement and emotional appeal.
- **Service value chain.** Based on research by academics Heskett, Sasser and Schlesinger, this concept identifies all aspects of a service company that contribute to earnings and growth. Properly used, it can improve profitability by focusing employees on the main functions of the business and discarding extraneous activities. It can be made relevant to employees at different levels in the organisation using the 'balanced scorecard' approach (see the Tools and Techniques appendix).

Leaders of large corporate businesses that need to improve the performance of a jumble of different service businesses will probably find portfolio planning tools are the most useful to them. The most famous of these is the Boston matrix (see the Tools and Techniques appendix), which was developed by the Boston Consulting Group, around the experience curve concept. An alternative portfolio technique is the directional policy matrix (see the Tools and Techniques appendix), which was developed by McKinsey for its client General Electric in the 1970s. Launched soon after the Boston matrix, as a result of the inadequacies with it, the directional policy matrix is more flexible because it uses criteria created by the management team themselves to evaluate their options. It also creates a healthy and enjoyable debate among leaders of the firm about their market and business strategies.

One final method of shaping a range of service businesses is based on the dynamics of different service offers. Are they supporting equipment or not? Are they scalable, process-based, services or high-margin professional services? A representation of these different forces, which is an adaptation of a proposal by McKinsey (see Auguste *et al.*, 2006), is shown in Figure 4.4. Debate among the leaders of a firm about where each service unit fits on these axes of different success criteria will produce clarity of purpose; it will focus the service business around clear strategic intent.

Each of these ideas is explained in the Tools and Techniques appendix but can also be found in any good corporate strategy book and applied relatively easily. They are also, probably, interchangeable. Purists would argue (probably correctly) that each has complexity and has a

Types of service business.

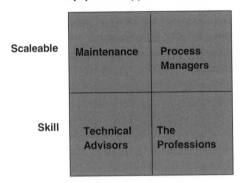

Figure 4.4   Using the forces of service success to clarify the shape of service businesses

different role in the formulation of a firm's strategy. Some would suggest that a large business developing a well-rounded strategy should cover them all. For the busy leader, though, any one can be used as a banner to rally the business around one clear purpose. This, in itself, will improve performance. Of overriding importance, though, is the achievement of clarity. Whether leaders use the 'experience curve', or a 'mission statement' or a 'service concept', the business needs clarity to thrive. It is more important to create a sense of mission than, for example, to create a neatly composed, bland 'mission statement'.

Hamel and Prahalad themselves said that their research showed there were several common components of good strategic focus. It:

- Captures the essence of winning.
- Is stable over time.
- Creates a sense of urgency.
- Develops competitor focus at every level through widespread use of competitive intelligence.
- Provides employees with the skills needed to work effectively.
- Gives the organisation time to digest one challenge before launching another.
- Establishes clear milestones and review mechanisms.

These characteristics have all been used in one of the most successful companies to have emerged in the Internet era, Google. With its vision of becoming the perfect search engine, Google has grown rapidly from a start-up to a global phenomenon without losing its clarity of purpose.

IN SEARCH OF PERFECTION AT GOOGLE

Global technology leader Google focuses on improving the ways people connect with information. It has a simple vision, to be 'the perfect search engine'. Its innovations in web search and advertising have made its website a top Internet property and the Google brand one of the most recognised in the world (ranked seventh in Interbrand's 2009 global brands survey, and valued at $32 billion). Founded in 1998 by Stanford PhD students Larry

Page and Sergey Brin, Google is headquartered in Silicon Valley with offices throughout the Americas, Europe and Asia.

Search is how Google began, and it's at the heart of what it does today. Co-founders Page and Brin named the search engine they built 'Google', a play on the word 'googol', the mathematical term for a 1 followed by 100 zeros. The name reflects the immense volume of information that exists and the scope of Google's mission: to organise the world's information and make it universally accessible and useful.

In little over a decade, there has been immense change to Google and the web as a whole. The first web page was created in 1990. Eight years on web pages numbered in the tens of millions and search became crucial. At this point, Google was a small research project at Stanford that became a tiny start-up; their search index used by a few thousand people, mostly academics.

Fast-forward to today and the number of people who use Google daily is in the hundreds of millions; the majority of its services being available worldwide and free to users as they're supported by advertisements. Importantly, billions of people now have access to the Internet via computers and mobile phones. So a child in an Internet cafe in a developing nation can use the same online tools as the wealthiest person in the world. Brin is delighted, 'I am proud of the small role Google has played in the democratization of information, but there is much more left to do.'

On a Mission

Google believes its 'user focus' is the foundation of its success. Certainly, the most effective and ultimately the most profitable way to accomplish its mission has been to put the needs of its users first. Offering a high-quality user experience has led to strong word-of-mouth promotion and traffic growth. Google's dedication to its users is reflected in three commitments:

- We will do our best to provide the most relevant and useful search results possible, independent of financial incentives. Our search results will be objective and we do not accept payment for search result ranking or inclusion.
- We will do our best to provide the most relevant and useful advertising. Advertisements should not be an annoying interruption. If any element on a search result page is influenced by payment to us, we will make it clear to our users.
- We will never stop working to improve our user experience, our search technology and other important areas of information organization.

As with its technology, Google chose to ignore conventional wisdom in designing its business. The company started with seed money from angel investors and brought together two competing venture capital firms to fund its first equity round. While the dot.com boom exploded around it and competitors spent millions on marketing campaigns to build brands, Google focused instead on quietly building a better search engine.

The word quickly spread from one satisfied user to another. With superior search technology and a high volume of traffic at its Google.com site, Google identified two opportunities for generating revenue: search services and advertising. Over time, these two business lines evolved into complementary networks, AdWords and AdSense.

Today, Google generates revenue (97% in 2008) primarily by delivering relevant, cost-effective online advertising. Google's ability to 'monetise' increased traffic largely relates to enhancing the end-user experience, including providing end users with adverts that are more relevant to their search queries or to the content on the Google Network websites they visit.

'The perfect search engine', says co-founder Larry Page, 'would understand exactly what you mean and give back exactly what you want'. When it began, Google was successful precisely because it was better and faster at finding the right answer than other search engines at the time. But technology has come a long way since then and the face of the web has changed.

Recognising that search is a problem that will never be solved, Google continues to push the limits of existing technology to provide a fast, accurate and easy-to-use service that anyone seeking information can access, whether they're at a desk in Boston or on a phone in Bangkok.

Ten core principles continue to guide all of Google's activities.

1. *Focus on the user and all else will follow*

    From the outset, Google has focused on providing the best user experience possible; taking care that every new initiative will ultimately serve 'you the user' rather than its bottom line. Google's homepage interface is clear and simple, and pages load instantly. Placement in search results is never sold to anyone, and advertising is not only clearly marked as such, it offers relevant content and is not distracting.

2. *It's best to do one thing really, really well*

    Google does search. With one of the world's largest research groups focused exclusively on solving search problems, it's been able to provide continuous improvements to a service that already makes finding information a fast and seamless experience for millions of people. Google's hope is to bring the power of search to previously unexplored areas, and help people access even more of the ever-expanding information in their lives.

3. *Fast is better than slow*

    'We know your time is valuable, so when you're seeking an answer on the web you want it right away – and we aim to please.' Google may be the only organisation in the world whose goal is to have people leave its homepage as quickly as possible. The average response time on a search result is a fraction of a second, and work continues on making it go even faster.

4. *Democracy on the web works*

    Google search works because it relies on the millions of individuals posting links on websites to help determine which other sites offer content of value. As the web gets bigger, this approach only improves, as each new site is another point of information and vote to be counted. Similarly, Google is active in open source development through the collective effort of many programmers.

5. *You don't need to be at your desk to need an answer*

    The world is increasingly mobile: people want access to information wherever they are, whenever they need it. So Google is pioneering new technologies for mobile services that help people around the globe to do any number of tasks on their phone. Android, a free, open source mobile platform, now brings the openness that shaped the Internet to the mobile world.

6. *You can make money without doing evil*

   Google is a business. Revenue is generated from offering search technology to companies and the sale of online advertising. To ensure it's ultimately serving all users, a set of guiding principles allow only ads to display: that are relevant where they are shown; which don't interfere with ability to see the content requested; and are clearly identified as 'sponsored links' to maintain the objectivity that users trust.

7. *There's always more information out there*

   Once Google had indexed more HTML pages than any other search service, its engineers looked to information that was not as readily accessible. From integrating new databases, to adding the ability to search news archives, billions of images and millions of books, Google continues to explore ways to bring all the world's information to people seeking answers.

8. *The need for information crosses all borders*

   Google's mission is to facilitate access to information for the entire world. Its search interface is available in more than 110 languages, offers people the ability to restrict results to their own language, and translation tools to discover content in languages they don't speak.

9. *You can be serious without a suit*

   Google is built around the idea that work should be challenging, and the challenge should be fun. Google puts great stock in its employees and embraces a culture that breeds innovation as one of its fundamental strengths. Despite its rapid growth, Google constantly seeks to maintain a small-company feel that promotes interaction and the exchange of ideas. The atmosphere may be casual, but as new ideas emerge in a café line, at a team meeting, or at the gym, they are traded, tested and put into practice with dizzying speed.

10. *Great just isn't good enough*

    Google sees being great at something as a start, not an end point. Through innovation and iteration, Google aims to take things that work well and improve upon them in unexpected ways. As important, it intends, to anticipate needs not yet articulated by its world-wide audience, and meet them with products and services that set new standards. 'Ultimately, our constant dissatisfaction with the way things are becomes the driving force behind everything we do' they say.

Summary

Google's mission – to organise the world's information and make it universally accessible and useful – remains as consistent as its focus on the users of search.

Google continues to think about ways in which technology can improve upon existing ways of doing business. New areas are explored, ideas prototyped and budding services nurtured to make them more useful to advertisers and publishers. However, no matter how distant Google's business model grows from its origins, the roots remain in providing useful and relevant information to the millions of individuals around the world who rely on Google search to provide the answers they are seeking.

BRAND EQUITY AS A SOURCE OF STRATEGIC ADVANTAGE

In a world dominated by luxury brands, celebrities and business icons, it is strange that technology companies so rarely grasp the remarkable strategic power of building a brand. Any service company will have developed a reputation (good or bad) by its approach to its work and the chatter of its customers afterward. This reputation can be a major influence on both the customer's likelihood of buying again and their willingness to recommend the business to others. In other words, the reputation of a business affects its future profits. Experience from numerous industries suggests that this reputation can be developed into a brand strategy that increases profits and customer satisfaction much further.

There are a number of issues that marketers must analyse and understand about their company's reputation or brand capability if they are to develop a coherent service marketing strategy: Do they, for instance, have any 'brand equity'? Can their business afford to invest in the reputation of their company in order to build an effective brand dynamic in their most important markets? Are the leaders supportive or sceptical of the value of brand investment? Is there any existing brand management capability?

Probably the first and most basic is the meaning of brand integrity in their market. As Chapter 5 will demonstrate, it is essential that the promises and expectations raised by marketing campaigns are properly reflected in the service experience. Does the way employees behave, the process through which the customers move, the environment in which the service is experienced and the technology used to deliver it, embody the brand promise? Does this happen every single time that people experience the service?

The second issue to understand is how famous the brand is and what it is famous for. Fame is a major ingredient in the success of brands. Yet many technology companies are inexperienced with brands and are reluctant to spend sufficient funds on marketing activity, which will make their brand famous. Those with significant consumer businesses, such as BT, Hewlett-Packard or Centrica, are better at this than their purely business-to-business peers, except where those peers have purposefully set out to use business-to-consumer techniques, such as at Accenture. They have demonstrated that this attitude to brand investment has to be changed if successful positioning is to be achieved. How strong is the company's reputation? How famous is its brand? Are the businesses leaders prepared to invest in the fame of their brand?

Third, marketers must identify a number of 'attributes' or 'brand values' that resonate with potential buyers. These will be either functional attributes which describe what the brand does (such as 'diagnoses', 'solves problems', 'advises') or characteristics that are judgements of value (such as 'integrity', 'quality', 'innovation' or 'excellence'). If this technique is to be successful, it is essential that the values are really experienced by the customers who use the firm's services. If not, the brand strategy will be merely a theoretical and meaningless corporate statement which is unconvincing to them or to employees. The firm should use properly conducted diagnostic research techniques with each of its intended audiences to understand how the service can be crystallised into a few realistic brand values. This is sophisticated and carefully engineered work, which must endure over years, and not a meaningless, changing wish list.

Many also set out to understand the financial value of their brand. In the past, marketers have found it hard to translate brand strategies into clear numeric data which wins the respect of their CEO, MD or finance director. Investment in many brand programmes, particularly corporate brand strategies, has therefore been based on little more than faith. As a result, worthy brand strategies have not got the investment they warranted and some non-brand-literate companies have not even begun to venture down this path. Yet, over the past two decades the accountancy profession has become increasingly focused on valuing 'intangible assets'. The international accountancy standards now require, for instance, that acquired brands are valued and recorded separately on the acquirer's balance sheet.

As developed economies have changed from a manufacturing to a service bias, many companies have found that intangible assets have grown in importance. From this point of view, brands are simply one type of intangible asset in a firm's array of assets and it is possible to use state-of-the-art valuation techniques in order to calculate their value. These include modern methods such as 'real option valuation' and 'royalty relief', which are more robust than some of the established, more judgemental methods. More importantly, it is possible to estimate the likely success of different brand strategies and to estimate the likely future brand value based on those strategies. In other words, it is possible to put real numbers behind brand strategy and calculate the likely return on investment. That makes it easier to understand the investment. However, to do that, it is essential to merge marketing and financial techniques in a profound way. Valuation and economic modelling can be merged with brand positioning tools to give a detailed, manageable route map for one of the company's most important areas of investment.

These issues, once properly understood, can be used as the basis of an effective brand strategy, as explained in Chapter 5.

CUSTOMER EQUITY

It is surprising how many organisations lie to themselves about who their customers are and the nature of their relationship to them. Many, for instance, claim to deal 'at board level' or advise 'the C suite', when, in fact, they deal at much lower levels. Others claim to have strong customer loyalty, with buyers who are reluctant to move to competitors. In some cases, though, this is more like apathy because the offer is not of sufficient importance for customers to bother to move elsewhere (in utilities, for example).

It is sensible to set out to understand exactly who buys what, where the revenue actually comes from and the likelihood of customers buying again. This propensity to repurchase is called by some accountants 'customer equity' because it is an asset, reducing, for example, cost of sales. A marketing plan cannot be started unless the marketer has an objective, clear-sighted view of the people who buy. How many are there? What revenues do they give and is there a means of understanding the profit per customer? Do any large customers dominate the firm's activities and is this a strategic vulnerability?

PORTFOLIO STRATEGY

Very few employees join an organisation that is completely new and has no customers at all. It is also rare to start with no products or services. So, a marketer must therefore understand the existing range of services offered by their firm to the market. How long have they existed? How profitable are they? How do they compare to the competition?

Care should be taken, however, when setting out to use any of the portfolio planning techniques taught in business schools and discussed earlier in this chapter. Many were developed for consumer products and in different market conditions to those experienced by service firms. The Boston matrix, for example, relies on the two axes of growth and market share, which may not be appropriate in many service markets. Similarly, many of the assumptions in product lifecycle analysis may not be appropriate. A detailed research project comparing the features of different services to customers' professed needs and competitors' offers may yield more useful insight into the firm's portfolio capabilities.

'GO TO MARKET' CAPABILITY

The marketer must take an objective view of their company's ability to go to market; the effectiveness of its sales and marketing infrastructure. The importance of developing marketing infrastructure inside an organisation is rarely emphasised by marketing writers. Yet, as discussed in Chapter 2, the maturity and influence of a company's marketing organisation develops over time. The effect of previous marketing experience creates expectations in a company of what is allowable, or do-able, as a current marketing strategy. For instance, business-to-business firms are often reluctant to conduct major TV campaigns. However, the advent of cable TV in hotels, 'narrow casting' in general and technology like YouTube makes it possible to communicate with business customers via digital advertising. To do so, though, the marketing director would have to argue successfully in the face of corporate scepticism. Having won the argument for the concept, they would then have to establish the internal management skills and the external agency infrastructure to execute it.

A similar process must be undertaken to build any new infrastructure for marketing or new forms of sales distribution. Marketing leaders have to convince senior colleagues of the value of the new approach and its associated capabilities. The investment needs to be weighed against other needs in the marketing stable and, perhaps more difficult, potential investments in other parts of the business. This will have both budgetary and political dimensions which can take time to skirt around. Once approved, though, funds need to be released, any suppliers chosen, talent recruited, technology deployed and processes set up. Only once these are in place can the work begin to flow through and new-style marketing programmes be executed. Whether it's direct marketing and database management, sales organisation, TV or PR, the establishment of marketing infrastructure is progressive; it takes time to evolve.

So, the marketing capability and expertise of an organisation develops as the firm becomes more market-orientated. It would be possible for a company to achieve things in the third year of marketing services that it would not be possible for it to attempt in year one. As more and more agencies, systems and people are established in the process of marketing, then the competence to make decisive changes in the market grows. Realistically, it is not possible to establish this infrastructure in the first few months of any new strategy due to limitations in the perspective of management, limited resources, inexperience and the capacity of the organisation. So investment in the development of marketing functions has to be prioritised. The pace of development of a marketing infrastructure, the acceptance of the role of marketing and its achievement is related to the evolution of an organisation and to the speed at which it can become market-led.

At the time of writing, some of the largest technology firms are entering their third decade of competitive service marketing. Having started with basics like packaged services and collateral, many have built progressive organisational competence in the marketing of services. Some have cooperated with leading academics to research new and developing ideas (like, for example, the co-creation of services). As many of the case studies in this book demonstrate, their ability to innovate new propositions, to launch service brands and communicate enticing concepts has progressively grown more sophisticated. They are beginning to rival, in service marketing, the sophistication of consumer brand marketing. Others in this sector need to take a hard-headed view of the organisational competence of their marketing and sales infrastructure in the service sector. Are the basics in place? Do elementary processes work? Does the marketing function really understand the concepts of service marketing so that they are able to design, launch and sell attractive profitable services?

HOW DOES THE COMPANY DEVELOP STRATEGY?

Finally, the service marketer should think about how their company develops strategy. The approach in many large, technical businesses is similar to that discussed in business schools and found in numerous academic books. Markets are examined, objectives set, research undertaken and programmes designed. This approach is 'procedural' (where a number of prescribed steps are followed to arrive at a particular point). It suits large, process-based organisations that need detailed investment strategies to be developed and large numbers to be coordinated. It particularly appeals to executives with a technical background because it breaks the vague and difficult task of strategy development into deceptively clear steps and activities. Yet it is not the only valid approach.

Some, for instance, adopt a 'functional' approach; where it is someone's job to draw up a well-presented and detailed strategic document. In others, strategy is not so logical or carefully developed. In businesses owned by one dominant leader or professional partnerships, there are not well-worked strategic plans or detailed planning timetables. The direction of the business can be more intuitive or ad hoc. The focus of executives in these organisations is on understanding the views of the leaders and determining the relevance to their area of responsibility. Strategic direction can be 'emergent' or 'extant'. It consists of a pattern of decisions by a dominant business leader, which are largely intuitive and often understood only in retrospect. These businesses are, though, frequently hugely successful at generating wealth. So this method of determining strategy is successful, despite being less clearly articulated.

These different approaches have different strengths and weaknesses; and none is ideal. Service marketers must, though, understand the approach to strategy development in their own firm. It is senseless to try to write a marketing plan which is more like a detailed book in an organisation that is freewheeling and erratic in its approach to business.

They also ought to be clear about the difference between the role of marketing strategy and the content of an operational marketing plan. Academic tomes have been written on the creation of marketing strategy and methods of marketing planning; plus the differences between the two. The first, like business strategy, creates a sense of direction. It is likely to include: market ambitions, portfolio changes, positioning, brand objectives and infrastructure projects. It can be developed as part of corporate strategy and can be articulated in a separate document. It gives a longish-term direction to marketing, a framework for decision-making.

Market planning, by contrast, is detailed and operational, giving resources and timing to marketing activities like campaigns and new service launches. In some modern companies their various marketing plans are little more than the detailed components entered into their campaign management systems. Both marketing strategy and marketing planning, though, must be conducted in keeping with the spirit, style and nature of the organisation. It serves no purpose to produce a detailed and academically well-rounded marketing plan if the organisation can do nothing with it.

SUMMARY

This chapter has explored a number of strategic issues that service marketers should examine in their own firm as they address their market. What is the strategic focus of the firm? Who are its customers and what does it offer them? How do those services match customer need and competitor offers? Does it have a brand or a reputation that might be built into a brand? Before seriously addressing the market, these issues ought to be analysed and understood in detail.

5
Creating and Positioning a Service Brand

INTRODUCTION

Brands are probably the most enduring and valuable entities that marketing skills have created for the business world. They differentiate an offer from competitors, protect against price erosion and increase margins. Whether they are product brands, service brands or corporate brands, they are amongst the most valuable, yet least tangible, of assets. Managed carefully, they can be the basis of sustainable competitive advantage by building strong, profitable bonds with loyal buyers. Yet, this can only be achieved by creating a branded proposition that stands out from the crowd and consistently delivers what it promises. Service marketers should use brand techniques to claim their space in the most important market of all: the minds and imaginations of their buyers. They need to position themselves in the service market, convincing their buyers that they are as good at creating relevant services as anyone else is at creating appealing products. With the right visionary leadership and political will, firms can create and manage a branded service proposition in such a way that it gives a clear direction to everyone involved, whether customer or employee. This chapter explores how to develop a successful service brand and to position it in a service market.

WHAT BRANDS ARE AND THE EFFECT THEY HAVE

A brand is an entity that engenders an emotional response from a group of human beings so that they pay more than they probably should for the purely rational components on offer. It is a changing, multi-faceted entity which creates a variety of impressions in the minds of different human beings through many different stimuli. It causes people to buy again and again, often at an inflated price, creating valuable equity. As the respected advertiser Jeremy Bulmore, a director of marketing services group WPP, said in the annual lecture of the British Brands group (Bulmore, 2001):

> Brands are fiendishly complicated, elusive, slippery, half-real/half-virtual things. When chief executives try to think about them, their brains hurt.

A brand is also a very valuable intangible asset comprising the goodwill of buyers and nurturing their future buying intent. Or, as Tim Ambler of London Business School has said, they are 'the future cash flow of the business' (Ambler, 2003). In 1996, for example, Professor Alvin Silk (Silk, 1996) estimated one of the world's most valuable brands, Coca-Cola, to be worth $39bn; whereas nearly 10 years later, Interbrand and *Business Week*'s estimate put it at $67bn – a remarkable increase in wealth. That 2004 survey contained a number of technology companies that were building a brand in services such as IBM (valued at $53bn in 2004 and $56bn in 2009), GE ($44bn and $48bn), Nokia ($24bn and $30bn) and HP (static at around $20bn). The changing value of this remarkable intangible asset is an addition to normal sales, revenue and profits.

Brands are among the most valuable and powerful propositions ever created by businesses. They differentiate an offer from competitors, create a price premium, enhance margins and encourage customer loyalty. Moreover, their effect is enduring. There are several brands in existence today that were created over a hundred years ago. Tables 5.1 and 5.2, for example, show simply the country and date of origin of a number of well-known brands that were created before 1900 and are still sold today. Some of these entities have been creating wealth by appealing, in a unique way, to a succession of human beings for several centuries.

Modern business executives listen to many conference speeches which assert that it is difficult to create substantial enduring value. There have even been ridiculous advertising campaigns which have claimed that 'the product you create on Monday is a commodity by Wednesday'. At the same time, business books routinely suggest that companies must continually innovate risky new products to survive. Yet the brands in these tables have commanded the attention of segments of customers and earned relatively high margins for many generations. Brand creation is, it seems, a powerful way to create substantial, profitable, long-term business. Marketers working for technology firms have a responsibility to their shareholders to convince their colleagues to invest in this remarkable approach.

Effective brands are most common in the consumer goods industries but are also increasingly found in both mass services (e.g. Virgin, McDonald's and Thomas Cook) and in professional services (e.g. Deloitte, McKinsey and Clifford Chance). They affect the price of every product or service these firms offer in every part of the world. They set expectations of quality and communicate subliminal messages of the firm's *raison d'être*. They are one of their owner's major intangible assets and should therefore be one of the prime strategic issues for leaders to address.

It is surprising, therefore, that brand management is not more generally recognised to be the powerful tool that it is. It has certainly had more impact on profit than many things attempted over the past years by different management teams. During that time companies have reached for changing management fashions (such as total quality management, process re-engineering, CRM or globalisation) as a means to create profit. It is astonishing that more have not invested resources and attention in this proven and enduring approach.

Perhaps one reason for this is that brand work involves such a wide spectrum of activities. At one end, creative agencies and consultancies help to design new images or new names. These projects attract public attention and, sometimes, criticism or even ridicule. At the other end, journalists like Naomi Klein (Klein, 2001) have challenged the ethics and integrity behind brand building. They have suggested that brands exploit workers and trick buyers, generating outrageous profits for unscrupulous owners. Moreover, the range of experts operating in the field of brand management proliferates by the day. In addition to professional brand managers in large corporations, there are strategists, design consultants and valuation specialists.

Despite the disparity of work, it is beyond dispute that a carefully designed image rests in the memory of buyers and helps them to choose products and services. Numerous firms have proved that, by managing that image carefully, a product or a service will appeal time and time again to a group of interested buyers. It becomes a familiar part of their life, giving them consistent benefits in their day-to-day activities.

As a result, they will pay a premium for the offer and develop a loyalty towards it. Over time, they become fond of these entities and, if they think about it, regard them as part of the landscape of their life. What starts as simple reassurance about quality or consistency becomes, on a deeper but hard to measure level, an emotional bond in a hectic modern lifestyle. Consequently, there are people who feel warmth towards a tin of paint, a sugar-filled drink or a pair of sports shoes. In fact, these items mean so much to them that they can be as upset and unforgiving if they think a favourite brand has been damaged as when a favourite soap character is killed off. The *Economist* said, for example, in 2001:

> Increasingly customers pay more for a brand because it seems to represent a way of life or a set of ideas. Companies exploit people's emotional needs as well as their desires to consume. Hence Nike's 'Just do it' attempt to persuade runners that it is selling personal achievement or Coca-Cola's relentless effort to associate its fizzy drink with carefree fun. Companies deliberately concoct a story around their service or product, trying to turn a run-of-the-mill purchase into something more thrilling. (© The *Economist* Newspaper Limited, London, September 6 2001)

Brands are part of the alchemy of marketing, a significant and important contribution to the creation of successful, enticing services. There is a clear correlation between developing organisational competence in marketing and professional capability in brand management; between effective brand positioning and enduring profit. This simple, yet hard to manage, strategy has created enormous wealth over the past century for both product and service companies with either a consumer or business emphasis. Yet it is hard for inexperienced firms to identify and manage all the components of an effective brand programme. As a result, many companies have ignored, to their detriment, the precious role that brands can play in the life of both companies and buyers. It is therefore possible that they might miss a very powerful, proven source of enhanced profit.

THE BASICS OF BRAND STRATEGY

Many technology companies do not have product brands and do not have a company name that commands a brand premium. They are therefore at a major disadvantage when trying to take a viable position in a service market because they do not have reputation, fame or allegiance that they might use to help convince people to buy their services at a premium. Nor do they have the competence to create and manage branded propositions or, usually, the political support from their leadership. Yet, convincing their buyers that they have moved to a service offer and positioning themselves effectively in the service market is crucial to the success of the whole venture. Without it they are unlikely to sell many services. They need to learn and apply the concepts of brand management as part of their service entry strategy if they are to be successful; they need to get the basics of branding established. These include two key concepts, brand integrity and fame.

Brand Integrity

The concept of brand integrity is behind the success of all brands. It means that the promise created for buyers by the brand name is always delivered when they use that product or service. Consistent delivery of an offer is the fundamental principle from which brands originated; often so obvious that it is taken for granted, unarticulated and not explicitly managed.

At the start of the industrial revolution most buyers could not rely on purchases of any kind to be consistent, reliable or safe. As in the undeveloped world today, a brand gave a much-needed promise of quality and delivery. For example, an advertisement from the 1880 *London Illustrated News* for Pears soap refers to the fact that the company had then been offering that product to the London market for over 100 years. (It is still a healthy brand today.) The message is not about differentiation, nor is it focused on identification with aspirations or health issues. It is about consistency. It has been an 'honest' product. And today's enduring brands became market leaders through consistent delivery to the expectations of buyers. Consumer brands have to deliver the promised function of the product, time after time, to be successful.

Table 5.1 Some product brands created before 1900

Product	Category	Country	Date	Product	Category	Country	Date
Aspirin	Pharmaceutical	Germany	1899	Woodbine (Wills)	Cigarettes	UK	1888
Renault	Cars	France	1899				
Bassett: All Sorts	Sweets	UK	1899	Lee Kum Kee	Food sauces	China	1888
Bugatti	Cars	France	1899	Kodak	Cameras	USA	1888
San Pellegrino	Bottled water	Italy	1899	Bulmers	Cider	UK	1887
				Smirnoff	Vodka	Russia	1886
Dentyne	Chewing gum	USA	1899	Coca-Cola	Drinks	USA	1886
Dewar's White Label	Whisky	UK	1899	Avon	Cosmetics	USA	1886
				Bacardi	Drink	Cuba	1886
				Whittard	Tea	UK	1886
Michelin	Tyres	France	1898	Glenfiddich	Whisky	UK	1886
Nabisco	Biscuits	USA	1898	Raleigh	Bikes	UK	1886
HMV	Entertainment	UK	1897	Lyle's Golden Syrup	Food	UK	1885
Chad Valley	Toys	UK	1897				
Hornby	Toys	UK	1897				
HP	Sauce	UK	1896	Andrew's	Liver salts	UK	1885
Shredded Wheat	Food	USA	1895	Fabergé	Jewellery	Russia	1885
				Dr Pepper	Drink	USA	1885
Lifebuoy	Soap	UK	1894	Bovril	Processed food	UK	1884
Barbour	Clothes	UK	1894	NCR	Technology	USA	1884
Andrew's liver salts	Pharmaceutical	UK	1894	UHU	Glue	Germany	1884
				Black and White	Whisky	UK	1884
Famous Grouse	Whisky	UK	1894	Jaeger	Clothes	Germany	1884
Hershey	Confectionary	USA	1894	Waterman	Pens	USA	1884
Mikimoto	Pearls	Japan	1893	Bulgari	Jewellery/ luxury goods	Italy	1884
Pepsi	Drink	USA	1893				
Halls	Cough sweets	UK	1893	Breitling	Watches	Switzerland	1884
Rowntree: Fruit Gums	Confectionary	UK	1893	Tiptree	Jams	UK	1883
				VAT 69	Whisky	UK	1882
				Cerruti	Watches	Italy	1881
Alfred Dunhill	Men's accessories	UK	1893	Rowntree: Fruit Pastilles	Sweets	UK	1881
Shredded Wheat	Food	USA	1892				
				Scott's	Porridge oats	UK	1880
McVitie: Digestive	Biscuit	UK	1892	Maynard	Sweets	UK	1880
				Hovis	Bread	UK	1880
Maxwell House	Coffee	USA	1892	Spear's	Games	UK/ Germany	1879
Ingersoll	Watches	USA	1892	Eu Yan Sang	Chinese medicine	Malaysia	1879
Wrigley	Confectionary	USA	1891				
Parker	Pens	USA	1891	P&G: Ivory	Soap	USA	1879
Player	Cigarettes	UK	1890	Quaker Oats	Food	USA	1878
Canada Dry	Drink	Canada	1890	General Electric	Consumer electronics	USA	1878
American	Tobacco	USA	1890				
Hovis	Bread	UK	1890	Ericsson	Technology	Sweden	1876
Nintendo	Entertainment	Japan	1889	Budweiser	Beer	USA	1876
Dunlop	Tyres	UK	1889	Tufts	Soda fountain	USA	1876
Sharwood	Spices	UK	1889	HP	Sauce	UK	1875
Schwartz	Spices	Canada	1889	Remington	Typewriter	USA	1874

Table 5.1 (*Continued*)

Product	Category	Country	Date	Product	Category	Country	Date
Southern Comfort	Whisky	USA	1874	Peek Frean	Biscuits	UK	1857
				Burberry	Clothes	UK	1856
Church's	Shoes	UK	1873	Spiller's: Dog Food	Pet foods	UK	1855
Levi Strauss jeans	Clothing	USA	1873	Louis Vuitton	Clothing/ fashion	France	1854
Chivers	Jams	UK	1873	Steinway	Pianos	Germany	1853
Horlicks	Drink	UK	1873	Aquascutum	Clothes	UK	1851
Stork	Margarine	Holland	1872	Singer	Sewing machines	USA	1851
Folger	Coffee	USA	1872	Bally	Shoes/leather goods	Switzerland	1851
Adams	Chewing gum	USA	1871				
Penhaligon	Men's scent	UK	1870	Jacob's	Biscuits	Ireland	1850
De Beers	Diamonds	South Africa	1870	Moss bros	Clothing	UK	1850
Campbell	Soup	USA	1869	Eno's	Pharmaceutical	UK	1850
Pillsbury	Bakery products	USA	1869	Oxo	Meat extract	UK/ Uruguay	1850
Tabasco	Sauce	USA	1869	Beecham's Pills	Pharmaceutical	UK	1850
Eno's salts	Pharmaceutical	UK	1869	Lobb	Shoes	UK	1849
Jack Daniels	Whisky	USA	1866	Bally	Fashion	Switzerland	1848
Nestlé	Baby milk/ confectionary	Switzerland	1866	Omega	Watches	Switzerland	1848
BASF	Chemicals	Germany	1865	Carlsberg	Beer	Denmark	1847
Bertolli	Olive oil	Italy	1865	Siemens	Electronics	Germany	1847
Bovril	Beef stock	UK	1865	Cartier	Jewellery	France	1847
Jacob: Cream Crackers	Biscuits	Ireland	1865	Maple's	Furniture	UK	1847
Fisherman's Friend	Lozenges	UK	1865	Pond's	Cosmetics	USA	1846
				Lindt	Chocolate	Switzerland	1845
McDougal's	Flour	UK	1864	Krug	Champagne	France	1843
Victory	Lozenges	UK	1864	Glenmorangie	Whisky	UK	1843
Robertson's	Marmalade	UK	1864	Bassetts	Sweets	UK	1842
Benedictine	Brandy	France	1863	Pilsner	Beer	Czech Republic	1842
Fray Bentos	Processed food	Uruguay	1862				
Bacardi	Drink	Cuba	1862	Whitman's	Chocolates	USA	1842
Rowntree	Chocolate	UK	1862	Royal Doulton	Pottery	UK	1842
Heinz	Processed food	USA	1861	Beecham's Pills	Pharmaceutical	UK	1842
Peek Frean: Garibaldi	Biscuits	UK	1861	Spiller's	Pet food	UK	1840
Otis	Elevators	USA	1861	Maker's Mark	Whisky	USA	1840
Swan Vesta	Matches	UK	1861	Pimms	Drink	UK	1840
Paneri	Watches	Italy	1860	Kikkoman	Soy sauce	Japan	1838
Peek Frean: Nice	Biscuits	UK	1860	Knorr	Soups & food stuffs	Germany	1838
Vaseline	Skin care	UK	1860	Hathaway	Shirts	USA	1837
Tag Heuer	Watches	Switzerland	1860	Tiffany	Jewellery	USA	1837
Chopard	Watches	Switzerland	1860	Procter & Gamble	Soap	USA	1837
Tate & Lyle	Sugar	UK	1859				
Canadian Club	Drink	USA	1858	Bird's	Custard powder	UK	1837
Peek Frean	Biscuits	UK	1857	Hermés	Fashion/luxury goods	France	1837
Smith & Wesson	Firearms	USA	1857				

(*Continued*)

Table 5.1 (*Continued*)

Product	Category	Country	Date	Product	Category	Country	Date
Lea & Perrins	Sauce	UK	1837	Asprey	Jewellery	UK	1789
Colt	Firearms	USA	1835	Waterford	Glass/crystal	Ireland	1783
Earl Grey	Tea	UK	1834	Schweppes	Carbonated water	Switzerland	1783
Bassetts	Confectionary	UK	1832				
Terry's	Confectionary	UK	1830	Veuve Clicquot	Champagne	France	1782
McVitie's	Biscuits	UK	1830				
Spiller's (later pet food)	Bakery	UK	1829	Bass	Beer	UK	1777
				Breguet	Watches	France	1775
				Mappin & Webb	Silversmiths	UK	1774
Guerlain	Perfume	France	1828				
Baedeker	Travel guides	Germany	1827	Breguet	Watches	Swiss	1773
Suchard	Confectionary	Switzerland	1825	Wilkinson Sword	Armaments	UK	1772
Clarks	Shoes	UK	1825				
Cadbury	Confectionary	UK	1824	Yardley	Cosmetics	UK	1770
Glenlivet	Whisky	UK	1824	Hennessey	Drink	France	1765
Macintosh	Clothes	UK	1823	Wedgwood	Pottery	UK	1759
Robinson's	Fruit juices	UK	1822	Fry's	Confectionary	UK	1759
Tetley	Tea	UK	1822	Guinness	Beer	Ireland	1759
Huntley & Palmer	Biscuits	UK	1822	MAN Group	Engineering/transport	Germany	1758
Johnny Walker	Whisky	UK	1820	Cinzano	Drink	Italy	1757
				Vacheron Constantin	Watches	Switzerland	1755
Beefeater	Gin	UK	1820				
Chubb	Locks	USA	1818	Wolsey	Knitwear	UK	1755
Armitage Shanks	Toilet ware	UK	1817	Axminster	Carpets	UK	1755
				Coalport	China/pottery	UK	1750
Remington	Firearms	USA	1816	J&B	Drink	UK	1749
Royal Doulton	Pottery	UK	1815	Villeroy & Boch	Tableware	Luxemburg	1748
Coleman's	Mustard	UK	1814	Chippendale	Furniture	UK	1740
Purdey	Firearms	UK	1814	Sèvres	Porcelain	France	1738
Crawford's	Biscuits	UK	1813	Blancpain	Watches	Switzerland	1735
Courvoisier	Cognac	France	1811	Remy Martin	Drink	France	1724
Heal's	Furniture	UK	1810	Martel	Drink	France	1715
Denby	Pottery	UK	1809	Cow & Gate	Dairy products	UK	1711
Colgate Palmolive	Toothpaste	USA	1806				
				Crosse & Blackwell	Soup	UK	1706
Pernod	Drink	France	1805				
Veuve Clicquot	Champagne	France	1805	Twining's	Tea	UK	1706
				Meissen	Porcelain	Germany	1705
Crombie	Clothing	UK	1805	Cutty Sark	Whisky	UK	1698
Chivas	Whisky	UK	1801	Dom Perignon	Champagne	France	1693
Greene King	Ale & beer	UK	1799				
Jim Beam	Whisky	USA	1795	Kronenbourg	Beer	France	1664
Sarson's	Vinegar	UK	1794	Haig	Whisky	UK	1627
Oban	Whisky	UK	1794	Bushmills	Whisky	UK	1608
Girard-Perregaux	Watches	Switzerland	1791	Beretta	Firearms	Italy	1526
				Lowenbrau	Beer	Germany	1383
Booth's	Gin	UK	1790	Stella Artois	Beer	Belgium	1366
Pear's	Soap	UK	1789	Weihenstephan	Beer	Germany	1040

Table 5.2   Some service brands created before 1900

The service	Category	Country	Date	The service	Category	Country	Date
Morrison's	Grocery	UK	1899	W. H. Smith	Retail	UK	1848
Bergdorf Goodman	Retail	USA	1898	Deloitte	Accountancy	UK	1845
Maxim's	Restaurant/ luxury goods	France	1893	The Co-op	Retail/banking	UK	1844
Sear's Roebuck	Mail order	USA	1893	Thomas Cook	Travel	UK	1841
Fenwick's	Retail	UK	1891	Heal's	Retail	UK	1840
Savoy	Hotel	UK	1889	Linklaters	Law	UK	1838
Slaughter & May	Law	UK	1889	P&O	Shipping	UK	1837
Westinghouse	Electronics	USA	1886	Tiffany	Retail	USA	1837
Sketchley	Cleaning	UK	1885	John Menzie's	Retail	UK	1833
Marks & Spencer	Retail	UK	1884	Harvey Nichols	Retail	UK	1820
Sainsbury's	Retail	UK	1882	Debenhams	Retail	UK	1813
Sears	Mail order	USA	1880	Claridge's	Hospitality	UK	1812
F. W. Woolworth	Retail	USA	1879	Citigroup	Banking	USA	1812
D H Evans	Retail	UK	1879	Clifford Chance	Law	UK	1802
Boots	Pharmacy retail	UK	1877	Hatchards	Bookseller	UK	1797
J Walter Thompson	Advertising	USA	1878	Norwich Union	Insurance	UK	1797
Liberty's	Retail	UK	1875	Lombard Odier	Banking	Switzerland	1796
Bloomingdale's	Retail	USA	1872	CMS Cameron McKenna	Law	UK	1779
Cable and Wireless	Communications	UK	1869	Gieves and Hawkes	Tailors	UK	1771
Saks	Retail	USA	1867	Christie's	Art auctioneers	UK	1766
KPMG	Accountancy	UK	1867	Lloyd's (TSB)	Banking	UK	1763
Marshall Field	Retail	USA	1867	Rothschild	Banking	UK	1769
HSBC	Banking	UK/China	1865	Sotheran's	Book sales	UK	1761
John Lewis	Retail	UK	1864	Hamley's	Toy shop	UK	1760
London Underground	Transport	UK	1863	Cox & King	Vacations/ tours	UK	1758
Lillywhite's	Retail	UK	1863	Coutts	Banking	UK	1758
H. Samuel	Jewellery retail	UK	1862	Dollond & Aitchison	Opticians	UK	1750
J P Morgan	Banking	USA	1861	Sothebys	Auctioneers	UK	1744
Macy's	Retail	USA	1858	Freshfields	Law	UK	1726
Banco Santander	Banking	Spain	1857	Fortnum & Mason	Retail	UK	1707
The Halifax	Banking	UK	1855	Bank of England	Banking	UK	1694
Standard Chartered	Bank	UK/Asia	1853	Barclay's	Banking	UK	1690
Le Bon Marché	Retail	France	1852	Pickford	Removals	UK	1690
Wells Fargo	Financial	USA	1852	Mitsukoshi	Retail	Japan	1673
Reuters	News	UK	1851	Lloyd's	Insurance	UK	1668
Moss bros	Retail	UK	1851	Thurn und Taxis	Postal services/ goods	Germany	1489
Pullman	Rail services	USA	1851				
Macy's	Retail	USA	1850				
Harrods	Retail	UK	1849				
PricewaterhouseCoopers	Accountancy	UK	1849				

Admittedly, expectations change over time and there is a need to keep pace. For instance, some food products reduced their sugar content dramatically in the last half of the 20th century. However, this happened step by step as consumers' tastes changed, moving in line with expectations so as not to lose the precious brand franchise.

Consistency in delivery is also vital for service brands. It is essential that the promises of the advertising and the expectations raised by promotion are properly reflected in the service experience. This means that the cocktail of service components (the way people behave, the process through which the customers move, the environment in which the service is experienced and the technology deployed) must embody the brand promise. Moreover, this must happen every single time that people experience the service. This is probably the most challenging and difficult aspect of marketing management experienced by any specialist in the field, for two main reasons.

First, the delivery of all services is unpredictable and variable. It is hard enough to translate research insights into product propositions. It can also be difficult to communicate the essence of a brief to design agencies and production people in order to create or change a physical product. Once done, however, the factory can normally continue to produce it without difficulty. It is much harder, though, to ensure that a service delivers its promise time and time again. Many mass services (in the airline, fast food and car hire industries, for example) set out to achieve this by sophisticated process design and the use of technology. They seek to 'industrialise' their service. However, this is more complex for business services, especially consultancy services, because they comprise sophisticated employees who do not wear uniforms or adopt any of the behaviours of mass services. The firm has to understand the common values that its professionals employ in their work and embed them in the design of their brand. Reverse engineering the brand in this way, around the client's experience, is the only way they can ensure that its promise is consistently delivered.

The second reason is scope and responsibility. Service delivery is normally the responsibility of the operations of the business but there are important components (HR, marketing and IT, for example) that are not usually under the direct management of a common unit. Brand specialists are unlikely to be given the power to affect all operations directly. They must work through influence and collaboration to achieve the desired experience for customers. In short, specifying, designing and managing consistent service to meet a brand position is both very important and very difficult.

For service businesses, then, brand integrity rests in the service experience. The people who deliver the service and the environment in which the service is experienced must match the expectations set by communications to the market. It is much more than designing a corporate logo, adjusting a strap-line or choosing a snappy name. When dealing with services, brand specialists have to focus on attitudes, principles and values that run through the organisation and which buyers actually experience. The emphasis of the firm should be on how the brand is experienced on a day-to-day basis and all of the components of the service must be integrated in one programme in order to reinforce the brand claim. Service marketers must begin their brand strategy with the integrity of the brand claim; the consistency between the service firm's external marketing and its customers' experience of the service.

Fame

Despite the concern of several modern commentators, fame and celebrity have been aspects of human society for a very long time. Actors, musicians, singers, politicians and even criminals are admired and followed for their public recognition as much as for a particular skill or achievement. Some are even famous for little more than continual appearance in the media; famous for being famous. Alongside favourite TV shows and magazines, these celebrities become a familiar part of day-to-day life. They are intriguing, beguiling and incredibly valuable as a result. People feel a sense of warmth towards them and tend to follow their own package of celebrity, magazine, style and soap opera.

Brands also occupy this arena. Their fame and their familiarity to different groups of buyers is part of their success. In the same way that a consumer might buy Burberry and be a fan of Tom Cruise, a business buyer might buy Pink's shirts, read the *Economist* and choose McKinsey. The wide knowledge in the general population of what these brands stand for enhances their appeal to those that buy them.

It is rather odd, then, that marketing professionals tend to dilute the importance and impact of fame by calling its use in brand work (rather meekly): 'brand awareness'. In non-marketing companies this allows leaders to under-invest in the creation and promotion of one of their greatest potential assets. In others, fame is a major ingredient in the success of their brands. Firms that have been successful in exploiting this phenomenon have invested, over the years, in marketing programmes aimed at making it familiar to as wide a range of people as possible. These might comprise a range of activities from brand awareness advertising, through product placement in feature films to high-profile sponsorship. All are aimed at the very simple objective of helping both the potential buyers, and the larger population, to remember the brand and to aspire to have its benefits. In other words, they set out to make their brand famous. In fact, research suggests that sales and revenue decline over time, as memory fades, when these programmes are stopped or paused in difficult times.

Aspiration is an important aspect of the fame and celebrity phenomenon. People mimic their favourite film stars or sports heroes because they want to be associated with the success represented by their lifestyles. This is often subliminal, but nevertheless very powerful. A young woman styling her hair like Jennifer Aniston or teenage boys dressing like British soccer star David Beckham are all associating with perceived success, and buy associated merchandise as a result.

Association with fame and success is also evident in business-to-business markets. The leader of a modestly sized business who wants a big four firm as their accountant, the director who stands on an awards platform with a well-known academic guru and the chief executive who basks in the glory of the latest acquisition led by one of the big merchant banks, are all reflecting this phenomenon to a certain extent. Even at the very highest levels of business fame, or brand awareness, plays its part. A chief executive might choose a consultancy because they trust an individual in it as a result of a long and healthy business relationship. However, they may regard the quality of the firm's brand as a mechanism to legitimise their decision, communicating to their staff that their enterprise will receive quality work.

The creation, maintenance and exploitation of brand awareness are, therefore, key components of brand strategy. To succeed, a brand has to become famous. Yet companies that are inexperienced with brands are reluctant to spend sufficient funds to achieve fame. This attitude has to be changed if successful positioning is to be achieved. They must be prepared to invest.

In addition to the two basic precepts of brand integrity and fame, there are three fundamental strategies that service marketers must manage in order to create value through brands. These are: distilling the brand essence; creating brand attributes; and bringing the brand to life through employee behaviour.

Distilling the Brand Essence

Each brand has a simple truth or 'essence' which it leaves in the mind of its intended buyers and its wider, admiring, audience. This is a promise that creates expectation and demand. For example, the essence of the Disney brand might be 'childhood magic', whereas the essence of an accountancy brand might be 'financial rigour'. In clarifying this brand essence, the firm seeks to understand the fundamental truth about its promise to its buyers and to build its presence in the market around this truth. This is sometimes called the brand message, the main idea communicated to the market about the offer, creating a reason to buy. Over time, this brand essence develops depth and different aspects as buyers use it. It creates a brand 'personality', to which buyers relate. One of the world's leading specialists in this field, David Aaker, defines brand personality as:

> . . . the set of human characteristics associated with a given brand. This includes such characteristics as: gender, age, and socioeconomic class, as well as such classic human personality traits as warmth, concern and sentimentality. (Aaker, 1996)

Brand personality is distinctive. Offers like Nike and Apple stand for something that is clear to analysts, buyers and employees. These brands are so rich and meaningful that different aspects of their personality appeal to different segments of customers, extending their effect. The customers themselves often interact with these entities as if they were people, developing, over time, a relationship which becomes part of the backdrop of their lives.

Brand Attributes or Values

It is possible to identify a number of attributes or values that will resonate with potential buyers. By making these values explicit and communicating them effectively to the intended customer groups, service marketers can make their brand strategy more effective.

Some of these characteristics, called hygiene factors, are essential if the firm is to compete in the market. They are likely to be strong in all competitors and discounted by customers. 'Motivators', on the other hand, are brand values that are meaningful to customers. They might include soft factors such as generosity; describing the tendency of a supplier to give insight and help which is not always charged but is part of a professional approach. On the other hand, they might be matters of style, such as the way relationships are conducted. Emphasis on these will, over time, differentiate the brand from competitors.

Academic researchers have uncovered three categories of brand attributes through their work, which seem both practical and useful (Darby and Karni, 1973). They include 'search attributes' (like name, price and product characteristics) which consumers can understand before purchase. The second group is 'experience attributes' (like fun, emotion and entertainment) which can only be evaluated after purchase and during consumption. A third group, particularly relevant to advisory businesses, is 'credence attributes'. Although these cannot be directly experienced or evaluated (and are more intangible than others), they are, nonetheless, very important. A number of corporate consumer brands have, for instance, been damaged by the discovery that their product is produced in child labour sweat shops. This damage to their credibility has affected profit and cash flow until repaired.

If this technique is to be successful, it is essential that the values are really experienced by the customers who use the firm's services. If not, it will be merely a theoretical and irrelevant corporate statement that is unconvincing to them or to staff. Service marketers should use properly conducted diagnostic research techniques with each of these audiences to understand how the service can be crystallised into a few realistic brand values. This is sophisticated and carefully engineered work, which must endure over years and not turn into a meaningless, changing wish list.

Reflecting the Brand in Employee Behaviour

As explained in Chapter 2, the employees of a service business are a very important aspect of its approach to market. Their day-to-day activities are, probably the most influential aspect of the brand. The language and behaviours of people employed by firms as different as Virgin and IBM are established; and people are only recruited if they will embrace them. The personality, dress and work style of the people employed by these organisations reflects the aims and objectives of their brands. The front-line employees of Virgin's businesses reflect the individualistic, modern wackiness of the Virgin brand as much as the consultants of IBM demonstrate its technical expertise and first-class business problem-solving.

Service marketers need to ensure that the promise of the firm's reputation and brand is delivered through its people every time the buyers experience its service. For instance, if the brand promises technical excellence, elite heritage, 'guru'-style insight or humour, then the people that buyers interact with must demonstrate these characteristics in their dress, language and behaviour. If not, the brand will be devalued over time, affecting earnings and margins. So, the people of the firm need to be chosen, trained and rewarded to reinforce its brand characteristics. Recruiters should be briefed to find behavioural and attitudinal characteristics (in addition to technical skills) which clearly reflect the organisation's brand values. These values then need to be reinforced through training, communication, example, performance appraisal and recognition.

Employees who understand and identify with the culture and objectives of their employer are more likely to give appropriate service and satisfaction to buyers. A mass service such as an airline will therefore make enormous efforts to make sure that staff are presentable, properly trained and able to take initiative within the scope of their jobs. This is more complex in a professional service, however, because the service is being carried out by people who are highly trained and also highly individualistic in their approach to life. They are self-motivated and committed to the technical areas of their work. They are extremely unlikely to wear uniforms or respond well to either over-engineered processes or unrealistic company propaganda. So, internal communications programmes to motivate such sophisticated employees to give appropriate service must be carefully crafted.

Specifying and influencing the human behaviour needed to reflect the brand is the hardest part of most brand programmes and it is particularly challenging if the firm needs to find a new position in the market. Leading service companies who want to change their brand perception find it a major task to motivate employees to adopt new brand values. Not only must they reposition their brand in a new, unfamiliar market, they must also embody the new brand values in people's behaviour as much as in any design or packaging. For most that have succeeded, it has been a decade-long journey of trial and error using world-class branding and HR techniques. Marketers who are handling the brand need to understand the attitude of their service people to their work and to the firm itself. They also need to understand how customers view these employees' behaviours. The new brand position and values should then be crystallised into specific human behaviours.

So, employees of the firm need to be an important focus of the brand manager, and need to be considered carefully in the design stage. When creating the brand essence and values, designers should derive their work from the attitudes of employees and customers' perceptions of them. Unfortunately, brand designers often start with a blank sheet of paper and build brand concepts around the leaderships' perceptions or aspirations. If, however, these are not dominant attitudes and behaviours at the customer interface, the brand will seem remote and irrelevant. It is better to start with a hard-headed audit of what is experienced by customers and make that the foundation of the strategy. Aspiration can be built on top.

Once a brand concept which is relevant to the service market is created, it should be tested on groups of employees to check that it is credible. Designed properly, this is an opportunity to unite the firm in a clear direction and motivate all the employees. Done badly, it will de-motivate and undermine the credibility of the leadership. The design and strategy should be adjusted in response to employee discussion groups as much as with customers. Once the concept is finalised, a detailed launch and communications plan to employees should be constructed; and this should be executed before it is launched to customers. If the firm has a thousand employees, there are more than one thousand opportunities every day for the brand to be communicated and, more importantly, demonstrated, to customers. This will affect their perception, causing them to buy again, develop relationships and recommend the firm to others. In short, it will increase revenue. An investment should therefore be made in a carefully crafted internal communications programme that is credible and sustained.

CORPORATE BRANDING: THE BRAND STRATEGY FOR SERVICE COMPANIES

One of the world's leading specialists in service marketing, Leonard Berry, concluded, after years of research into service industries, that there is a major difference between the way product and service brands should be created and managed (Berry, 1988). A product company can create a brand, which has its own presence in the market and is independent of its owners. When it is properly managed, buyers respond to the proposition itself, incorporating it into their purchase habits and returning again and again. The corporate entity behind the product proposition can be irrelevant to them. Brand house Unilever, owns, for example, brands as different as: Magnum ice cream, Vienetta, Dove soap, Lipton tea and Slim-fast. In these product brands it is the image and performance of the brand itself which carves out a position in its market. For other brands, though, the owner's name does become associated with the product brand (e.g. the Mars bar, Heinz ketchup and Kellogg's cornflakes), but the owner is often unseen and irrelevant to the consumer's interaction with the brand.

The dynamic with service brands is completely different, however, because the emotions engendered by the buying process are different. As we have explained, services (from a mass consumer service to customised or sophisticated consultancy services) involve a process through which the buyers and users must move. People must surrender themselves to the service provider and this yielding of control creates anxiety (which increases with the importance of, or unfamiliarity with, the service). As a result of this anxiety, service buyers reach around the service proposition itself to seek emotional reassurance from the entity in charge (without being aware that they are doing so). As a result, the great service brands tend to be corporate brands (e.g. Virgin and IBM). As Berry says: 'service brands should be the firm's name and should not be individualised'. This is represented in Figure 5.1.

This has implications for many aspects of brand development and naming strategy. For instance, a product company can organise itself so that brand management is handled by a division of specialists, sometimes under a director of brands. While being integral to the health of the company, the brand group can be managed as one function within it. However,

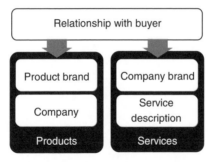

Figure 5.1 The difference between product and service branding

brand management for a service company is about dealing with the corporate brand, the name and reputation of the company itself. It involves a different set of stakeholders, including the firm's leadership, and a mix of responsibility. The corporate name might, for example, be the responsibility of a corporate relations director as opposed to the marketing director. Competitive positioning and brand essence for a service company involve the whole firm and are more complicated to handle; especially if the firm is inexperienced in, or unsupportive of, brand approaches.

Naming strategy is also different because the names of individual services have to be simple functional descriptors. Examples might include: business class from Virgin versus business class from American Airlines; or managed services from IBM versus managed services from Fujitsu. It is the corporate brand that gives a promise of style and difference to the service category. Branded service propositions or made-up names are unlikely to be successful.

One company which found this out to their cost was a UK retail bank called 'Midland' (now part of HSBC) in the 1980s. Its research showed that the population's attitude to banking was changing. It hired an experienced marketing director from the consumer goods sector who created a range of new bank accounts for young professionals and other customer groups. Called 'Vector' and 'Orchard', they were extensively researched and launched with millions of pounds-worth of advertising. They were impeccably designed using state-of-the-art consumer product techniques and were intended to be stand-alone offers in the market. However, they failed to draw in enough new customers and, after an embarrassing waste of funds, the bank 'returned to traditional banking'. Later, a new corporate brand, First Direct, was launched on the basis of the lessons they learnt through this failure. Similar attempts to 'productise' services or create an independent range of named services have been attempted and, generally, failed in the computer and consulting industries. Service brands tend to be corporate brands.

All of these generic principles underpin the success of brands. They are the fundamental basics which need to be in place to create an effective brand strategy. This is an essential, preliminary step towards creating a brand, and that, in turn, is an important component in the creation of enticing and profitable services.

BEYOND PETROLEUM

On 14 April 2009, BP celebrated its centenary. After 100 years in business, it is now one of the world's largest energy companies, providing its customers with fuel for transportation, energy for heat and light, retail services and petrochemical products for everyday items. Its 2008 revenues were $365 billion, from activities in about 90 countries through 96,000 employees.

Chairman Peter Sutherland described 2008 as an unprecedented year of dramatic change in the company's annual report. 'In the space of a few months, we went from a record oil price of more than $140 per barrel, and BP reporting two consecutive quarters of record profits for the group, to a recession in most of our major markets.'

Despite this, BP remains in a strong position, with strong assets including an established and enviable brand position in an increasingly difficult energy market. But this was not always the case. In 1997, BP's chief executive accepted the causal link between climate change and human activity; a year later, in response to plunging oil prices, BP triggered a wave of consolidation across the oil industry by merging with or acquiring a number of companies. In 2000, it therefore recognised that its brand needed refining and repositioning.

Growing Concerns: Energy Security and Climate Change

With both energy demand and carbon emissions rising, the world needs every sustainable, affordable energy source available. World energy consumption increased by 2.4% in 2007, the fifth consecutive year of above-average growth. Consumption could be 45% higher in 2030, with non-OECD countries contributing 90% of that energy demand growth according to International Energy Agency (IEA) estimates. Further estimates predict the world's population will grow from 6.7 billion today to 9 billion by 2050, with people in China and India moving from a rural to an increasingly urban way of life.

All available primary energy sources will be needed to address the twin challenges of energy security and climate change. Fossil fuels will play an invaluable role, but there is an urgent need for transitional incentives to bring forward the development and deployment of low-carbon technologies.

As a company with its legacy in oil, BPAmoco (briefly the company's name until 2000 before the rebrand) appeared to have a short shelf life as it focused on a fossil fuel that was perceived to be damaging the environment and shortly to run out altogether. BP claimed that it was committed to:

- Exploring, developing and producing more fossil fuel resources to meet growing demand.
- Manufacturing, processing and delivering better, more advanced products.
- Enabling the material transition to a lower carbon future.

But this was not the perception, and so the company realised it was time to take action and reposition itself in a changing market.

Repositioning the Brand

In 1999, British Petroleum merged with Amoco and then acquired the Atlantic Richfield Corporation Aral, Veba, Burmah Castrol and others. The new company sought to position itself as transcending the oil sector, delivering top-line growth while remaining innovative, progressive, environmentally responsible and performance-driven. BP wanted to demonstrate to key opinion leaders, business partners and its 100,000 employees worldwide how it intended to go 'beyond petroleum'.

BP intended to position itself as a new type of global energy company – one that confronts difficult issues like the conflict between energy and environmental needs and takes action beyond what is expected. The goal was to engage BP's extremely diverse employee population in this transformation, unifying them under the new brand in the process. Four characteristics (innovative, progressive, environmentally responsible and performance-driven) became the brand values and shaped the company's visual identity.

A new visual identity was designed to break with tradition. It was unlike any other energy identity, setting BP apart in the market and communicating more about the company's future than about its heritage. To reinforce this effect, all elements of the company's identity changed.

BP is now the company's global brand, appearing on production platforms, refineries, ships and corporate offices the world over, as well as on solar products, wind farms, research facilities and at retail service stations. The brand is summed up by the phrase 'beyond petroleum', since the company recognises that meeting the energy challenges of today and tomorrow requires both traditional hydrocarbons and a growing range of alternatives. Beyond petroleum is both what the company stands for and a practical description of what it does.

To bring the brand to life, BP created two key elements of its visual identity – a 'wordmark' and a logo. The wordmark (bp) was designed in lower-case letters to reduce the 'imperialism' of the brand, breaking with the company's traditional associations. The Helios logo accompanies the wordmark, named after the Greek sun god and symbolising diverse forms of energy. The colours of the Helios, white, yellow and green, suggest heat, light and nature. It is also a pattern of interlocking shapes: a single entity created by many different parts working as one. This was particularly relevant as the new identity was created after a series of mergers and acquisitions, and so the new brand worked to unite all the heritage companies and employees into a single company with a common vision and identity.

Once new brand and identity was developed, it was time to create a campaign that communicated the new BP both internally and in the external market.

The Brand Campaign

In July 2000, BP launched its new positioning. Brand champions ran an internal change programme using leadership communications, brand toolkits, chat room promotions, CEO satellite broadcasts, town hall meetings and celebrations around the world. Externally, BP worked with Ogilvy to produce an advertising and PR campaign, including a video news release of the new BP Connect retail service station, managing a press tour of the prototype with major media.

Launch-day media coverage included top-tier placements in print and broadcast, emphasising the new logo and the effort to 'go beyond'. BP's global workforce responded well, and in the month following brand re-launch, 76% were favourable to the new brand, 80% were aware of the four brand values, and 77% believed it was credible for BP to go 'beyond petroleum'.

After the initial launch, BP polarised opinions, dividing people into those who were impressed by the company's new approach, and those who saw the new brand campaign as 'greenwashing'. So, BP's communications strategy was developed further, to give a voice to real people about their energy concerns. The company responded using simple and straightforward language about how it was tackling those issues. BP recognised that it couldn't solve these issues alone, but was not sitting on the sidelines – this was embodied by the company's 'it's a start' language.

All BP executions aimed to demonstrate how the company goes beyond petroleum, both through its investments in alternatives, and how its policies and actions break with the conventions of the oil industry of the past.

Today, BP is constantly reviewing the brand positioning and how it is communicated through regular research. It has to remain relevant and support both BP's current business and future aspirations.

Results

BP's repositioning campaign won *PR Week* 2001 'Campaign of the Year' award in the product brand development category, and later a 2007 gold Effie from the American Marketing Association. Between 2000 and 2007, BP's brand awareness went from 4% to 67%. In Interbrand's 2009 Global Brands Survey, the brand was ranked 83rd in the world, up one position on 2008, and was valued at $3.7 billion.

But the repositioning exercise has not been without its critics, and BP stands accused of 'greenwashing'. *Fortune* magazine's Cait Murphy commented, 'Here's a novel advertising strategy – pitch your least important product and ignore your most important one ... if the world's second-largest oil company is beyond petroleum, *Fortune* is beyond words.' Despite this, a Landor Associates survey of consumers in 2006 found that 21% of them thought that BP was the greenest oil company, followed by Shell at 15% and Chevron at 13%.

PRACTICALITIES: HOW TO CREATE A BRAND WHERE NONE EXISTS

It is possible for firms of any size to create a brand that customers prefer, and pay a premium for, over many years. It is possible for service marketers to advise their boss how to create an entity which will earn a price premium, be distinct from the competition and garner a high margin over many, many decades. There are a number of steps toward this which, although they sound straightforward, are profound and difficult to achieve if a company has no brand experience.

Step One: Decide which buyer groups to focus on and concentrate on them to the exclusion of others

Clear customer segmentation (see the Tools and Techniques appendix) and deep understanding of human motivation are not academic or irrelevant theoretical ideas. They are the practical foundations for effective brand strategies. Brands appeal to the beliefs, needs and aspirations of people. They need to be built around a clear, unique human insight. Technology marketers have frequently failed to convince their colleagues and their leaders to invest in choosing a distinct customer segment. Technology companies tend to settle on broad, meaningless groupings (like 'SME' or 'EMEA') and, worse, try to continue to reach all groups. Even after an extensive, well-crafted segmentation project, senior teams can be reluctant to grasp the implication of concentrating on the most attractive segments: declining to serve the others. Many seem to want to wistfully cling on to all possible chances.

Yet the experience of those who are developing differentiated and exhilarating services in this sector demonstrates that they cannot even begin to position their brand effectively if they are not prepared to invest time and effort in understanding the human beings that make up a distinct group. Even in business-to-business markets, the start of effective brand development is human insight.

Step Two: Invest in understanding the rational and emotional needs of the group of buyers

Marketers need to invest in understanding the customers whose attention they want to attract. They must commission good-quality research, as discussed in Chapter 3 and detailed in the Tools and Techniques appendix. This will be needed to identify and validate customer segments as much as to concept test the emerging brand strategy.

Step Three: Develop a brand strategy

The company needs to create a specific strategy by which it will develop and handle its brand. They might adopt, for example, a 'monolithic' brand strategy. This means that the company will have only one brand, which will be reflected in all its 'public face' (from building livery to the behaviour of its people). This allows the firm to invest in one intangible asset and to communicate that one entity through all points of contact with buyers. To succeed, all identity and communication pieces must reinforce the values and image of this master brand.

'Sub-branding', by contrast, is the creation of many different brands. These might be offers in their own right or linked to either the company brand or a generic brand. While completely stand-alone brands are unlikely to be successful for service companies, a form of sub-branding that is open to them is to apply the brand to organisational groupings. By applying the corporate name to different business units, the firm signals that the subgroup is operating its business with a similar style and approach as the corporate entity. For example, a group of businesses competing in business strategy could be branded as: 'KPMG Advice' or 'McKinsey Strategy' or 'IBM Business Insights' (all fictitious). Each signals a different approach to consultancy subliminally communicating a different 'flavour' of service, based on the main brand's essence. This form of brand architecture can support different organisational groups and their position in different markets.

The brand strategy needs to articulate: the intended segment of customers, the positioning, the rational and emotional promise, brand values, any strap-line and the managerial plans to ensure that this strategic intent is actually experienced by customers in the service.

Step Four: Design the brand

The success of a brand rests on the shorthand it creates in the minds of buyers, so a clear, unique proposition needs to be designed to represent it. A professionally managed design process needs to be used for this which is not as straightforward as it sounds. First, a designer needs to be employed to create a colour scheme which reflects the mood and style that the firm's leaders want to achieve. For instance, a company with a long heritage may want a 'classic' style, whereas a high-tech service management firm may want a modern image, or a management consultancy may want an open-minded, fresh approach.

Colour gives this subliminal message. A designer should first produce a palette of colours representative of a 'mood and style' which reflects the firm's objective, such as the Helios created to bring BP's repositioning to life, using white, yellow and green to represent heat, light and nature. There is added complexity if the firm is international, because colour gives different subliminal messages to different cultures. For example, to British eyes pure white means freshness and cleanliness. To Japanese eyes, on the other hand, it is the colour of mourning. The firm must take on board the subliminal message of the intended colour scheme and not just the aesthetic appeal to its leaders.

The design scheme must cover all of the 'public face' of the firm in every physical manifestation of its presence. This encompasses such things as: signs on buildings, design of reception halls, letterheads, fax headers, business cards, invoice formats, email, presentation slide format, the website, proposal documents, report covers, conference appearances and so on. Some organisations even apply it to staff briefcases, PCs, vehicles, toolboxes and other items regularly seen by customers. Large firms often have to undertake an extensive audit of all the points of contact with their public and are frequently surprised by the extent of the project. A useful technique here is a 'contact audit' (see the Tools and Techniques appendix), which ensures that all points of interface with the public are identified.

A designer needs to apply the colour scheme to representations (mock-ups) of all these items. These must be tested in discussion with customers and employees to gauge their reactions to the design before being finalised. Some companies are hesitant to put unformed propositions about their own business to customers. However, experience suggests that customers are often willing to help a worthy supplier to find their way and the risk of launching an unpopular or negative design is worth the much lesser risk of imposing on a customer's time a little.

Very often a firm uses a 'strap-line', a short statement intended to communicate the essence of the brand, to reinforce the thrust of its approach. For example, at the time of writing, Orange uses 'Together we can do more', BP uses 'Beyond petroleum', Nokia uses 'Connecting people', bmi uses 'Better for business' and Eurostar uses 'Little break, big difference'. A strap-line should reflect the brand essence. If its claim isn't truly reflective of the firm's approach, then neither staff nor clients will see it as having any relevance and it will have no effect on the brand's asset value.

It is absolutely essential that any words associated with the brand or names are checked before implementation. This might seem to be common sense but many fail to do it well. There are numerous chances to make mistakes in this area and they can be very costly. Choosing a name, for instance, where the web address (URL) is already owned by someone else, or means something obscene in a foreign language, is more common than is usually admitted publicly. Choosing a strap-line that will not translate into other cultures is also very common. Before the project is finalised all names should be trademarked, URLs tested and translation into important cultures of the world checked.

It is a huge task to ensure that all contact points with customers in all parts of the firm are subject to the redesign. Once implemented, however, it is essential that someone is responsible for controlling the integrity of the design. There will always be a reason why employees feel they need to adapt a colour, a piece of design or a strap-line. This should be resisted at all costs as it undermines the subliminal message of the brand and damages the financial value of the asset. Even a slight change in colour will create an impression of confusion and diversity in a customer's mind once a number of pieces of material are produced differently in different parts of the world.

Current practice is to develop an internal website where all standards and materials, together with an explanation of the importance of compliance, are set out in full. Managers across the firm can access and download materials from the site. This helps ensure that all aspects of brand design are produced in accordance with the overall design scheme. It is absolutely essential, however, that leaders at all levels of the firm reinforce this necessity. If they do not, they are damaging a valuable intangible asset of the firm and are being negligent.

Step Five: Test, launch and communicate it

Brand communications must be integrated into the mix of communications that the marketer crafts to influence the generic market and specific customer groups. Sometimes it will be an objective in its own right, with a dedicated campaign, and at other times it will be a component of a campaign. Yet it must always be a specific ingredient of communications policy and planning if it is going to earn substantial margins over a long period of time. This is explored alongside other communication tasks in Chapter 9.

PRACTICALITIES: REPOSITIONING AND BRAND EXTENSION STRATEGIES AS A MEANS OF ENTERING SERVICE MARKETS

So, how should technology companies that already have strong brands or a company reputation reposition themselves in a service market? How can they convince existing customers, or potential new buyers, that they have viable services to offer?

Consumer product companies have been very successful at repositioning product brands. Cosmetics companies extend one brand into ranges of products and confectioners into different types of chocolate bars and ice cream. Buyers seek a familiar taste or style in the new category of product. Examples include the Mars bar launch into ice cream and the extension of the Dove range of personal hygiene products from one soap.

Some service companies have also succeeded at penetrating markets through similar brand repositioning. Richard Branson's Virgin Group has moved from entertainment to airlines to trains and into many other retail areas, like communications services. Each move had a sound business reason (normally based on shaking up a stodgy, cosy market), but each time the brand took existing buyers into that market and picked up others en route. For instance, the early Virgin flyers were young business people who had learnt the group's style by experiencing its entertainment before they started their career. They also moved with it when they were ready to invest in financial services. Yet, with each move, the brand picked up new customers too, such as in the story of Virgin Media. It is a classic example of successful repositioning.

This is no small task and the start of this risky process is a clear understanding that the brand exists in the most important place of all: the imagination of its customers and potential buyers. Years of successful product business and rewarding transactions for both the business and its buyers have created a valuable heritage; a brand equity. The goodwill towards the company might have slightly different meaning for each buyer, but each nuance is encapsulated in its company reputation. It is this that the firm needs to move progressively into the service market.

This manoeuvre needs to be undertaken carefully, taking customers with the strategy and picking up others en route. Moreover, under no circumstances must it endanger the existing business or brand franchise. A poorly executed programme is likely to cast doubts on the competence of the firm. It will cause customers to wonder if the company can still run its existing technology business effectively.

The leadership needs to demonstrate that it is applying its normal business approach, resources and skills to the service market. It might have been innovative or respectful or fun or thoughtful in its technology ventures. It needs to be the same in services. For example, at the time of writing, the Finnish mobile manufacturer, Nokia, is moving into services. As its established reputation is for excellent design and fashionable products, it needs to approach services in the same way.

Once again, there are several steps to successful brand repositioning in a service market, which may benefit from external support if there is little expertise internally to carry them out.

Step One: Undertake a brand audit and identify transferable brand values

The marketer needs to understand the strengths of the company's brand and the franchise it holds with four distinct audiences: its existing customers, its employees, city investors and the general public. A 'brand audit' (see Tools and Techniques appendix) is a research process that determines this. It identifies the image that each audience has and reduces these to identifiable values. It also highlights any areas where there may be inconsistencies. These range from items where the design conflicts with perception to operational issues such as flaws in customer service.

The marketer must use this data to identify brand values that might be a basis for a new position in the service market. For instance, the firm might be seen as 'responsive,' after years of listening to customers and responding to them. Or it may be perceived to be 'high integrity' or 'high quality'. All of these are values that meet service needs. Any negatives that are identified by the brand audit should be confronted honestly. If they are major perceptions or impediments, they must be confronted by the future strategy and by clear, effective communication plans.

Step Two: Identify the intended brand position

Perceptual maps (discussed in Chapter 3 and detailed in the Tools and Techniques appendix) can be used to clarify an intended market position. The maps can be used to determine the likely number of customers, the most relevant value proposition and to anticipate the likely reaction of other competitors. They can also be used to work out the service style which is appropriate for the firm's strategic position. Perceptual maps can also help with internal communications of the new position. This highlights any gaps between what the firm believes and what the buyers actually experience. It enables the firm to create internal communications and education programmes to bridge the gap.

Step Three: Create a transition strategy

There are a number of ways in which a brand can be moved in a service market. The first is a phased transition. If the firm has a strong brand, it can create its new service business by association with that brand. For instance, a product company could create a separate division, which specialises in services. IBM used this strategy as it moved substantially into services. As a result of nearly a decade's investment, 'IBM Global Services' is now one of the world's largest services businesses and has a healthy brand reputation. The juxtaposition of the company name with services signalled to buyers that the approach and values of the firm were to be applied to the service market. The heritage, quality and expertise of the world market leader could be experienced in the service market.

A second transition strategy is the dramatic move. This is best achieved through acquisition or joint venture. A product firm is unlikely to have the competences to successfully set up and run a new service business very quickly. So, a method of minimising risk while setting up expertise quickly is through purchase or mutual partnership. Each has its strengths and weaknesses. Acquisition is an immediate commitment. It gives the firm the ability to create its own direction with the new service business from day one and, as importantly, signals its intention firmly to its customers. It can change the brand name immediately or it can phase the change. When, for example, IBM bought PricewaterhouseCooper's vast consultancy business it changed the name immediately, signalling the intent to integrate this organisation into its core service business.

A third transition strategy is to build on gradual organic growth. A manufacturing firm is likely to have some service business based on, for example, maintenance work. This can be used as a basis to build the service business. However, it is usually difficult to finally convince the market that the business has become a service company, even when it is the dominant nature of the business.

Step Four: Launch and communicate the brand

Whatever transition strategy is chosen the firm must invest in communicating the new brand. This involves serious investment in convincing the market that the firm now has a viable service brand. One of the best known service brands in the technology sector is, for example, Accenture. Yet its fame is no coincidence. When, in 2000, it was first set up from the then accountancy firm Andersen, it was launched with an advertising and sponsorship budget of $100 million. Since then, it has consistently invested in its brand, communicating its 'High Performance Delivered' promise through association with sports celebrity Tiger Woods, and most recently underlining its credibility with the campaign 'We know what it takes to be a Tiger'. This campaign demonstrated how the company draws on its work with clients in numerous industries around the world to help transform the businesses of new clients. The communication strategy must match the transition strategy though. The communications plan to match a dramatic move, like an acquisition, must be shaped very differently from that supporting a phased transition.

Step Five: Measure the brand change

The firm needs to know the view of customers in the intended market of its brand relative to competitor brands. As the brand is such an important asset, it is sensible to set up measures of its health in the marketplace. To do this, it is normal to establish some form of brand tracking survey. This can either be done by buying into one of the many brand-tracking surveys run by large marketing firms, or by creating a proprietary survey. The best surveys use some form of conjoint or trade-off technique to break down the elements of the brand and the way it resonates with customers relative to competitors. These surveys allow the firm to adjust the strategy in light of changing customer views and competitor actions.

Virgin is one of the companies that understands the value of branding and measures its brand performance carefully, particularly in competitive markets such as that of mobile telephony and home communications, where it has established a strong position in a relatively short space of time.

LAUNCHING VIRGIN MEDIA

Almost 10 million customers choose Virgin Media in the UK today, which makes it one of the largest residential broadband providers in the country; the UK's largest mobile virtual network operator, and its second largest home phone and pay TV provider.

In cabled areas, its advanced fibre optic cable network reaches 12.6 million homes, delivering high-quality TV with pioneering 'on demand' content as well as broadband services up to 50Mb. In addition, Virgin Media offers ADSL broadband and telephony throughout the country, delivered over telephone lines.

But it's not all about size. Virgin Media is a company that understands people as much as technology. In other words, it's a typical Virgin company (in fact the largest Virgin branded company in the world with revenues just over £4bn in 2008), even though it was created from six very different companies as recently as 2006.

Creating a New Virgin Company

In November 1999, Virgin Mobile launched as the world's first mobile virtual network operator, listing on the London stock exchange in July 2004. In March 2006, ntl and Telewest, the UK's two leading cable companies, announced the completion of their merger (which also included Flextech Television, responsible for channels such as Living, Challenge and Bravo; Virgin.net, an ADSL-based national Internet service provider; and ntl:telewest business, which provided business-to-business IT, data, voice and telephony products and services). On 4 July 2006, ntl Telewest completed its acquisition of Virgin Mobile.

At this point, planning began to relaunch the group of companies as a Virgin branded company. The newly merged executive board wanted a relaunch within 3 months, and a new managing director of brand, Ashley Stockwell, was hired to lead change workstreams across the company. He quickly reset expectations on the executive team that a brand launch like this one was about more than a logo, and as such would take much longer than they had planned. The relaunch would have to cover more than 15,000 staff, 100 buildings and a fleet of over 5500 vehicles.

Coming from the Virgin Group, the new MD understood his first job was to make sure that everyone inside the new organisation knew what it meant to be a Virgin company. A cultural change programme was designed to brief all staff on Virgin's core values:

- value for money;
- competitively challenging;
- innovative;
- brilliant customer service;
- good quality;
- fun.

All of the companies involved had very different cultures, so the merger and creation of a new single culture under the Virgin brand was a critical success factor. Luckily, most people were thrilled to represent the Virgin brand, and welcomed Virgin as a distinctive and attractive company for which to work.

Perhaps more challenging was the need to consolidate legacy systems and processes across the new company. Members of the newly formed brand team collaborated with workstream leaders on a number of change projects to make sure the customer experience would deliver the Virgin brand promise. For example, in the billing workstream, a number of legacy systems were consolidated with the aim of simplifying bills so that customers would quickly be able to understand their bill and have charges for all the services they used in one place.

Brand team members reviewed all services and propositions to customers, again making sure that they represented the Virgin brand and values; and were meeting customers' expectations around great service and value for money.

As the relaunch date loomed, an internal teaser campaign began with the words 'Start the revolution'. All the usual internal communications channels were used in addition to more guerrilla tactics, such as stickers appearing behind toilet doors. For managers, the campaign was titled 'Lead the revolution' and involved a special programme of workshops to brief them on how to lead in a Virgin company. This was followed up with the do-it-yourself 'Revolution takeaway' in a Chinese takeout-style bag, containing tools and props to help managers work through with their teams what being Virgin meant in their area of the business.

One key message that all managers received was that expectations internally and externally would be high with the creation of a new Virgin company. Known for coming into markets and shaking things up on behalf of the consumer, everyone expects change overnight when a Virgin company launches. Stockwell made it clear to all that the launch was more of a journey which would take the company 3–5 years to complete.

The Birth of Virgin Media

The combined company relaunched as Virgin Media in February 2007. Virgin Media was a brand new entertainment and communications company offering consumers a great choice of the latest products and technology, outstanding customer service and great value.

Internally, the launch campaign built on the teaser campaign, with the rallying cry 'Start the revolution'. On launch day, everyone received a 'Party in a box' containing bunting, streamers, balloons, coasters and a shot glass, along with a little red book that explained what it meant to be a Virgin company.

Externally, a teaser campaign led up to the launch on 8 February 2007, using advertising, direct marketing and press coverage to announce the arrival of Virgin Media and demonstrate with both the look and feel, and the tonality of the communications, that this really was a new company built from the best of the companies that merged to create it.

Just as he had done many times before for other Virgin companies, Richard Branson took part in a publicity stunt on launch day. He spent the 8th February in a glass box in Covent Garden, London, alternatively using his mobile, broadband and TV to spread the word about Virgin Media, and receiving celebrity visitors through the day.

A special package was sent to existing customers letting them know that Virgin Media was now their supplier, summarising what had changed, and how the changes would benefit them. For most existing customers, the visible change occurred when they turned on their television or went online. Where once they saw an ntl or Telewest home screen, they now saw the new Virgin Media electronic programme guide, and the old ntl and Telewest portals were replaced with virginmedia.com.

The old Virgin Mobile stores were rebranded and began to sell the other three services now available to customers. Signage on all other premises in the estate was changed over a two-month period. Perhaps the most challenging external rebrand was on the fleet of vehicles, which took longer to change simply because it was a working fleet and needed to be tackled gradually to avoid any drop in customer service.

Repositioning the Brand as it Grows

When Virgin Media launched, its focus was on announcing its arrival and its points of difference, to challenge competitors in the market. The competitors were tough and had deep pockets with which to fight back once Virgin Media launched in their industry. With customer churn a vital industry performance indicator, it wasn't long before things became very competitive and frenetic, with everyone competing for more customers at the expense of creating real value. For example, everyone was leading their advertising with price messages, and then with the technological features of the offer – benefits were hard to spot and so customers increasingly treated the services on offer as commodities.

Virgin is used to this type of competitive response from larger, more established players, having roughly 20% of its major competitor's resources in most of the markets in which it competes. And so, while its early offers to customers were around better value and led with price, over time the company began to reposition itself around its other differentiators. In April 2009, Virgin Media began to establish a more premium position, emphasising its great service, the entertainment aspects of its offers and focusing on benefits. The message to customers was clear: Virgin was different and worth paying more for.

Today, the brand team remains at the heart of the business, looking after corporate identity, core brand look and feel, tone of voice, championing the customer's experience and delivering the brand via internal communications. The team regularly gets involved in new product development, for example taking the lead on the design of a new modem to support the new ultrafast 50Mb Internet access, the UK's fastest broadband service, in December 2008.

Creating Value the Virgin Way

At the heart of Virgin's ethos is the idea that if you look after your staff, they will look after your customers, and your shareholders will be happy. This has certainly worked at Virgin Media.

Brand tracking measures show a steep climb for the brand since it launched in February 2007: *Marketing* magazine's Best Loved Brand survey showed a leap from 53rd in the UK to 6th in just 12 months, and a recent *Readers Digest* survey showed Virgin Media on a par with the BT brand; quite something in just two years. The sector's key performance indicator, customer churn, is significantly down from the ntl Telewest days. Virgin Media also uses the 'net promoter score' (NPS) to keep track of how it is doing. The company tracks NPS through every part of the customer journey, and can give scores to individual agents, giving them specific feedback on what they're doing well and what they need to improve.

As more people join the company, it remains just as important to share what it means to be a Virgin employee, and this is reinforced in the recruitment process, induction programmes, training and performance reviews. When he can, Richard Branson will call in to see staff around the country, listening in on their conversations with customers to keep in touch with what they expect from Virgin.

He also gets involved in helping customers use the services on offer, creating 'webisodes' that can be viewed online, which explain how to get the best out of the services available.

So, from being the first people in the UK to offer customers TV, broadband, phone and mobile, Virgin Media sees the future as bursting with fresh entertainment and communication possibilities. Their ongoing mission is to bring customers all the excitement possible and make their digital place the brilliant place it should be.

A NOTE ON POLITICAL WILL AND LEADERSHIP VISION

So why hasn't brand strategy been more fully adopted by different companies? Unfortunately, all the famous branding exercises look good in retrospect. Before Nike emerged as a leading sports brand, people bought gym shoes or sneakers. Teenagers (and their parents) would have been astounded to be told that they had to pay over $100 for a pair of 'running shoes' and would wear them to go out with friends. Similarly, before Intel, microprocessors were just 'chips'. It took visionary leaders to invest in the development of those brands, to invest in brand strategies that turned a near commodity into a value proposition. The leaders of firms like Intel and Virgin Media had to put their personal political capital behind the risk of a commitment to brand.

Truly adopting brand management involves both visionary leadership and a change of organisational emphasis. The brand, once created, should give direction to everyone in the business. However, large firms do not typically have the political commitment to alter the balance of power radically in their internal operations in order to achieve this longer-term benefit. They often have to be driven there by relentless market forces, going through traumatic management change en route.

Smaller companies, on the other hand, can be daunted by the power of better-known consumer brands like Coca-Cola or Nike. They forget that many successful brands (like DELL, Google or even Microsoft) were built from scratch by business leaders with very modest initial resources. However, whether brand success has resulted initially from vision, or luck, or the ravages of the market, the steps needed to succeed are well known. If the firm's leadership is committed to creating effective services, robust plans need to be put in place and brand strategy ought to be part of those plans.

SUMMARY

Effective brand management is a powerful tool and an important aspect of service marketing. It communicates the firm's approach to customers, engenders loyalty, distinguishes the firm from the competition and contributes to the firm's success in the service market. It can only succeed, however, if forcibly backed by the personal political capital of the firm's leaders.

6

Innovation and New Service Design

INTRODUCTION

Technology firms are rightly proud of their capital infrastructures, the expertise they offer their customers and the investments they make in R&D. Enormous efforts are made to create new products and to refine existing ones; and some place similar importance on services now that they make up a large proportion of their business. Leading companies like IBM and GE stimulate innovation in their service portfolio, while some consulting firms with a technical emphasis, like Accenture, harvest 'thought leadership' from their projects with clients as a source of new services. But many do not take a thoughtful approach to innovation, and are surprised to find that well-established tools and techniques exist, which can turn these assets into competitive, appealing services. Services can be created, developed, improved, marketed and withdrawn, just like any other offer. Called New Service Design (NSD), this approach can also revive services, like utilities, which are in danger of becoming undifferentiated commodities. This chapter discusses innovation and service design. While it offers a comprehensive and practical guide to the service design process, its emphasis is on ways of crystallising these approaches into lively, appealing and valuable service propositions.

SERVICE VISION

In London and other capitals of the world, Absolut vodka has set up 'ice bars'. They are made out of ice shipped from Russia and kept below freezing. Visitors put on heavy clothing to visit the bars for short intervals where they drink vodka from ice tumblers. Yet, for this service experience and a tiny shot of vodka, they pay nearly the same as the price of a bottle of Absolut vodka.

In the 'male grooming' market, the manufacturers of razors have invested in different ways to produce razors which give a closer shave. They have differentiated their product by producing razors that have two, three, four or even five blades. There is now a wide range of choice from electronic and battery to 'wet' razors and disposables, forming a highly competitive, international market which it is extremely difficult to successfully enter. Each produces a new price point for the products, but the price per shave is very low in most countries. Despite this, New York's Malka and Myriam Zaoui have created 'the art of shaving', a premium service focused on shaving. In addition to a range of innovative shaving products, the company offers a shave with a wet razor in its Manhattan-based 'Barber spas'. Its 'Royal shave' combines master barber services with aromatherapy knowledge to produce a shave at a cost that is many times the price per shave of a man's own. It is sold as a unique and special service experience, which, at the time of writing, is being franchised across America.

In the airline market of the 1980s, passengers had a dreary time. Each airline's service was as bad as every other, providing stodgy food from harassed staff, with awful track records in luggage safety. It is even reported that some staff referred to their passengers as 'SLC': self-loading cargo. Industry research at the time showed that frequent flyers would tolerate one airline for a while, then switch and switch again until they rotated back to the original carrier. Into this market came Colin Marshall from the ocean cruise market as MD of the recently privatised British Airways; a government-owned engineering-led service. Backed by his chairman, Lord King, he set out to revolutionise air travel through a massive customer care programme. Called 'putting people first', it changed the fortunes of BA's service and had a dramatic impact on its profits. BA shot to market leadership and became 'the world's favourite airline'. Rivals, like Jan Carson of SAS, copied the approach but, whatever they did, their 'me too' strategies failed to dent BA's lead. Until along came a young pop entrepreneur with so little business experience that he called his ventures 'Virgin'. Yet, he knew the young, upwardly mobile executives who were beginning to make headway in their careers because he had made a fortune entertaining them. His vision was for a different, entertaining, even sexy, service experience and, over two decades later, a large number of people are still buying it.

In China at the end of the 20th century, no-one skied despite there being the right topology and weather not too far from the capital. Yet a group of entrepreneurs recognised that an ambitious, successful and young workforce was hoovering up the trappings of achievement. Chinese people became, for example, the world's leading buyers of luxury goods and began to follow global celebrities. They set up the Nanshan ski village just over 60 km due South of Beijing. With 12 trails, snow sledges from Canada, snow groomers from Austria and a range of sports, it is, at the time of writing, one of the most advanced facilities in the country aimed at an estimated three million skiers. This restless, changing society also pays for service experiences.

In business-to-business services, a young Victorian insolvency specialist, William Deloitte, set out to audit (for the very first time) the then 'space age' industry of trains. His vision was for a 'financial reckoning', which led and assisted new, technologically advanced businesses. The vast accountancy and consultancy firm that still bears his name today is still seen as a pioneer and thought leader in this field. When Professor James McKinsey started his consultancy firm, his vision was of 'management engineers'; professionals who used a 'scientific process' to solve business problems rather than push a particular service. His mission was:

> To help clients make distinctive, lasting, and substantial improvements . . . and build a great firm that is able to attract, develop, excite and retain exceptional people.

His firm is now one of the most respected and distinctive in the business world. It frowns on the word 'sales', and counts the leaders of the world's most powerful businesses amongst its alumni.

Walt Disney had a service vision which led him to create theme parks that used the skills of the theatre to give an outstanding service experience to families. So did the founders of Marshall Field, Harrods, Marriott, Hilton, Starbucks and Macdonald's; as did the politicians who encouraged IT firms in India to offer outsourcing on the world stage. They all proved that it is possible to create an attractive, profitable and unique service which appeals time and again to a group of customers. They have shown that people are hungry for experiences and revel in branded services. Just as a consumer might say that they are a *Vogue* reader and shop in Harrods, they are as likely to say that they have been to Disneyland and flew there with Virgin. There is profit and enduring success in the creation of enticing, enjoyable, branded services.

There is no reason why the service of technology companies, should not be just as enticing, distinctive and profitable as these branded services. If this industry moves from being technology and engineering led then utilities, telephone services and computer services can be just as attractive. It is the job of marketers to create appealing offers for their customers on behalf of the owners of their firm and there are a number of aspects common to these successful service businesses that technology marketers should take on board; all discussed in this chapter.

One example is that the provision of any given service must be 'holistic'. This means it is not only important to design all the basic components of a service, but to ensure that they work together in such a way as to create an integrated experience which meets all the buyer's expectations without interruption or aberration. Integrated service is a little like a theatrical performance. In a ballet the components of the production are prepared and rehearsed so exactly that, at a particular beat in the bar, of a particular piece of music, the point of a ballerina's toe will come down at a particular point on the stage which will be lit by a particular light. Each component of the ballet – whether music, stance, dress, gesture or set – is integrated together to give a seamless (and wondrous) experience for the audience. In the same way, the components of services need to be identified and integrated so as to make the service experience enjoyable and complete.

The key components of service, which need to be integrated in this way, are:

- Brand and image.
- Environmental design.
- Access to the service environment.
- Access into the service process.
- The behaviour of people.
- The process through which the customer moves.
- The technology used to support the service.
- The product or physical components that are part of the offer.
- The means of measurement of the service.

Marketers must not only decide the means of delivery of each of these items, but, more importantly, they must plan how they will all be integrated. Focusing on these features and their integration means that excellent services can be planned from the start. The concept of service component integration ensures that marketers can create unique offers which give buyers choice and enable suppliers to make a profit. For some this will be through trial and error, driven by a demanding, intuitive and visionary leader who has a clear idea of what they want to achieve and will not be satisfied until it is realised. For others it will be through the steady application of appropriate techniques to a market insight or an innovative idea. While, for others, it will be hammered out in the half-world of a 'skunk works', sheltered from politically hostile opponents. Each of these is an expression of 'new service design'; the application of innovation to services and the pursuit of unique, appealing experiences.

COMMON MISTAKES IN SERVICE DESIGN

Unfortunately, there are a number of mistakes that occur again and again when technology companies set out to launch new services. In the same way that some new products do not sell, there are many services that do not get off the ground, or do not make as much money as they could, due to mistakes by their marketers. The most common mistakes include the following.

A Lack of Differentiation

Due to inadequate design and marketing, many services look exactly the same as those of competitors and, as a result, price becomes the basis of choice. This is particularly true in utilities, where a generic service, offered to all (the true meaning of a utility) is the heritage of the company. With the deregulation of many of these markets (such as gas, water, electricity and travel), the competition for customers that has ensued has been mostly based on price. There is little evidence that the service offered by these massive organisations is substantially different for different customer groups or from their competitors.

Allowing the Service to Become a Commodity

A commodity is a product or service that is not valued by potential buyers. They see it as a necessity and are not prepared to pay what they consider to be a high price for it. Yet, products and services only become commodities because suppliers allow them to. It is quite possible to turn commodities (like maintenance and voice telephony) into value propositions with a well-thought-out approach. In the past 30 years, for instance, a relatively small group of companies have created wealth and differentiation with the ultimate commodity: water. Bottled 'mineral' water is not always as healthy and pure as it is claimed or perceived to be. The filtration process of a relatively small private company, however big their brand, is tiny when compared to the massive accumulated investment that nations have made in their water supply. The items that modern science extracts from public water systems are remarkable, and private water companies cannot possibly keep up. One study found, for example, that nearly 12% of food poisoning cases in England and Wales could be attributed to bottled water (Evans *et al.*, 2003). Nevertheless, this industry has made a fortune for its shareholders by implying that their product is healthier and by convincing restauranteurs to 'nudge' customers into purchase by saying 'would you like still or sparkling water?' If this industry can create a value proposition out of water, technology marketers can change the perception of so-called commoditised services.

A Poor Understanding of Buyer Needs

Often people do not know what they need or want. They might, for instance, be unaware of all the technical possibilities that a technology or an industry might offer. It is not unusual for customers to only use a fraction of the capabilities of even their most familiar modern tools, like their mobile phones or personal computers. Just as important, they may be unaware of, or unable to express, their own unarticulated needs. So their vision will be limited and they will rarely be a reliable source of innovation or valuable new ideas, unless specialised research techniques are used to stimulate their imagination. If, for example, researchers at the end of the 19th century had asked people if they wanted 'a black drink, full of caffeine and sugar' they would probably have reported very little demand, and Coca-Cola would not have got off the ground. But once the proposition was placed in front of potential buyers in a way that was attractive, the most successful marketing proposition of the 20th century was born. Research that can uncover such possibilities needs to be explorative or observational.

One disservice that the gurus and proponents of customer care have done is to imply that, by simply asking buyers what they want and meeting or exceeding those desires, firms will engender loyalty and create profit. Yet, if people generally do not know what they want and cannot envisage new propositions, their vision will be limited to improvements on an existing service. They may ask for it to be provided faster or cheaper (a course which will drive a supplier progressively out of business), but it is very rare that they will suggest creative new insights. So, if suppliers take a technical, superficial approach to the analysis of their customers' needs they will be misled. There are a range of needs beyond the purely technical, such as the true benefits sought by customers and the underlying emotions or unarticulated needs. An insight into these might open up a totally new approach, such as that developed by Orange with its per-second billing offer once it realised the dissatisfaction caused to customers by paying for time they had not used via 'per-minute billing'.

Over-reliance on Industry Reports

Another common mistake is to rely too much on industry-produced reports as a means of gaining market perspective. In many technology industries, there are specialist research companies who dedicate themselves to tracking and analysing buyer and supplier behaviour. This has been particularly true in the IT sector with its proliferation of analysts, such as Gartner and IDC. In some cases, analysts even report on specific companies (e.g. Vodafonewatch). Very often these research companies are staffed by technical specialists from the industry conducting trend analysis based on previous interviews and analyses. As a result, there is a danger that industry research reports are too generic and used by internal management or sales people to justify a preconceived notion, so that industry mediocrity becomes a self-fulfilling prophecy. Few are likely to be the source of radical innovations.

The 'One-off' Service

This phenomenon seems to occur when a need is expressed by an important customer to an ambitious or forceful senior executive. The supplier may never have addressed this issue before but, no matter, a project team is drawn together of various functional experts from across the organisation. It then creates a unique answer to that customer's needs, supported by the senior executive but often with limited further exposure. As it is an individual answer to a very specific need, it is highly customised, demanding both effort and costs; sometimes it barely scrapes a profit. Once successful, though, it is presented at management meetings across the firm as a potential for the future portfolio and seen as a 'future direction for the company'. However, this potential relies on the company's ability to replicate or industrialise services. If there is no established process to turn ideas into replicable services, the costs of turning such highly customised projects into firm-wide offers are prohibitive, and the attempt fails without the momentum and backing of the maverick senior executive.

'Over-claim'

The sales literature of technology firms routinely claims that their service is 'leading-edge', 'world-class' or 'the best'. Yet there is rarely any objective measure or justification for this claim. In fact, the service is often the same as that of its peers. Many exaggerate the benefit of their service and create cynicism amongst their buyers.

The Lack of a Meaningful Proposition

A good number of service offers, particularly in business-to-business markets, are unclear and complex. In many instances the supplier fails to create a simple value proposition. A mistake that is often made is to suggest that the firm is a 'full service provider', or that it can do anything. To the customer this appears as a lazy lack of clarity, which implies no particular skill or emphasis worth paying for.

A 'Product-led' Approach

Most industries have evolved through the sales boom of a growth market. This is a period of powerful natural demand, when it is only necessary for the suppliers to keep improving on the features of established offers. During this phase the industry is supply-driven and 'product-led'. This successful behaviour can lead to mistakes though. Markets and needs change as industries mature and firms which are primarily production machines are exposed to huge costly errors if they do not adjust after such a fundamental shift in their market. As industries reach maturity and customers have more choice, marketers need to take a 'market-led' approach, differentiating their products for different segments of buyers and against competitors. Service companies can have a similar 'product-led' mentality. Marketers pushing services which have not been adjusted to meet changing market needs are just as likely to fail.

Servility rather than Service

Customer orientation and service values are vital to the success of maintenance organisations. The ethos of large 'support' businesses is to solve difficulties as fast and as well as possible; to do what customers, and their representatives in the organisation, say. They must respond quickly and well to customer needs. However, years of support or, worse, giving away the service to sell technology, creates something akin to servility, and unthinking subservience is not appropriate when designing an appealing service. It undervalues expertise and business value, leading to discounting and, eventually, declining revenues.

Clearly, for companies to create real value out of the services that they are designing, they must introduce techniques, skills and practices to minimise these mistakes. They must ensure that their business becomes progressively more competent at NSD.

THE ROLE OF INNOVATION AND SERVICE DEVELOPMENT

It is sometimes difficult for engineers to think of their company's expertise as a competitive service that can be researched, designed and packaged into an offer independent of their technical products. To many executives in technology firms, it feels as though their services are merely an enabler of the real business. In the past, many have been sceptical that their customers would spend their money on vague, ethereal services rather than practical, high-tech products. While those in utilities can find it hard to think of their massive public facilities as a competitive service, which can be presented to different publics in a way which entices them to pay different amounts for different versions. But when their company's expertise is carefully crafted for segments of customers, wrapped in a brand, presented in conjunction with other capabilities and is accessible through a process, it can become a competitive, sexy and appealing service offer.

Technology companies offer services that are based on an infrastructure, a technology or a network. The performance, development and stability of that infrastructure affect the nature and the content of those services. Very often there is a core service ('communications' in telecommunications companies, 'support' in maintenance businesses and 'power supply' in utilities) and opportunities for added-value services, as illustrated in Figure 6.1. Unfortunately, it is quite common for marketers to reach first for 'added-value' services to raise their revenue

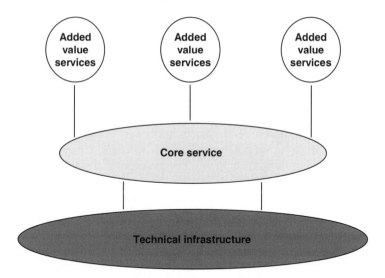

Figure 6.1   A representation of technology or network-based services

line, neglecting or dismissing the perceived value of the core service or the stability of the technical infrastructure. Yet, the perceived value of the core service is irretrievably linked to the technical infrastructure through which it is delivered and any 'added-value services' are dependent upon the perceived value of that core service. If there is little perceived value for the core service, there will be little value from any added-value services. So marketers should set out to hone and reposition all levels of service.

As the service sector has grown, so has the work by a myriad of firms to generate revenue by enhancing their core service or creating new services. Businesses in the service sector have experimented with NSD and explored related techniques. Some software companies, like Microsoft in Europe, have, for example, used their software development process to craft services; ensuring that they are fit for market through a rigorous process. Others have used service blueprinting and 'lean' techniques to improve customers' experiences of their services. At the same time, theorists and researchers have teased out many of the success factors. So, there is a growing understanding of the importance of NSD and which aspects of it are most likely to work effectively. The experience of these specialists should, at least, be considered by marketers in technology companies who are assessing their capabilities and trying to introduce new service offers. It is likely that, if they do so, the chances of success with new offers will be increased and less will be wasted.

There are several motives for the development of new services. It may be that the existing services are becoming obsolete and need to be replaced by new ones with better appeal to buyers. New service ideas may be launched simply to increase revenue or to take advantage of new opportunities. Clearly, if a company spots a gap in the market, it will seek to take advantage of that opportunity. Some may be created to stay ahead of competitors and others to use up spare capacity due to troughs in demand (Air Miles are a good example). Others are an attempt to overcome commoditisation. If buyers have lost a sense of value due to bad pricing practices by the supplier, services can be repositioned and re-launched with a new pricing regime and greater margins. Finally, the management team may be concerned about risk reduction or seasonality. New services might be launched or companies bought to balance a portfolio reliant on one service. Whatever the motive, however, service marketers ought to become as skilled, systematic and professional in new service design as they are in, say, marketing communication.

There are a number of distinct areas where they can apply these techniques. NSD can, for instance, be applied to the core service; even if it is a long-standing, boring public utility. Many service companies have invested resources, processes, training and skilled people into the systematic improvement of the core service that they offer to the market (e.g. gas supply to homes, rail travel and even water supply). They are moving down the 'experience curve' so that, over time, costs systematically decrease and enhancements are made to the service. This is a natural evolution which can be helped by explicit, planned NSD techniques. In fact, an NSD approach is particularly important if a company's core service is a commodity and it needs to turn its offer into a value proposition to stop the erosion of profits. It allows marketers to create a new perception of value for the core service. They can create different versions of the core service for different segments of buyers and introduce little innovations that enhance the existing service, and improve its perceived value over time. Through the application of creativity to the core service it is quite possible, then, to turn a commodity into a value proposition. Services offered by banks have been rescued from crisis and commoditisation in this way and the same techniques can be applied to technical services.

NSD can also be used to create added-value services. These might be:

- An annuity service, such as a maintenance service.
- A support service, such as break-fix repairs.
- A proposition wrapped around a product sale, such as a warranty.
- A service component of a product sale which now needs independent value, such as a consultancy or planning service.

They maintain the value of the proposition in a mature market, increase profit and enhance the relationship with customers. They also create knowledge of future needs through more regular contact with customers. So, it is possible to create new service propositions that have their own value and succeed in the market. They may be delivered electronically, through people or via physical channels. Either way, it is clear that the proactive creation of service, and techniques to make this effective, are an important competence that technical service companies need to foster.

INNOVATION IN SERVICE BUSINESSES

In 2006, IBM launched a massive international advertising campaign called: 'Innovation that matters'. The objective was to suggest that they had the means to help corporations innovate in all parts of their organisations ('structurally, financially and operationally'). Whatever their own intentions, the effect was to, once again, put emphasis on the word '*innovation*' in business circles. Since then, innovation has become a topical subject with books, seminars, 'summits' and conferences dedicated to it. Leading consultancies have published research and 'white papers' on innovation needs. Although this might seem a little faddish, there is no doubt that service businesses across the world are investing time and resources in innovation because owners and customers are demanding fresh approaches, particularly after the economic difficulties faced by many in the worldwide recession at the start of the 21st century. It is also clear that innovation management is at the heart of the successful differentiation achieved by organisations in other industries and is beginning to create real differences between the offers of different service firms.

There are well-researched projects and many anecdotal stories about the need to bring innovation to businesses. They suggest that innovation in all parts of a business, from strategy and management to marketing and sales, is vital in fast-changing markets; that sclerosis can destroy value. Creativity and innovation are particularly vital to a firm that needs to bring a range of new products and services to market. In fact, some researchers suggest that the ability to innovate new products and services is the key to survival for some companies operating in a number of mature or fast-changing markets. But what is innovation? Is it practical and manageable or does it just occur in a maddeningly haphazard way?

Actually, innovation is a management process; the mechanism by which an organisation captures, cultivates and exploits initiatives. It is different from creativity itself, which is the source of initial insight or ideas on which innovative new services are founded. In recent years a number of technology businesses have adopted successful methods of harvesting creative insight, in addition to ensuring that creativity is stimulated in the first place. They vary enormously but can be grouped into several generic approaches.

Innovation Founded on a Formal Process

The proactive creation of new products (NPD) has been established practice for many years in a number of different businesses selling physical products. Leading organisations in markets where new product innovation is a critical success factor, like consumer products, have highly sophisticated processes, which are managed at a senior level in the firm. They aim to present many innovations to a fast-changing market, knowing that some will be successful. As a result, NPD processes and concepts are well established and known to provide demonstrable value. Research has shown, for example, that a new product design process in manufacturing reduces risk of failure; also, that successful innovation is both costly and risky, but an NPD process reduces these costs. The factors that contribute to the success of new product design include:

- Senior management involvement and control.
- A clear and managed new product design process.
- Superiority over existing products.
- Investment in understanding the market.
- The proficiency of marketing operations.
- Degree of business/project fit.
- Effective interaction between R&D and marketing.
- A supportive management environment.
- Effective project management.

These factors are now widely recognised. As a result, many companies in many sectors have in place clear and formalised new product creation processes, portfolio management techniques and dedicated product managers. But how recognised and understood is the same process in service industries?

In 1991, Canada's Ulrike De Brentani (De Bretani, 1991) set out to resolve this question. He conducted research designed to apply the 'conceptual and research paradigms that have evolved from studies of new manufactured goods to services'. He concluded that the strategic issues facing service organisations were similar to those for product companies. While NPD technologies and the detailed steps included in the process vary, the underlying notions behind their use do not.

Two decades later, many leading service organisations now use a formal innovation process. They also create as many good ideas as possible and then reduce the number of ideas by careful screening to ensure that only those with the best chances of success get launched.

This occurs most commonly in companies that create and industrialise consumer services because the high volume of their business demands careful process design and technology deployment. As a general rule, those services that are high volume, low margin and easily reproducible can more easily be developed using a rigorous development plan than those that are highly customised (like consultancy or other professional services). American Express, for instance, has a highly developed process with strict controls over eight development 'gates'; while at the other end of the scale, even the venerable British law firm, Allen & Overy (which specialises in highly customised advice) has an 'innovation committee', which reports to its leadership team on programmes to capture and exploit creativity. It has successfully sponsored the launch of a series of IT-based services that, at the most sophisticated level, automate a number of their clients' processes.

So, the general concept of an 'NPD process' is applicable to service design, although in many cases certain elements are added or heavily modified. The normal approach is to set up a subcommittee of the leadership team. Prospective innovators are given a budget by the committee to complete initial feasibility. They are then required to return to the management group for further funds to go to the next phase. The committee reports to the main leadership team on the progress of all new concepts.

Figure 6.2 outlines the typical steps in a formal NSD process. It starts with the creation of ideas. Many service firms develop an attractive idea without first stopping to consider if there

A typical NSD process.

Step 1 Analysis: business backdrop and buyer analysis.
Step 2 Idea generation.
Step 3 Prioritisation against firm's criteria.
Step 4 Detailed component design and process blueprinting.
Step 5 Creating the value proposition.

Quality check: are the unique considerations of services thought through?

Step 6 Create the concept representation.
Step 7 Research: Focus groups with clients to test the concept.
Step 8 Write the business plan.
Step 9 Trials.

Figure 6.2 A typical NSD process

are alternatives which will have greater appeal to the market. This tendency is exacerbated if there is pressure to generate funds. It is sensible, however, to stimulate ideas from staff and customers deliberately before putting effort into new service creation. Some technology firms, such as Orange and Fujitsu, now use the concept of an 'advisory board' of customers to help at this stage.

Once a list of potential ideas has been created they need to be prioritised. This should be based on criteria related to the objectives of the firm, the need to generate more funds from customers and the need to develop a new area of expertise. The criteria must be specific to the firm and practical.

The process then moves on to the design of the detailed components of the new service and the translation of those into a value proposition. Finally, each aspect of the service, including marketing materials, sales process and service delivery, is summarised into a detailed business plan, which will test the viability of the service, through financial rigour and research. As shown in the case studies, this systematic approach was set up by Interoute as it grew and used by BT as it developed a new consultative service to help its business customers understand their options for reducing carbon emissions.

INTEROUTE GROWS THROUGH CUSTOMISING ITS CORE SERVICE

Interoute is a fast-growing international, communications network which offers 'business-class voice and data services at an economy price'. Its products and services include: band-width, virtual private networks, high-speed Internet access and transit, managed hosting, communications services and media streaming. Founded only in the 1990s and privately owned, it has 57,000 km of fibre and 59 data centres. It operates in 24 countries and 90 cities. The firm's service was, initially, as a niche player in the international telecommunications market. The fact that it has customers that are serious players in that market (Sprint, BT, AT&T, Deutsche Telecom and China telecom) demonstrates the quality of its service, technology and reputation.

It started as a basic 'bearer' network offering network capacity to other suppliers. This is regarded as a 'wholesale' business because other telecommunication suppliers bought extra network capacity. Yet, although it was successful and respected, the leadership of the firm were dissatisfied with remaining a commodity infrastructure service. They set a 10-year strategy to reposition the service of the firm. They started by offering 'DIY' technology to telecommunications experts but have climbed through 'collocation' and 'hosting' to 'application management'. The long-term vision of the firm is to be a player in the emerging 'cloud computing' marketplace, where computing applications will be provided as services which are remotely sourced in internationally hosted data centres. In order to do this, the company had to learn the skill of crafting its core service to different customers to improve perceived value.

The company has deliberately set out to move from a generic infrastructure service to a higher-value, tailored communications service. It identified buyers in individual companies who were either chief finance officers or chief technology officers. It then worked through the products, features and benefits that would appeal to each person in each organisation and tailored its offer to each company. The results were dramatic. Between the period 2004 to 2008, there was a compound average growth rate of 67% and 'EBITDA' quadrupled.

Over the same period, the percentage of its customers who were 'corporate' grew from 8% to 51%. The company has penetrated the corporate sector very effectively by reconfiguring its basic, core service to different corporate buyers. It has customers in the public sector (e.g. the European Union), financial services (ING and Morgan Stanley), services (Hilton and Yahoo), retail (WE and Chopard) and manufacturing (Ford and Siemens).

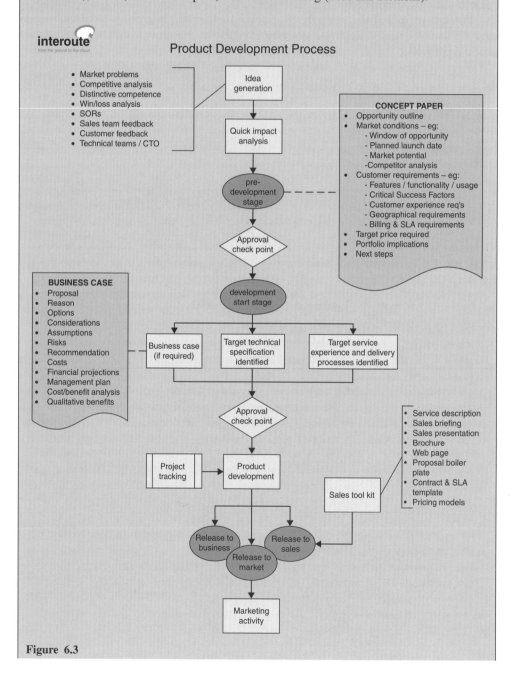

Figure 6.3

The respected industry observer, Gartner, recognised the effectiveness of the firm's strategy when it said:

> Interoute is pursuing a focused strategy based on extensive infrastructure ownership. It has been expanding its footprint principally through acquisition of long distance and metropolitan fibre assets, such as CECom covering central European countries. Principally a provider of wholesale services to other operators, Interoute also offers a growing range of services to enterprises including MPLS VPNs, Internet access, Ethernet and bandwidth services.

Tony Rogers, corporate product marketing manager says:

> The customisation of our core service to different customer groups has been one of the main components in our success. We are now formalising this into a more structured NSD process upon which we can create viable and robust services. We also see this as a tool to create added value services for different customer segments.

The process, shown in Figure 6.3, starts with idea generation similar to many others. It then, though, undertakes a 'quick impact analysis' of the ideas, to prioritise those which might move quickly into the market. The process then moves through similar 'gates' to other NSD processes, undertaking research and constructing business cases at the appropriate time.

As the firm has grown from a fairly small start-up with a small-firm ethos, this more structured approach has become necessary. Rogers says:

> People have seen the effectiveness of this more thoughtful approach and now understand that, as we grow and take on more significant corporate customers, we have to be more thorough and structured. If not, quality might not be so good and money might be wasted on mistakes. We have, though, tried to preserve the small company entrepreneurial spirit which made our services distinctive in the first place.

Interoute's combined network assets now represent one of the largest and most advanced voice and data networks in Europe, and are still growing fast thanks to the tailored strategy. It is now regarded as the operator of Europe's 'largest next-generation network'. In June 2009 the company won a business innovation award at the Global Telecoms Business Awards, for innovation in the International Service Delivery category. The award was for its 'Ground to Cloud' service delivery platform. This gives its customers the ability to take basic fibre services and add their own management layer or buy fully managed services including Interoute's monitoring and reporting tools. In giving the award, the organisers said:

> For those enterprises looking to a partner to take on the management of their critical IT infrastructure Interoute's cloud approach to services removes the complexity of building, managing and changing their infrastructure.

Innovation through Smallness

Other, firms emphasise smallness as a means to innovation. For instance, in a recent speech, British entrepreneur Sir Richard Branson attributed part of the success of his Virgin organisation to smallness. His policy, he said, was to break units in two once they had reached 'around thirty' people. He had found that it maintained entrepreneurial spirit and kept overheads low. This approach is founded on a story which has become the stuff of business legend. In the 1960s, the American multinational 3M nearly missed a major opportunity. Its Dr Spencer Silver had produced a 'low-tack reusable adhesive' (glue which didn't stick well). He saw the potential to create, what became, 'Post It' notes. Unfortunately, the concept was rejected by all. However, he did not give up on the idea but began to manufacture samples, which he gave to the board's secretaries. As they asked for more, word spread and demand grew until the amateurish manufacturing capability for the notes took over his supportive boss's office. The company eventually accepted the innovation, which became one of its most successful products. Then, as a response to the fear that its large, bureaucratic structure might miss a significant product opportunity because of its short-sightedness, it broke down into small cells. It claimed that this was responsible for a number of subsequent innovations.

The technology industry has seen enormous innovation through smallness. Not only have some large companies sought to create entrepreneurial growth through small business units, but the character of places like Silicon Valley has facilitated interaction between capital investors and numerous start-ups. Leading, creative technologists have grown used to the fact that tiny start-ups can grow into massive organisations like Google.

Innovation through smallness has also been the reason for the success of many consultancies. Even in the largest firms, success lies in their individual practices, which, like cells in a living organism, evolve and respond to changes because they are run directly by a partner. As an owner of the business they feel freer than corporate executives to take their own course and refine their own offer. In fact, some leading practices ask partners to write and then implement a 'business plan' which sets the direction for their practice; especially if in a completely new field of operation. There is even evidence of partners developing a practice in defiance of the wishes and policies of leaders; sometimes to great effect. So, fostering an innovation by dedicating a practice to it is common in consultancy. Typically, though, this is intuitive and not part of structured 'innovation management' policies or any managed process. Experience suggests that marketers can improve the success of this approach if they make explicit a mechanism to identify, nurture and prioritise developing ideas.

Outsourcing Innovation Management

There is a view that creativity occurs less often in large, bureaucratic and process-dominated organisations and that innovation management rarely manages to exploit it. Some product companies have, therefore, managed creativity by using an external agency; innovation through outsourcing.

In the early 1970s a young advertiser, David Crayton, had the European decorating company Polycel as a client. He was in discussion with them about a marketing campaign when they began to talk about a huge mistake. They had tried to move into paint and, because they were a glue company, were about to discard the resultant 'stuff'. It was far too white and sticky; but the main problem was that it dried as a rubbery skin. It could not be used as gloss paint under any circumstances. The company asked David if he could use his creative skills to save some of their investment. He went back to his agency and used his team to work through 'features', 'applications' and 'benefits'. They came up with the idea that it might be, in a country with old housing stock, 'a treatment for ceilings which did not need painting'. It was eventually launched as 'Polytex', becoming a successful product for the company. David went on to found 'Crayton, Lodge and Knight', a leading creative agency in this field.

These agencies, which came to prominence in the 1970s, are known as 'new product development companies'. They take a brief in the same way that advertising agencies or research companies take briefs from marketing departments. These briefs contain requirements to produce new products, offering benefits to new markets for whatever reason. The famous cream and whisky drink, Bailey's Irish cream, was, for instance, created in this way. The then owners of the drink commissioned it from a creative agency because they had spare manufacturing capability in their Irish plant. The original brief, it seems, contained a requirement to produce an alcoholic drink product which contained 'typically Irish components' like cream and whisky. These suppliers have specialist processes to explore the target market, to design the offer and to specify manufacturing and packaging. They return to the company a researched concept which can be taken, through a detailed business plan, into manufacture.

Exploiting the 'Diffusion of Innovation'

As discussed in Chapter 3, the diffusion of innovation is a recognised social phenomenon which contributes to the maturity of markets; a mechanism whereby societies learn new concepts and adjust to them. It involves word-of-mouth, education and subtle forms of publicity. One sociologist, Everett Rogers (Rogers, 2003), says of it:

> This information exchange about new ideas occurs through a conversation process involving interpersonal networks. The diffusion of innovation is essentially a social process in which subjectively perceived information about a new idea is communicated from person to person.

So, the acceptance and growth of an innovation in society occurs through dialogue amongst a network of interrelated people. Its diffusion depends on relationships and conversation.

Businesses can use this phenomenon to develop new services and marketing concepts. There is, for example, a specific type of innovation diffusion which occurs in the services industry. Consultancies which specialise in high-end customised services, like McKinsey, usually undertake complex projects to solve unique client problems. However, if their practice meets a problem several times, the type of engagement will be given a name and a new concept is born in the industry. Other suppliers will then take up the offer, creating processes and tools to handle it. It moves down the 'industrialisation' line, picking up volume but losing margin. Eventually the approach becomes commonplace, is captured in software tools, is trained in professional academies and undertaken by customers themselves. It becomes more of a commodity. Offers like: double-entry bookkeeping, portfolio planning, CRM, customer experience management and process re-engineering have followed this course; and it is seen throughout the service industry today. Yet, different offers suit different firms at different stages in the commoditisation journey. A high-end strategy firm like McKinsey pioneers new concepts, whereas one of the IT-based management consultancies, such as Accenture, might be better suited to volume, process-based offers. Marketers can identify potential innovations for their practice if they have a clear view of the concepts that are appropriate to its competences and by adapting ideas from others as they diffuse across the sector.

In his definitive study on the use of this concept in technology companies, Geoffrey Moore (Moore, 1991) pointed out the difficulties that companies have in bringing true innovations to market. He emphasises the need to take different approaches in marketing strategy and technique as diffusion progresses. He said, for instance:

> ... our marketing ventures, despite normally promising starts, drift off course in puzzling ways, eventually causing unexpected and unnerving gaps in sales revenues, and sooner or later leading management to undertake some desperate remedy ... the point of greatest peril in the development of a high-tech market lies in making the transition from an early market dominated by a few visionary customers to a mainstream market dominated by a large block of customers who are predominantly pragmatists in orientation.

The differences in the pace of development of high-tech services like broadband over optical fibre (originally proposed in the 1980s) and either the Internet or iPhone apps shows that this level of sophistication in innovation marketing is just as relevant in services marketing.

Innovation through 'Thought Leadership'

Thought leadership is one of the most successful marketing strategies that the world has seen. It is a vast, influential and diverse range of activities, which, at its best, produces systematic, iconic work like the *McKinsey Quarterly* and the *Harvard Business Review*. The worst, though, has been mocked by the *Economist* magazine as 'Thought Followship' because it is frequently just the erratic, whimsical jottings of consultants in different disciplines trying to 'market' themselves between projects; some of it is not leadership and some of it isn't even clear thought. Yet the power of this approach to amplify an excellent reputation and draw in clients is undeniable.

Thought leadership works because ideas that resonate have powerful influence on business leaders. In certain circumstances, it generates the business equivalent of urban myths. For instance, most management audiences will recognise:

- '50% of my advertising budget is wasted, I don't know which 50%.' Or
- 'it costs more to recruit a new customer than retain an existing one.' And
- 'the average dissatisfied customer tells 13 other people if they experience bad service.'

Yet, these were manufactured ideas promulgated by people trying to market a service and circulated amongst management networks to gain influence. Few know their source, context or relevance. In a busy job where executives do not get much time to think, they make decisions, policy arguments and presentations with these propositions clustered in the back of their mind. The ideas become *de facto* strategies and, as a result, they have wide impact on businesses, creating interest and demand in the firms that promulgate them. Ideas which have stood out from the crowd (like: the millennium bug, shareholder value, the dot.com revolution, CRM, customer loyalty and many others) have had remarkable impact. Many go through a clear evolution.

In the 1990s, for example, a thought called 'process re-engineering' came to prominence and is still familiar to many in technology companies today. It can be credited, it seems, to a consultancy called the Computer Science Corporation of America. One summer they met with people from M.I.T. and came up with an idea to dramatically improve organisational performance. Middle management could be cut out of an organisation and replaced with computers if top management streamlined the operational processes of their business through a clear rational approach. They then produced a book (*Reengineering the Corporation*) and the idea began to take off. Then the computer industry got in on the act because it realised it could sell computers through re-engineering projects by introducing systems analysis techniques. Some IT marketers in other companies took the case studies they had used as examples of the effectiveness of the previous notable idea ('Total Quality Management') and re-packaged them as 'Process Re-engineering', because it would sell computers. They also presented at conferences and wrote articles as if it had been happening for years; adding to the cacophony of voices. Finally, the financiers became involved in the idea. They were convinced of it by the professions and started to mark down the shares of companies who had not 're-engineered'. This prompted yet more companies to take on the idea, improving the performance of some but damaging others.

At the height of this fashionable idea two things happened. Firstly, academics started getting good research budgets to investigate using proper methodologies and substantiating the concept. (Now most business schools tend to teach 'process management' as part of their 'operations management' curriculum.) Secondly, sceptics wrote articles like 'Has process re-engineering damaged corporate America?' These iconoclasts make money by challenging ideas and suggesting that they are not a panacea that will change the whole of business.

So the steps to effective thought leadership appear to start with the choice of an idea. This can come from deep personal knowledge of an industry sector or a marketing team's prioritisation of candidate concepts. More often, though, it is an adaptation of an important or popular subject which has momentum and publicity. However, all who succeed with this approach have to put a degree of investment behind their work to gain appropriate impact. It can be surprisingly cheap to employ a journalist to ghost write a book on a subject, but the marketing of ideas needs sustained attention from a supplier if they are to be associated with them and exploit them as a source of services revenue.

'Thought Leadership' is both a source of innovative new services and a marketing tool that helps technology sales. Leading consultants like Accenture, IBM Global Services and Deloitte have skilled staff dedicated to this function. What starts with an idea and a marketing campaign can soon have consultants and leading practitioners dedicated to it. It becomes a 'practice' in its own right, earning, sometimes, millions of dollars in revenues. Leading service businesses have had, for example, practices focused on: 'CRM', 'lean service' and 'shareholder value'; each innovative ideas in their own time.

Innovation through Imitation and Adaptation

Another valid form of innovation is the adaptation of emerging ideas; when, for instance, a supplier adopts and adapts a new idea, product or service, which has been created and pioneered by a peer in their industry. As new concepts occur, they are often pioneered by small, flexible businesses or by notable individuals. However, it is when organisations with a huge footprint take them on board that they have substantial impact and a new market is formed. So it is a natural evolution and a source of real profit to identify emerging ideas and adapt them. Some would argue that this is not true innovation because it is not completely new. Yet, famous strategy professor Ted Levitt argued, as far back as 1966, that this was a valid form of innovation, which was less risky than blindly pursuing the entirely new idea or offer (Levitt, 1966). He said:

> We live in a business world that increasingly worships the great tribal god: innovation; lyrically hailing it not just as a desired but necessary, condition of a company's survival and growth. This

highly agitated confidence in the liberating efficacy of innovation has in some places become an article of faith almost as strong as Natchez Indian's consuming faith in the deity of the sun. Man creates gods according to his needs. Significantly, the businessman's new demigod and the Natchez's more venerable and historic god make identical promises. They both promise renewal and life.

Yet before all our R&D energies and imaginations are too one-sidedly directed at the creation of innovations, it is useful to look at the facts of commercial life. Is innovation all that promising? More important, how does a policy of innovation compare in promise to more modest aspirations?

What is needed is a sensibly balanced view of the world. Innovation is here to stay, it is necessary, and it can make a lot of sense; but it does not exhaust the whole of reality. Every company needs to recognise the impossibility of sustaining innovative leadership in its industry and the danger of an unbalanced dedication to being the industry's innovator. No single company, regardless of its determination, energy, imagination, or resources, is big enough or solvent enough to do all the productive first things that will ever occur in its industry and to always beat its competitors to all the innovations emanating from the industry. (Reprinted with permission from *Harvard Business Review*, Levitt, Sept/Oct 1966)

Innovation is not only about completely new ideas but also about the sensible adaptation and improvement of concepts pioneered by others. The executives of Britain's remarkably successful supermarket, Tesco, are renowned for seeking, adapting and trying their competitors innovations. IBM originally moved into computers (and later the PC) through imitation; as did Virgin in air travel (after Freddie Laker) and Ericsson in mobile phones. This is not to argue that the services or products that result are not unique or appealing. These companies adapt concepts without infringing copyright or patents; making them their own and producing a unique new value proposition.

Innovation through the Co-creation of Service

Creativity tends to occur in all businesses as they interact with customers. As local people (particularly in large, diffuse businesses) respond to needs, solve developing problems and tackle the unintended consequences of central policy, they frequently find creative new answers to needs. This is particularly common in service companies because of the involvement of human beings in the interaction between the two sides. Unfortunately, these creative answers to local problems are, like fast-blooming wild flowers, frequently lost. So, some companies put great stress on capturing local innovation across the operations of their company and communicating it across the whole firm. Some marketing VPs, for instance, see part of their role as identifying and legitimising innovation that occurs in the field. They stress the need to set up mechanisms capable of industrialising it and replicating it across their company. In many cases, this leads to them concentrate on mechanisms by which they can create innovative ideas with customers.

The 'co-creation of services' is the practice of developing new services through collaboration between companies and customers. Innovation and value is created by the firm and the customer, rather than being created entirely inside the business. Co-creation not only describes a trend of jointly creating services but also a movement away from customers buying products and services as transactions. It leads to purchases being made as part of an experience, so it is very relevant to the creation of the attractive, experiential services that are the goal of this book.

It is now common for firms to take ideas from their customers about adjusting the way in which they deliver products to them. They try to get customer input into the way they enjoy the service, in fact some create online communities, exploring and understanding how customers use the offer and then seeking their ideas on how to improve delivery. This is a way of improving the customer's experience without affecting the product itself. It is also a way to bring the customer inside the business, helping them work with the brand on the parts of the experience that matters to them most.

Co-creating service, however, goes beyond this; it might, like Amazon, allow customers to select and change their delivery times. Many airlines, for instance, allow real-time self-service for their customers, letting them choose and change their seat, meal and even their flight time. In some business-to-business cases, customers work with suppliers on a deeper level of engagement to create entirely new offers. The suppliers then bring these into their portfolio and offer them across their business. As the case study demonstrated, this was one of the by-products of the Orange listening programme.

VALUE PROPOSITIONS

A value proposition is an offer to customers which meets their buying criteria at a price that they regard, however illogically or unfairly they form that judgement, as 'value for money'. Contrary to popular opinion, price, or more accurately cheapness, is not a buyer's prime consideration. Sometimes they are willing to pay more for a particular item and sometimes they deliberately choose an 'expensive' offer. It depends on the value of the offer to them.

Some new services fail to make money, not through poor planning of the service concept or its components, but because it is misrepresented in the marketplace: it is not a clear, simple, appealing value proposition. The marketing and sales literature is merely a description of the various components and features of the service. To succeed, it is essential that the raw elements of the service are turned into a proposition to which buyers can relate. It would be ridiculous, for example, to describe a fast food outlet as follows: 'Our service contains people in a carefully designed uniform who prepare a limited range of food very quickly using the latest technology to cook; it is both hot and ready to eat very quickly'. Buyers would be bored by the time they reached the end. Yet this is exactly what many marketers, particularly in business-to-business markets, do with their services.

One fundamental reason for this failure is that the creation of a value proposition is frequently fragmented. Engineers will create the components of the service, an accountant will agree pricing, different sales people will work out benefits to customers, marketers will summarise the offer and pull all the details together, and an advertising agency will create a one-line summary of the proposition. As a result, a haphazard and unclear proposition is put to the market that fails to gain traction.

This crucially important function should be managed within the firm. An attractive proposition, which summarises the offer in clear terms, should be created and put to the market. The FMCG companies, like Gillette for example, have long excelled at this ('three blades, fewer strokes, less irritation') and marketing teams in leading technology companies are tackling it. They will often hold, for example, a 'value proposition' workshop to clarify the promise to customers before any offer goes anywhere near the market. Many will even use this approach with individual sales deals to significant customers. They think through the components of a dynamic model shown in Figure 6.4 below.

The components are:

• A clear understanding of the buyers the value proposition is aimed at, including their rational and emotional needs.

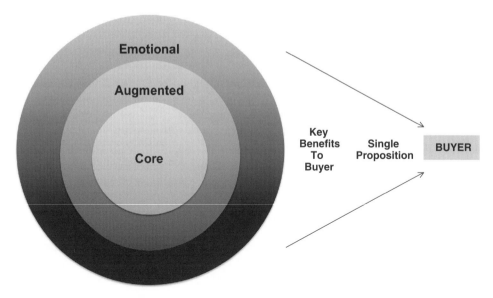

Figure 6.4   A structured route to a value proposition

- An integrated view of the tangible and intangible components of the offer, and their value to the buyers.
- The benefits of the proposition.
- An understanding of the differentiation or 'unique selling proposition' of the offer.
- A short market-based description of the offer, sometimes called the elevator pitch.

Where an offer can be tailored or customised to different segments of customers, the value proposition should be too, as demonstrated by BT in their development of a carbon impact assessment and the benefits it could bring different types of customers.

CREATING SUSTAINABILITY PROPOSITIONS AT BT

In 2008, BT was a co-editor of the SMART 2020 report published by the Global e-Sustainability Initiative (GeSI) and The Climate Group, which found that smarter use of ICT could help reduce global CO_2 emissions by 15% by 2020. Embarrassingly, analysts confirm that their sector, information and communications technology (ICT) accounts for 2% of global carbon emissions, equivalent to the much-maligned aviation industry. They even predict that it will eventually overtake the aviation industry in coming years. Clearly, a more sustainable approach to ICT is a must for all our sakes.

Yet BT has long been committed to sustainability and environmental issues which date back to the early 1990s. It has built one of the most flexible workplaces in the world, leading to a saving of over 38.6 million euros a year on travel costs, making staff up to 20% more productive through agile working, and increasing staff retention (96% of women return to work after having children, for example), saving millions in recruitment and retraining costs. So far, the company has achieved a 43% global intensity emissions reduction by becoming more energy efficient and increasing its use of renewable energy. In the UK, it reduced CO_2 emissions by 58% compared to 1997, with 41% of the company's energy coming from renewable sources.

BT has a well-established corporate responsibility team who continue to focus on the company's positive impact on society and the environment. It held the 'number 1' position in the Dow Jones Sustainability Index's ranking of telecommunication companies for eight consecutive years between 2001 and 2008. So, in 2007, BT Global Services commissioned far-reaching research to explore the most pressing issues for business leaders. Three strong themes were identified and adopted by BT as a result: collaboration, risk and resilience, and sustainability. The time was right to take the internal expertise BT had on sustainability out for the benefit of its clients.

Within BT Global Services, a sustainability practice was established, with the remit of creating and incubating sustainable propositions for the benefit of customers. 'BT has world leading sustainability credentials and has integrated its best practice and experience into its commercial offerings, which are assisting organisations to reduce inefficiencies, operating costs and emissions' said Ninder Takhar, head of marketing and propositions.

Sustainability

The practice defined sustainability as 'the transfer of benefits from one period of time to the next', and adopted the Bruntland Commission's thinking of sustainability as environmental, economic and social. Once in place, the team used Harvard's 'eight bubble' model of innovation, adopting the thinking that many ideas lead to several potential propositions, which in turn lead to one or two strong future businesses.

Internal research showed that while 35.5% of UK companies had a sustainability strategy, the remaining 77.6% were either aware but not compelled to act, or aware but did not know where to start. So, the first solution to be developed (in May 2007) was a carbon impact assessment (CIA), to help companies begin and continue their sustainability journey. It went through a gated development process that began with an initial 'playbook', comprising: an outline of the concept, the customer need it addressed, the potential revenues it would deliver, and the targets that could be set and measured for its success. Once the concept was through this first gate, a 'market requirement definition' document was developed to further clarify the size of the opportunity for BT. Next came a 'product definitions document', describing in detail the design of the new service and how it would be delivered to customers. Finally came the product launch readiness gate, involving sign-off and approval from all internal stakeholders of the new proposition, an essential step for the service to ever see the light of day.

BT analysed its customer and prospect base carefully before creating costomised value propositions for sales teams to take to market. Desk-based segmentation was run from an offshore base in India, which put several hundred companies through analysis. It included: their Board's commitment to sustainability, corporate responsibility commitments, stated targets, and publication of a CSR report with or as part of the company's annual report. These factors were weighted in terms of their 'importance', and each company scored, to create a 'dashboard' of targets based on their CSR sophistication. Red denoted a company that had barely started thinking about sustainability, amber was one that had embarked on the journey, recognising the importance of the issue to their business, and green represented a sophisticated company that had already made great advances in its working practices.

A generic proposition was created, promising customers that BT could help them 'build a sustainable organisation'. Beneath this umbrella, the less sophisticated companies were targeted with 'what will you do about the carbon reduction commitment?' This highlighted impending legislation to tax companies on the basis of their carbon emissions which would force them to reduce emissions year on year. For this group of customers, a 'quick start' assessment service was offered, to identify the main areas in which they could make reductions.

For more sophisticated customers, tailored value propositions were created, offering in-depth diagnostics for specific areas of business that the customer themselves recognised needed improving. Examples included creating more energy-efficient data centres to store their information, or establishing a more agile workforce through better collaboration and use of communications technology.

As with all propositions, BT needed to demonstrate its differentiation and answer the question, 'why BT?' To do this, the company talked about its focus on how people work, such as the impact of their commute to the office, and not just on IT hardware. It was backed up by BT's own experience and credentials in this area. Most of the other businesses offering carbon impact services considered only the hardware implications, or were niche consultants who had never implemented a carbon reduction solution of scale for their own businesses.

With such a lot of noise in the market, many people were cynical of suppliers 'greenwashing' to join the sustainability bandwagon. So, all of BT's thinking at this stage was tested externally through industry analysts, who advised on the relevance and differentiation of BT's propositions and the language used.

To get these propositions out to market, BT created an international campaign within the umbrella message of 'Bigger Thinking'. The campaign included internal training and handbooks to support customer-facing staff. It used conference platforms and a general thought leadership approach to sustainability to get the market's attention (recognised thought leader Dinah McLeod was now on board and leading the sustainability practice for BT). Finally, it included sales training, collateral and customer presentations, for sales and account teams to explore the issues directly with their customers and tailor the value proposition.

As is so often the case, the real challenge to the campaign's success lay in engaging the internal sales team. Used to selling very large-scale solutions such as data centres or networks, BT's sales and account managers struggled to see the relevance of the carbon impact assessment, either to their customers (many of whom had not expressed a need for it) or themselves (why sell a solution valued in the thousands of pounds when you could sell and get commission on one worth millions?)

Despite this potential obstacle, in the Carbon Impact Assessment's first phase it achieved its financial targets as set out in the 'playbook' and 'market requirements' documents. But there was more to come.

To drive further innovation, the strategy, marketing and propositions division brought together an advisory board of leading thinkers from around the world, including the ex-President of Costa Rica, who had already helped his country face an energy crisis and rebuild it using a more sustainable energy philosophy. This board challenged BT's approach, taking the initial success of the carbon impact assessment and using it as the basis for a much more ambitious vision of how BT could help companies, economies and societies in their drive to sustainability. The board set out a vision comprising: intelligent transport systems; smart utility metering; local revitalisation through use of communications technology in place of commuting; intelligent buildings; alternative energies; smart grids; carbon reduction, capture and storage. Following this momentous meeting, the sustainability practice went away to rethink their role and the scope of BT Global Services as a business.

With this bigger challenge on the table, the practice considered how they could help BT achieve such a huge ambition. In what was simply called the 'Ambition Plan', the practice highlighted the importance of all of BT's competencies and services, pointing out that these were where the real delivery of sustainability would take place. It reviewed the role of both the practice and BT Global Services as a 'thought provoker' and consultancy, identifying which of BT's other services were most relevant to support customers on their own sustainability journey. Rethinking the whole of BT's portfolio in this way, from a sustainability point of view, also resized the opportunity for BT from millions to billions of pounds.

Today, the practice focuses on embedding sustainability into the propositions across the whole of BT's portfolio, and views the carbon impact assessment as a consultation that helps to identify for customers which of the other propositions are best suited to their needs.

McLeod lists a number of lessons learned as a result of this journey. The first is about not 'boiling the ocean'. With any innovation, and particularly one based on a worldwide trend such as sustainability, the challenge often lies in clearly defining what to focus on. BT started small with the carbon impact assessment. It then faced an enormous challenge as the scale of ambition for sustainability was opened up by the advisory board. The re-scoping of the role of the practice them allowed the team to clarify how the new proposition sat with the rest of BT's portfolio, refocusing their activities with the sales and account managers.

The difficulty in engaging with these teams led the practice to simplify their messages and the proposition as time went on, and use case studies as proof points wherever possible. Proof in terms of impact on the customer's bottom line was vital, even with sustainability as the main value proposition. Differentiation also proved to be important, as in a market such as this there is room for so many players. BT's differentiation was built on their own experience of making ICT work more flexibly and sustainably over a number of years.

A final lesson BT learned was that in addressing a global issue like sustainability, with such huge potential environmental and social impact, the tone of communications needs to be humble. In spite of all their experience in implementing their own sustainable working practices, there was still so much for BT and others to learn about sustainability and its effects on the world. The practice concluded that all they could do was talk about the journey that everyone is on together.

A very important part of the perceived value of a service is the price that customers are charged. Important though this is, it is still only one aspect of service design and only one of the features that customers will consider. No buyer in any part of the world ever bought any offer on price alone. Buyers look for a cluster or mix of features which serve their needs and aspirations best. Price only becomes a determining factor if all the elements of the various offers are exactly the same. In practice, there are various approaches to pricing.

Cost-plus Pricing

This, the most common, involves estimating the time to be taken on each element of the project by each type of skill deployed by the firm. In reality, the cost of that time can be based on either a percentage mark-up on the cost of employment (so, the market for human capital determines value) or the firm's judgement about value.

Value Realised

This is sometimes called 'shared reward'. The supplier receives part of their payment if certain criteria, agreed with the customer, are met. Generally this applies to service firms that are focused on achieving efficiency and cost-savings for their customers. It was famously pioneered in the outsourcing market by EDS (now part of HP) taking over the parking ticket processing for a city in America. EDS was able to increase the proportion of parking tickets and fines collected, and as a result shared in the additional revenues that flowed to the city. Interestingly, this is also applied the other way (in terms of 'shared risk contracts'), whereby if an outsourcer does not deliver the benefits promised, they incur a penalty on their fees.

Competitive Approaches

Some price on the basis of what they think is appropriate in the market, based on what their competitors will charge. The drawbacks to this are serious. Firstly, they are unlikely to know exactly what others will charge and risk cutting prices too far. Secondly, they are unlikely to know the competitor's cost base, and may cut prices to a point that profits are eroded because they have a higher cost base than their competitors. Thirdly, the offer is unlikely to meet the customer's view of value and could become a commodity over time.

One of the most valuable and powerful applications of NSD, though, affects the perceived value of the core service; the pricing of basic services like maintenance or power supply. It can turn commodities of this kind into value propositions and transform the fortunes of massive

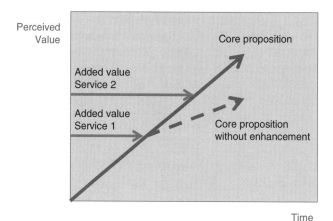

Figure 6.5 How to maintain the value of the core service
Reprinted with permission from *The Journal of Marketing*, published by the American Marketing Association, G.L. Shostack, April 1977.

organisations. Through a creative device, research and some form of innovation process, the supplier identifies improvements to the core service which will appeal to different segments of buyers. These are launched as a 'premium' or differentiated service, for which an extra price is charged. Eventually, as perceptions change, these fall into the core service and replace the basic offer, just as is the case with some products.

The basic car offered by a manufacturer establishes value expectations in the market. It is a benchmark which defines the features for which specific groups of buyers will pay extra. A cluster of features above this become the 'GTI' or 'Supra' model. In order to make the package aspirational, the supplier puts one feature that ought to be in the basic package into the elite version. Over time, car buyers begin to expect a different package of features which they think should be included 'as standard' with the prime offer of a car. The enhanced features become the standard offer and a new benchmark becomes the norm. To the modern car buyer the bundle of features offered by the car industry some 20 years ago (i.e. a low-performance engine with no sun roof and manual windows) would be unacceptable. Over time the core offer has increased in value through the introduction of added-value features.

This management of features is highly sophisticated and requires an institutionalised approach to NPD or NSD, which normally develops as a result of market trauma. There is no reason why, though, marketers should not initiate this approach as, say, a major, five-year strategy. A sustained programme to reposition the value of a network's core service is likely to be much more valuable than the creation of a myriad of 'added-value services' or numerous marketing campaigns to prospective customer groups. It does, though, need the progressive institutionalisation of NSD capability, similar in services to that which companies like BMW have achieved with manufactured products.

DIFFERENTIATION AND HOW TO ACHIEVE IT

In many demanding and difficult markets, suppliers have learnt to thrive by creating a truly unique offer, which commands the attention of their customers. In luxury goods, for example, the world's leading brands have learnt to create wealth from the perceived differences of, say, perfumes. The problem is that this looks sensible and obvious in retrospect. But perfume is really just fragrant water. There is no logical reason why Éstée Lauder should be perceived to be so different and more desirable than, say, Yves Saint Laurent. Yet, the world over, people flock to these different, desirable offers.

Other suppliers have succeeded in creating difference between business offers. The business class travel on Virgin Airways is perceived to be different to American airlines for instance. Whereas, in management consultancy, customers who see no significant difference between support or outsourcing contracts will understand the very different stance of, say, McKinsey, Deloitte and Accenture. Accenture has achieved this by concentrating on its theme of 'high performance delivered'. Its advertising, using Tiger Woods, conveys the message whereas its internal processes and training reinforce it. McKinsey, by contrast, felt at the start of this century that they had drifted a little from their prime emphasis; which they summarise as 'inspiring leaders'. They instituted internal global messages, training and support systems to reinforce it.

In fact, executives in technology companies have been trying to create the 'magic bullet' of differentiation for some time; and there are signs that it is beginning to happen. In mobile telephony, for instance, the different brand emphasis amongst European operators is leading to different emphasis in their service quality and the packages they offer. O2 has opted to concentrate on entertainment, aligning itself with the appropriate packages of services on a mobile and its stake in the massive O2 entertainment complex on the banks of the Thames in London. Orange, by contrast, has emphasised 'participation', Virgin mobile 'fun' and Vodaphone 'the individual'. If differentiation can be built amongst the services built on a mobile infrastructure, there is no reason why it cannot be built amongst the suppliers of any other service that relies on a technical infrastructure.

The reason many technology firms find it difficult to differentiate is that they start in the wrong place with the wrong assumption. Many assume that all customers want is the best products or services at the cheapest price. They then look to their own firm to try to discover ways in which they might be different to competitors. However, differentiation really begins with the needs of customers.

History shows that those who have succeeded in creating viable, profitable differences in other industries have done so through a systematic understanding of the needs of their buyers. Profitable differentiation starts with their desires and, contrary to popular belief, they will not all want the cheapest offer. They will want a mix of features and price, which represents value to them. As discussed in Chapter 3, some seek a features-rich offer with a high price and some a basic offer at a low price. The likely output of this understanding, reflecting buyers' needs and requirements, is represented in a perceptual map (see the Tools and Techniques appendix). The buyers' ideal purchases will be scattered around a line of value in the market (the line on which suppliers can achieve long-term position). Suppliers should choose where they want to focus and what they want to emphasise.

In fact, one important test of whether they really want to differentiate is choice; which customers they will not serve. Executives in technology firms seem to be very reluctant to say no to any group. They will do segmentation or value research, decide whom to focus on and then continue to service the entire market. Yet, to achieve differentiation, the customers to whom the company say no will define it.

CATEGORISATION AND ITS IMPLICATIONS FOR NEW SERVICE DESIGN

'Category management' is well known in the consumer goods industry as a way of producing and distributing products to buyers. It has been very successful in creating new markets, new products and lines of profit. Producers think of their products and display them in a category that is relevant to buyers, such as 'convenience health foods'. 'The response of' buyers leads to further ideas and product innovations.

The same concept exists in the service sector as a guide to spotting opportunities. The method of categorising a service affects service design, guiding the type of service to be created and identifying gaps in the market. It can therefore give competitive advantage and is a key strategic issue which ought to be defined by marketers as a framework for design work. There are a number of ways of categorising different services. These include the following.

Customised Services (Sometimes Called 'Solutions') versus Industrialised Services

When most executives think about the services of technology companies, they tend to think in terms of high-end customised services such as outsourcing and consultancy in business-to-business markets. In consumer industries, however, services are the subject of intensive efforts over years to streamline them and make the process of delivery to a mass market as efficient

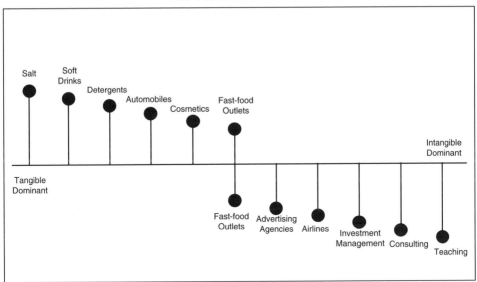

Figure 6.6 The goods/services spectrum

as possible (e.g. utilities and airlines). This was called the 'industrialisation of services' by Theodore Levitt (Levitt, 1976). The margins, approach to market, degree of client engagement and personnel used vary enormously between these two extremes.

Product-based Services versus Pure Services

Some services are adjuncts to product propositions and are therefore intimately tied to the value, development and pricing of that product, such as central heating boiler guarantees and maintenance. Other services are unique propositions in their own right which will stand alone in the marketplace, such as energy efficiency assessments. This was described by Lyn Shostack in her groundbreaking article (Shostack, 1977) and is illustrated in Figure 6.6.

A service supplier can use this as a tool to develop new services and manage its range of offers. For instance, in the early 1990s the IT industry was comprised mainly of product companies with associated technical support skills. Their advice and repair services were built around products, represented by the left-hand side of the diagram. While many industries make enormous efforts to simplify their propositions to buyers, the IT industry at the time was complex and tended to pass this complexity on to its buyers. A myriad of suppliers were therefore able to create healthy businesses by moving further to the right-hand side of the diagram and providing independent service offers, based on advice and assistance.

Technology-based Services versus People-based Services (High Tech versus High Touch)

Some services (self-service check-in at the airport) comprise technology through which customers are served. Other services are reliant on the skills of people. In some cases, the difference is based simply around the buyers' preferences. In others it is linked to market evolution. A supplier in a market where service by people is the norm might introduce a technical innovation whereby customers can undertake some work themselves. This leads to choice in the market. Some services are performed on behalf of a buyer, whereas others provide an infrastructure whereby they can perform the service themselves; such as printing their own boarding pass for a flight rather than collecting it at the airport. In initiating this, though, the supplier needs to plan the use of the technology, the impact upon the service process, how clients will be trained to use the technology and how this training will be spread to the total customer community afterwards.

The latter two points are very important. Customers need to be shown how to use the technology. People have a reluctance to change and to adopt new technology and this reluctance needs to be managed. Simple brochures or unattended demonstration equipment will not work. This needs to be non-intrusive, led by people who are competent and, above all, preserve the dignity of the customer. The introduction of technology in this way will reduce costs but will also increase perceived quality, because clients will be in more control of their service experience. However, the converse strategy is also effective on some, rare, occasions. In a market where suppliers have become remote or the service automated and commoditised, a premium supplier can gain advantage by introducing people into the service. To be successful, the supplier's brand must be high quality and aspirational and, in addition, the service must be high priced.

A Membership Relationship versus No Formal Relationship

This classification suggests differences according to the relationship of the buyers to their supplier. Membership may range from formal paid-for inclusion in a club (e.g. frequent flyers) to an emotional attachment to the grouping or community at whom the service is targeted. Other, project-based, services (such as consultancy) may be accessed once, when there is an immediate need, and then rarely or never again.

PRACTICALITIES

The Techniques of Component Design

Just as living entities have detailed and connected sinews wrapped around skeletons, so products and services comprise components which can be identified, adjusted and mixed to appeal to different groups of buyers. These components might be physical, technical, artistic, human or emotional. Product designers use design and planning systems to adjust product features because manufacturing or packaging processes need detailed specifications if they are to change; and so should service designers. Volume services in particular need equally detailed specifications.

Unfortunately, as services are intangible, some suppliers launch an ill-defined or poorly designed idea because of a lack of internal competence and often use the excuse of market urgency. Such poorly planned initiatives rarely succeed. Experience shows that suppliers need to think through the components of a service and ensure that all relevant processes are in place before launch. In short, service companies need to engage in proper component design; a task for which detailed and recognised techniques are available. The three main techniques are: 'features analysis', 'blueprinting' and 'molecular modelling' (see the Tools and Techniques appendix).

Features analysis, for example, suggests that every proposition has three levels of features. The 'core' feature is the heart of the proposition that buyers are seeking. It is often difficult to define but obvious once found. One management consultancy decided that it was in the business of 'problem solving', for example. 'Augmented' features are those items (such as materials or processes) which are the methods by which the service is delivered into the market through that supplier. Finally, and most importantly, are the emotional issues, which need to be built into the proposition. These are often a major factor in the customer's view of quality and value, and are often neglected by technologists and engineers.

Another well-developed technique is called 'blueprinting'. Every service that is offered to a market contains a process through which buyers move. This can be mapped and designed from the customer's point of view, plotting out all the elements of the experience they will receive. The service process must direct and encourage customers to receive the benefits they are seeking. It must be designed to help clients and to allay any fears they have. Any hiccup in the process or lack of forethought will cause them to become disaffected.

Finally, 'molecular modelling' recognises the mix, in some propositions, between physical and service content, tangible and intangible components, if you like. It is undertaken in a similar way to features analysis.

Portfolio Management

Product portfolio management is a much discussed aspect of marketing theory. It is argued that, in order to stay competitive, a business must continue to offer an up-to-date and broad range of products in a relevant way to a particular market. If a company is offering a range of products that are, for example, all reaching the end of their life, then its survival is threatened; a high-risk position. Strategists argue that new offers should be created before this dangerous situation is reached. Logic would imply that such a strategy is relevant to service businesses and that a regular competitive review of the range of offers would benefit the firm. However, as discussed in Chapter 4, this is predicated on a dubious concept, the product lifecycle (see the Tools and Techniques appendix). Marketers should be cautious about coming to conclusions that their service is 'mature' or 'dying'.

The 'Boston matrix' (see the Tools and Techniques appendix) is a good example of why marketers need to be educated, careful and thoughtful in the development of portfolio strategy. Many are surprised to learn that it was not based on the product lifecycle but around the concept of 'the experience curve'. Nor was it intended to analyse individual products and services but a portfolio of businesses. It was an attempt to give corporate strategic planners a way of evaluating different business units in different markets. A business which has just started operations would be a 'question mark' because the skills of the management to improve on its competence would be unproven. A company that had established itself in the marketplace and was beginning to thrive could be considered to be a 'rising star'; more established companies are 'cash cows' (producing profit but unlikely to achieve dramatic growth) and those in decline are 'dogs' (candidates for withdrawal).

At face value, the Boston matrix would appear useful to the leadership of service firms who own a range of businesses which each focus on a single service or market. The varying core competencies of these different businesses in their different markets may allow them to be managed as if they were product businesses specialising around their centre of expertise.

However, its applicability is still not clear-cut. For example, one of the fundamental assumptions of the Boston matrix is that market share is a measure of success. This is based on the experience of consumer product industries in 1970s America, where the likes of Coca-Cola and Pepsi, or Unilever and Procter & Gamble, fought over tiny percentage increases in share. However, market share is not necessarily a critical issue for some service businesses. In some cases, such as those serving the public sector, a large share makes it more difficult to win work. Some large businesses have a procurement policy to share their work out among suppliers and not rely on one single supplier (such as that adopted by supermarket Sainsbury following the failure of their single supplier outsource to Accenture).

Another well-known portfolio technique is the 'McKinsey/GE matrix' (see the Tools and Techniques appendix). This tool has been used in various service organisations to help them prioritise resources and approach new markets. In each case it has proved to be a useful tool for creating viable, practical service strategies. It is particularly powerful at creating a consensus among executives during the ranking of criteria and the rating of individual practices.

A further method of developing a product portfolio is 'line extension'. This categorises products according to a 'range' or a 'line'. It is completely independent of the product lifecycle concept and leads businesses to apply progressive innovation to their business processes. Designers look for applications based around the competence that produces them, and then try to apply them to new markets or different segments within a market. Theorists argue that the range should have 'width and depth'. 'Width' refers to the number of different lines offered, while 'depth' refers to the assortment of offers. Again, this has been successfully used in service industries.

A drawback of recognised portfolio planning techniques is that they do not help managers to think through the product/service mix of their propositions to market. However, the model created by Lyn Shostack in Figure 6.5 can be a particularly useful and practical method of planning the strategic direction of a portfolio. Changes in the strategic direction and positioning of the company, or changes in demand, can be incorporated into this diagram and used to plan the structure of the company's 'proposition portfolio' (i.e. the mix of products and services).

Should there be 'Product Managers' for Services?

Product and programme managers are very familiar to technology companies. These are normally senior executives who are dedicated to the management of the features and benefits of a product and its success in the market. Product managers often have worldwide responsibility for the creation of new products, the adjustment of existing products and the deletion of products from the range. They are invariably people who have either a technical background and/or a deep knowledge of the organisation. They are trained in new product creation techniques and have responsibility for the business plan of the product, for understanding the requirements of buyers and for the bottom line profitability of the whole range. Clearly if the creation of new services is important to organisations, they should consider creating dedicated 'service managers' who have equal weight and responsibility. Yet in larger technology firms, engineers or consultants will often be responsible for the resources to deliver services to customers. Despite the fact that services are frequently at the heart of the business, NSD techniques are often rudimentary and the role of product managers for services non-existent.

The role of a 'new service development manager', modelled on the product approach, would be to:

- create new services;
- adjust the features mix of existing services;

- withdraw old services;
- prepare and manage the business plan for a particular service;
- be responsible for the profit of the service being brought to market;
- establish all processes necessary to launch the service successfully and generate revenue from it.

Despite the lack of product managers for services, the industry has evolved an even more significant role. Some create 'service champions', senior individuals able to give senior executive sponsorship and responsible for the spread of a new service concept through the organisation. These tend to be senior executives with weight in the organisation who take on the role of 'political champions' for the success of new service ideas. This role is necessary because of the difficulty of selling new service concepts inside an organisation, particularly one that is engineering-led; they need to get buy-in from many different functions and individuals. A service champion also needs the skill to encourage the members of the new service project team to perform at their optimum level and achieve the NSD objectives. He or she must lead the team and smooth over disagreements and, particularly, give support to the new service concept at all levels in the organisation.

One difficulty in NSD is how to achieve the necessary level of integration between headquarters functions and field groups, particularly if the latter are independent or remote units in different nations. Ideas generated in the field need to be drawn together and coordinated by an individual with an international perspective. They should gather teams from across the firm to develop each idea if they are to establish it as an accepted part of the company's service range. Service champions have to overcome any dislocation between the local and global perspective.

Measurement of NSD

Finally, marketers ought to establish criteria by which the new services that are created are considered to be effective. The measurements put in place by top management give signals to the organisation as to what is important and what gets priority. If the creation of new services is important to a company, appropriate success measures must be created. De Brentani's research (De Brentani, 1991) identified a number that are used in American and Canadian firms:

- sales performance;
- competitive performance;
- cost performance;
- 'other booster' (i.e. how it affects other sales and costs).

SUMMARY

Understanding the management, structural issues and techniques involved in developing new services and breathing new life into existing ones should be of far greater importance to technology firms than it currently is. The processes and concepts behind new service development have been shown to provide demonstrable value. It can teach firms valuable lessons about how to turn the acknowledged expertise of their service delivery managers and consultants into a much more lucrative competitive service offer, whether with the firm's core service, added-value services or brand new offers. Getting to grips with this important part of the business should be high on the marketer's agenda. It is not at all easy for anyone to create an offer which is distinct and different from others; but it has been done many times before. In consumer products, for instance, British Potter Josiah Wedgwood (1790) and American entrepreneur Henry Heinz (1890) managed to create unique offers in businesses which were harsh, cheap commodities and in pre-industrialised, war-ravaged economies. In consumer services, leaders as different as Marshall Field (1910) and the modern-day Richard Branson have been able to create distinct, unique offers for which buyers were happy to pay more. And, for the sake of doubt, this has also been achieved with sophisticated business services. From banks like Rothschild and Coutts, to consultancies like McKinsey and Accenture, educated business leaders have been able to create and maintain distinct, valuable offers. If they can make themselves wealthy by crafting unique service offers, so can marketers in technology firms. Commoditisation, cheapness, erratic cash flow and price slashing are not, by any means, inevitable. It just takes a little forethought, application and durability to achieve genuine differentiation through innovation.

7

Selling Services

INTRODUCTION

The subject of selling has been covered in many courses and books over many decades. The objective of this chapter is to focus on what is different about selling services in technology firms and the relationship with service marketing activities. It covers the sales techniques that have proven most effective in these businesses, and how to apply them. It discusses, for instance, the lifecycle basis of selling services, as opposed to a product. The sales process restarts each time a product is sold but, when a service is sold with it, it is likely to be required for a number of years and, over time, the revenue it generates is likely to exceed that from the original product sale. So, while product sales tend to be short-term and transactional, service sales are longer-term and more dependent on the relationship built between buyer and seller.

Selling is not an erratic, ad hoc activity, nor is it best undertaken by a loud-mouth in a suit that shouts just as loud. It is not exaggeration, lies, false claims or pressurising people to buy things they do not want. All of these constitute poor salesmanship, which causes bad word-of-mouth, terrible reputation, law suits and, eventually, business failure. Sales is a credible business function with particular skills, processes and concepts that need to be as rigorously applied as those in other specialities. In fact, an understanding of how to apply the relevant sales principles to different sales environments is essential to revenue growth. This chapter looks at different sales processes, structures and, specifically, the techniques for selling services in technology firms.

HOW PEOPLE BUY

It is difficult to sell effectively without understanding how, why and when people buy. This is not, however, as straightforward as it sounds. Human beings are erratic, unpredictable creatures, driven by both their rational and emotional natures when making decisions; particularly, when buying anything (whether for work or personal life). Sometimes they approach decisions carefully, collecting as much data as possible before they commit to a purchase; at other times they return to trusted suppliers or brands and sometimes they buy, whimsically, on impulse. Although they sometimes have fixed ideas of what they want, it is just as likely that they are unclear on a particular occasion or cannot articulate their needs.

Considerable research and experimentation has been conducted to understand buying motivations and behaviours. Much of this is useful in giving pointers to practical sales approaches or market opportunities. Yet it is dangerously easy for a firm to institutionalise a half-baked view of buyer needs or to ignore them because it has succeeded by creating and pushing certain products or services in the past.

The Consumer

When consumers shop (whether physically, by telephone, by catalogue or through the Internet) they may have a cluster of needs in their minds, which are not necessarily refined into a fixed product purchase. They might only have an intuitive, confused mix of ideas or preferences

Figure 7.1 How consumers buy

and look for the proposition that best fits them. Their thinking process is broadly represented in Figure 7.1. This is, though, an inprecise representation because the intensity and duration of these different stages differs according to their experience, understanding and view of the purchase. Also, people are not necessarily linear or logical; they may miss out or mix up many of these stages.

First they tend to become aware of a need. This may happen quickly (perhaps prompted by the breakdown of an old domestic appliance) or through growing awareness (like an ever-closer house move and the need to arrange a new broadband connection) or, perhaps, over time (say, by the gradual ageing of a gas boiler). They will then think through what to buy, when and how to buy it. They might, for instance, research an involved purchase through the Internet or by visiting showrooms. Once they have refined their search they will discuss the potential purchase with the seller before buying. Sales people can increase their chance of success if they are able to influence their customer's thinking at the planning or evaluation stage. Nevertheless, experienced sales people are able to take people who have made up their mind on a competitor's service back to their evaluation and needs, matching them to their own proposition. Understanding, listening and matching needs to service benefits or outcomes is therefore essential to the sales process.

For instance, suppose a mother wants to buy a home broadband service to meet the needs of her whole family as they move house. She might be very familiar with the latest offers from seeing newspaper advertisements or receiving other forms of publicity. She might, though, savour the experience of spending a morning online, browsing through the offers of the leading suppliers. Or she might be too busy to either read adverts and direct marketing or to do her own online research. However, she is likely to have in mind a number of different attributes that her new service must include. She might consider: the style of equipment provided, the speed of connection, scope of package (TV, telephone and broadband), brand, contract flexibility and price (all represented in Table 7.1).

These attributes vary in nature and importance according to individual preference, occasion and culture. Different people will describe them differently and rank them in different orders of importance. Interestingly, contrary to popular belief and even in a recession, 'price' will

Table 7.1 Home broadband: a fictional example of home broadband service features

Features
Equipment style
Speed of connection
Package scope
Brand
Contract flexibility
Price

Table 7.2   Home broadband: a fictional example of graded home broadband service features

Features	Ideal	Acceptable	Unacceptable
Equipment style	Sleek 'hub'	Modern, discreet router	Large router
Speed of connection	50MB download speed	20MB download speed	Less than 20MB
Package scope	Phone, broadband and HDTV	Broadband and TV	Broadband only
Brand	BT	Virgin Media	TalkTalk
Contract flexibility	Totally flexible	Six months fixed	One year fixed
Price	£20 per month	£30 per month	£35 or more per month

generally be ranked after several other features. Despite what many sales people say when they lose a deal, no one buys products or services solely on 'price'. They don't go 'looking for a bit of price', but an offer which meets their needs. There will always be anywhere from two to four service attributes that will be more important than price. However, if all of these more important attributes are exactly the same as competitor services, then price becomes the only difference. So, if suppliers do compete on price, it's because they haven't understood the cluster of needs that are in their buyers' minds, and created a distinct offer that meets those needs. It is their own fault.

Using a research technique called 'zones of tolerance', created by America's Valerie Ziethaml (see Ziethaml and Bitner, 2003), each of these attributes can be broken down into components which are 'ideal', 'acceptable' and 'unacceptable' to the woman wanting to buy the broadband service (as shown in Table 7.2).

During the evaluation phase of the purchase, it is likely that the buyer will come across a number of different choices that meet these needs. Several will be candidates to fit her varied criteria, despite being very different mixes of features. Some will have few features at a low price, and some will be features-rich at a high price. Some will have components that the buyer has not thought of before, stimulating curiosity or desire. It is quite possible that a well-packaged mix of features could entice the woman to pay more than her original 'unacceptable' price.

The most successful sales person is likely to be the one who has delicately elicited these needs and suggested the offer from their own company that best meets them. Of course, they do not have Ziethaml's objective and systematic research techniques available for each shopper, so they must (through technique, experience or intuition) learn ways to understand their customers' preferences. They have to listen, observe and diagnose before they 'sell'.

Neil Rackam's (Rackham, 1995) famous and exhaustive research (he spent 12 years studying 35,000 sales transactions) emphasises the importance of listening and questioning in sales success. His SPIN model, which has been at the heart of much sales strategy in Western companies for several decades, codified four categories of questions: 'situation', 'problem', 'implication' and 'need-payoff' questions. Once the customer's situation is clear, the sales person scopes the problem with the customer and then identifies the implications for the customer if that problem isn't resolved. Finally, the sales person will establish the benefits of resolving the problem and create an urgent need to do so in the mind of the customer.

However, there will often be other influencers on the purchase decision, which complicate the sales process. The woman's partner and children will have a view on the most appropriate broadband service, with other preferences such as free WiFi access, or security options. This is called the decision-making unit (DMU) and is shown in Figure 7.2. Sales people must

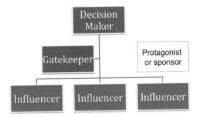

Figure 7.2   The decision-making unit

understand the roles and importance of the different people in this DMU if they are to succeed. For instance, used car sales people are taught to watch any new family coming onto the sales forecourt before approaching them. Their behaviour, and their answers to a few simple questions about what they are looking for, will reveal who is the actual decision-maker, who holds the purse strings and the degrees of influence of other members of the family in this particular purchase.

Business or Organisational Buying

One of the main differences between organisational buying and consumer buying is the more formal recognition of the role of different individuals in the business DMU. This affects the sales cycle because the sales team needs to talk to and nurture more people; and governance rules in Western companies seem to be increasing this method of buying. For their article in the *Harvard Business Review* two researchers (Trailer and Dickie, 2006) surveyed over 1200 companies. They found that the individual 'economic decision maker' was being replaced by:

> committees and multiple layers of approval, all equally important to the decision to move ahead. This is partly why the length of the average sales cycle keeps increasing.

They found that companies reporting a sales cycle of seven months or more had climbed from 18% in their previous surveys to over 25%, and those under three months had dropped from 50% to under 42%.

Suppliers have to become knowledgeable about the roles of these different people, the weight given to different purchase criteria and the sources of information used to come to a decision. There are 'influencers', who research or contribute to the evaluation of the proposed purchase and at least one 'gatekeeper', normally a reasonably senior person who presents the project to the decision-maker or a specialist, such as a purchasing manager. A 'sponsor' is an employee who is particularly well disposed to the supplier.

There are clear differences in the way different organisational buying groups behave in different sizes of company. For instance, the influence of the chief executive is likely to be more dominant in smaller companies and, because they have limited expertise, they are likely to filter information through trusted advisors or business networks. In large firms though, the buying of some services, such as consultancy or managed communications services, can sometimes be individual with little involvement from purchasing specialists, although this is decreasing as corporate governance and risk management become more important to shareholders and policy-makers. The situation is complicated by changing business strategies, by the network of relationships between firms, by the interactive nature of dialogue, by governance requirements and by the influence of the formal purchasing function (particularly in government or other public sector purchases).

Service providers can be particularly troubled by purchasing managers when they first encounter them. These specialists, often professionally trained, are employed to get the best value for their organisation. Contrary to popular belief, though, they do not just want the cheapest price. In fact, some deliberately cut out the lowest (and the highest) priced suppliers in any buying process because of concerns over quality of delivery. They will have a clear set of criteria by which they will evaluate and decide. Inordinate effort must be made by sales people to understand those criteria and how their offer is perceived to measure up against them.

Some academics have suggested that business buying is more rational and objective than consumer sales because of the existence of objective evaluation processes and the buying function. Yet emotion still plays a part in business sales. People at work make emotional decisions as much (some would argue more) than logical, objective and analysis-based decisions. A chief executive may want to buy a consultancy project to write a business plan from a particular firm because he fears acquisition and wants to use the brand of his supplier to reassure shareholders. A business buyer may be worried about the impact of a purchase on their budget, or their political standing after choosing a poor-performing supplier. Above all, though, as executives develop in their career they rely more and more heavily on the beliefs, opinion and intuition that they have developed through long experience in their industry. As they become more senior they deal with more strategic and less clear-cut issues, and so rely on their judgement (or 'gut instinct') even more. This means that, despite governance and rational processes, senior business leaders take intuitive and emotional decisions and must be influenced at that level. Experience suggests that suppliers who understand these emotional drivers of business decisions are able to charge higher prices than those who do not.

Figure 7.3 represents, very broadly, the typical steps in the business buying process. The first two stages are frequently undertaken inside the firm when assessing straightforward products such as paper, chemicals or Internet servers. Executives will determine the business needs and set policy. If they need information on suppliers, then the sources they refer to are: representatives of their firm, word-of-mouth among peers, industry networks (e.g. contacts in other firms or industry associations), directories, press releases, brochures and Internet searches. Each of these has different influence in different markets. For instance, research has shown that, in business-to-business markets, personal contacts with colleagues, including peers in other companies, is the most influential source of information whereas direct mail has virtually no effect. One group of academic researchers (File *et al.*, 1995) showed that word-of-mouth was rated as 'very important' by chief executives of US companies buying consulting services. Boston-based IT Services Marketing Association regularly reports the same phenomena through their surveys entitled 'How customers choose'. In their 2009 survey, they found that buyers are three times as likely to speak to their peers as use any other route to get information on suppliers who can help them with IT solutions. That is not to say that other information sources (directories, advertising, professional registers) can be neglected. Buyers

Figure 7.3   Steps in organisational buying

often use them early in the process to build up a short-list, or to find alternative suppliers, or to gain a perspective on favoured suppliers.

Some service suppliers have changed their sales strategy because of the need to influence the first two stages of the business buying process. Their experience is that, if they can influence these formative stages, the content of any request for proposal is richer and more favourable to them. Their experience suggests that suppliers who limit themselves to responses to tenders lose out on deals and potential incremental services. However, the people involved in their customers' strategy and planning processes are likely to be in different parts of the organisation and more senior than those they usually deal with. They are also more likely to weigh alternatives and seek answers to problems as they formulate policy. As a result, these first two stages can become the consultative and diagnostic stage of service projects focusing on the project scoping, pricing and contractual process. In short, they can develop into consultancy purchases. So a number of technology firms (like Nokia, IBM, Unisys and Ericsson) have moved from simply responding to tenders to a business model, which advises on the development of strategy, by creating a consultancy arm. They have, in their words, 'moved up the value chain' and the development of these consultancy skills has ensured the health of their businesses.

Although there has been detailed academic work, research and models developed about organisational buying behaviour, research into the organisational buying of services was a relatively neglected area until the 1990s. That decade saw the rise of outsourcing, facilities management companies, call centres and an explosion in professional services; particularly in management consultancy projects built around huge systems and change management initiatives. These firms needed to understand the buying criteria of their customers, so demand for research and concepts applicable to this field grew, as did the number of third-party intermediaries who built their businesses around interpreting buyers' needs and running procurement processes on their behalf. There is now a broad range of experience, concept and research available to thoughtful service marketers in technology companies.

THE SALES STRATEGIES OF TECHNOLOGICAL SERVICE COMPANIES

Different approaches to creating sales structures and skills are seen in technological service firms. They include the following.

Attempting to Train Product Sales People to Sell Services or to Turn them into Consultants

A number of service companies are adjuncts to product companies or have a heritage of product sales. It is natural, then, for them to try to get product sales people to 'sell services' or 'advise on the structure of the deal' (to consult). These are two distinct approaches and some flounder because they confuse them.

The first, 'getting sales people to sell associated services' is normally a cultural change requiring discipline and training. It may be that the firm has grown by selling products and that associated services (maintenance or financing, for example) have been neglected. A first step is to check that all the customers who have bought products have the necessary support contracts and the administrative arrangements to pay for them. This may sound simplistic but a surprisingly large number of first-rate companies have given service away 'free' to help the price of product sales. Two academics (Colletti and Fiss, 2006) found, for example, that the sales people of a:

> multinational software company ... were consistently giving away the time of its billable professional services staff to secure new sales contracts. On the surface, this wasn't a bad idea. However, it ran against the corporate strategy of focusing on top-line growth at a time when there were few opportunities to increase profitability through cost cutting. (Reprinted with permission from *Harvard Business Review*, Colletti and Fiss, July/Aug 2006)

Sales people may have been so focused on selling product that they have discounted the value of support services or, much worse, not included them in the sales process. If so, little contractual documentation will exist for their customers' service needs. Even those organisations that have discovered and resolved this obvious problem have to conduct regular audits to ensure that their controls continue to work. A second step is to consider rewarding sales people on total income from a customer. For instance, if a product costs x, associated services can amount to at least another $2x$ over its lifetime of use. When sales people realise this and are incentivised accordingly, they can quickly develop skills to sell associated services. This change alone can revolutionise behaviour towards services.

The second approach, 'getting people to advise on the nature of a deal' in a business-to-business environment is a 'consultative sales' approach. It requires sales people to engage in conversations with customers earlier in the sales process. The sales people attempting this need to develop their listening skills more, perhaps learning advanced consultancy techniques to help the buyers to diagnose a problem and subtly propose answers to it. They also need to learn how to relate to more senior, thoughtful people. Some find this easy to do because consulting techniques merely put words and methods around listening approaches that they have learnt intuitively through years of sales experience; but many find they are not comfortable dealing with the vagaries of different strategic choices.

Although there are similarities in the sales techniques of traditional field sales operations and consultancy, there are also some very real differences (see the summaries in Tables 7.3 and 7.4). These differences must be taken into account if a successful consultative-sales approach is to be taken to the customer.

It appears that it is arduous to move product people into selling service, and experience over the long term has shown it is rarely successful. A number of companies have tried and failed. In their article on network sales for example, two academics (Üstüner and Godes, 2006) reported the experience of Hewlett-Packard's computer services organisation during the 1990s:

> The unit had planned initially to penetrate new accounts by satisfying the (essentially commodity) hardware reordering needs of purchasing managers and then use those initial contacts to open the door to the IT department where it could sell more complex hardware. HP planned to leverage those contacts to build relationships in the c-suite in order to sell lucrative consulting projects. This foot-in-the-door plan failed ... had HP spent more time evaluating the underlying prospect networks, they might have foreseen that their client contacts would not necessarily help them move higher in the organisation. (Reprinted with permission from *Harvard Business Review*, Üstüner and Godes, July/Aug 2006)

The most notable example of a revised approach was IBM as it moved more forcibly into services. It first used its account managers to handle service sales. Yet it eventually had to turn around its business, laying off a large number of its then 300,000 employees and recruiting more with different skills. It found that very few of even these highly skilled and very well-trained sales people were able to make the transition to selling concepts and

Table 7.3   Similarities between product sales and consultancy

Product sales and consultancy
Payment by results crucial to achieving sales and revenue
Cash flow management critical
Brand is an important issue
Good-quality collateral is important
Characteristics of people who generate revenue

services. In the end, IBM developed a strategy built on its knowledge of sectors and used that to penetrate the global services market. This led them, ultimately, to the massive acquisition of PricewaterhouseCoopers' consultancy division, which contained people whose whole career had been spent in consultancy and who had contacts at senior level in leading organisations.

Table 7.4   Differences between product sales and consultancy

Product sales	Consultancy
Regional sales	Industry knowledge
Closing technique	Problem diagnosis
Awareness building	Reputation building/demand pull
Product knowledge	Industry knowledge
Sell benefits of products	Diagnose needs & solve problems, involving elements of strategy
Response to tender once needs clear	Discussion earlier in sales process
Contact at middle & senior management level	Contact at senior & top management level

Teaching Service Managers to Sell

A number of companies have tried to use maintenance technicians, engineers and managers from their service division to handle the relationship with important customers. The logic of this approach is based on the fact that purchasers of products with a long sales cycle will be visited more frequently by service people than sales people (indeed, service people may even be constantly 'on site' with customers). If a customer normally replaces their product every three years, then contact with the sales organisation will be infrequent; whereas service people are, at least, likely to be visiting annually and could schedule more frequent visits while attending to other customers in the area.

This approach has hidden costs. It will affect, for instance, the performance of the service division, increasing the time they spend with customers, reducing their visit ratios. A more subtle drawback is that, although they have a service ethos, they are often unable to adopt a consultative approach to their customers and have limited sales ability. It has been attempted successfully at document outsourcer, Xerox. But a carefully developed programme to give the service managers the tools and techniques needed to up- and cross-sell was an important factor in its success. So, this approach can result in a long, arduous training and recruitment process, and can take several years to engender an effective sales ethos. As a result, some companies have had to draw back it.

XEROX GLOBAL SERVICES: HELPING SERVICE MANAGERS SELL

Xerox Global Services (XGS) is the world leader in document outsourcing services, with more than 20 years' experience and over 15,000 business professionals in 160 countries. XGS is one of three main business units within the Xerox Corporation, a $17.6 billion technology and services enterprise that helps businesses deploy smart document management strategies and find better ways to work.

An Opportunity in Waiting

Having always been renowned for its approach to sales training and development, XGS became aware of considerable limitations in the way that its on-site services were being promoted. An audit confirmed that only a portion of accounts were promoting Xerox services within their sites and almost none had a realistic plan for their promotional activities. Of the sites that did have promotional material in place, significant inconsistencies in Xerox messaging and branding were found. In addition to the obvious impact on the Xerox brand image, they were also missing an important opportunity with a captive audience (the existing client base) to 'up-sell' and 'cross-sell' Xerox services.

An extensive web-based survey of Xerox operations employees validated these concerns. For instance, employees felt constrained by: time (45%), skills (38%) and affordability (28%). Overall, the survey confirmed the need for an easy-to-use, centralised marketing support programme for operations people.

Since XGS marketing thought that it can be up to five times cheaper to keep an existing client than it is to attract a new one, they felt strongly that both revenue and ROI could be quickly and profoundly affected by designing campaigns to existing clients; by ensuring that consistent marketing messages were seen by existing clients on a regular basis. The resulting 'Xerox Site Marketing Programme' was developed to provide easy access to professionally created, on-message, on-brand marketing materials that could be customised by operations employees in a matter of minutes. The programme was also designed to bridge the gap that existed between the Xerox operations and sales teams assigned to support the same account.

The programme achieved revenue growth in its first year of over $600M from existing clients. So, it has established itself as a valuable resource to Xerox on-site managers around the globe.

Innovation in Action

The concept of promoting Xerox services inside a managed services account, or branding Xerox's on-site presence, was not a foreign one to the account teams. What the Site Marketing Programme did was revolutionise the process by which account teams marketed to their sites by providing a web-based interface to dynamically create a marketing plan. This included the strategies and tactics most appropriate for that client site; and the ability to produce customised marketing collateral that supported the plan.

Xerox operations employees can now create an unlimited number of Site Marketing Portals (typically one for each client or client location under their management). Each is initiated by answering questions from the portal's four-part assessment tool about their specific client environment (services offered, industry, contract length, applications, awareness levels, communication methods available, client objectives, etc.). This information creates the content on that company's Site Marketing Portal and generates, interactively, a customised, end-to-end marketing plan for the client and that Xerox team to jointly implement.

Every tactic outlined in the site marketing plan is present in the portal. It covers all aspects of on-site services promotion; from generating demand to strengthening business relationships within the company. New content and capabilities are added regularly, ensuring that Xerox operations personnel have the latest tools available to help their clients receive value from their services. Customisation of templates can be done online through a simple web interface, which generates high-resolution files that can be printed on-site (or produced off-site) and delivered directly to the client's facility.

Users have access to both static, but customisable, content and artwork (including a capabilities brochure, team directory, posters, wide-format standing signs, tent cards, calendars, open-house event save-the-date postcards, invitations and fliers, e-blasts, company newsletter articles and ready-to-use surveys). It includes links to additional Xerox resources such as existing sales collateral, flash executable files and whitepapers.

The Site Marketing Programme accommodates three branding options for all customisable templates. Users can choose a Xerox-only brand, a client-only brand, or a co-branding option, providing the flexibility for the account team and client to select the option that is best aligned with their partnership and marketing strategy. Once the branding preferences are selected, the end-user never has to worry about logo placement on the individual collateral pieces, which are all generated automatically to correct size and placement. A facility that not only increases speed in production, but ensures the materials meet brand standards for both Xerox and its clients.

Although originally intended to be a post-sale marketing tool, the Site Marketing Programme has also proved to be a differentiator in a pre-sales situation. The benefit being the capability to promote the services chosen by the client from day one of any new contract. So, 'pre-sales' external collateral was created about the programme, and included in many new tender responses.

Implementation

After seven months of planning and development, the Xerox Site Marketing Programme was introduced with a pilot in Xerox's largest North American accounts. A full-scale US-based launch, in April 2007, was followed closely by a Canadian and Developing Markets launch; and a European pilot the same year.

Much consideration, collaboration and modification was required to create a successful programme, with real attention paid to areas such as content, language translation, client-specific brand requirements, naming conventions and system administration. To be accepted as a standardised global marketing tool, input was needed from each of Xerox's three operating companies (North America, Europe and the Developing Markets Organization). In addition to the initial survey to determine market needs, a 'global community of practice', comprising key stakeholders from each region, met bi-weekly. A repository of existing content was established and regionalised as appropriate. The need for several local versions of the portal soon became evident, with the need for one mirror version of the English portal to be created in each supported language.

A robust administration site was established to provide access to metrics including (but not limited to): the number of registered users, the number of marketing plans created, the number of customisable pdf templates downloaded and then printed on-site, or centrally. At a moment's notice, Xerox was also able to track: the user frequency, page popularity, page visits by hour, day, month and user registrations by country and date.

System management initiatives included a 'Problem Resolution Log', for documenting and tracking the status of all problems, and a 'Site Marketing Hotline' that provided dedicated telephone and email support. As the global marketing 'quarterback', the XGS marketing team handled all problem resolutions and content updates required for both the programming and financial aspects of the programme.

A challenging aspect of the programme was the inability to 'track beyond the metrics'. While it was easy to produce reports and spreadsheets that provided data on the number of portal activities, Xerox was originally unable to track the use with clients. A number of measures were put in place to address this:

1. An inspection plan that tracked client-facing site marketing activity which was supported by senior executives.
2. Identifying and communicating to senior leadership the new revenue signings associated with accounts that have marketing plans versus those that do not.
3. Instituting a 'Best Practices' library to document and share how the site marketing materials were being used in client sites, and the resulting metrics.

Among the metrics reported, the number of marketing plans created stands out as their most significant measure of programme success. The presence of a documented marketing plan inside an account signifies that collaboration has taken place between the Xerox operations and sales teams, and their client. It means they have identified and agreed upon the most appropriate ways to generate awareness, and utilisation, of the Xerox services provided to that client.

The site marketing plan had also enabled account teams to execute the plan and penetrate other departments within the account that had previously been untapped (including the client's marketing department). This opened the door to deeper client relationships, revenue growth and protection against competitive entry.

Enhancements to the Site Marketing Programme since its launch include the addition of new content and innovative tools. Improvements have been made to the marketing plan, such as: the ability to prioritise the suggested marketing tactics, to edit tables and charts within the plan and refined reporting capabilities. XGS intend to enhance the look and feel of the portal, add even more content (both static and customisable) and launch internal contests to increase awareness and encourage additional use of the tool.

Account Growth Achieved

Xerox made a commitment to ensure that their brand personality and messaging was communicated accurately and consistently, inside their existing client base. The Site Marketing Programme has enabled this objective to become a reality.

Within a year of launch, the Xerox Site Marketing Programme had reached 100% of its $600 million original target from within the existing client base. At the click of a mouse, Xerox account teams across the world now have centralised access to all the strategic and tactical marketing tools they need to help increase the awareness and usage of Xerox services. It also helps them promote future innovation within client accounts.

CREATING A 'SERVICE' SALES FORCE

Most of the world's developed economies now have a thriving service sector and the businesses in these sectors often have sales organisations or consultancy arms to generate revenue. So a number of technology companies have recruited new consultants or sales people from successful service businesses. These specialists have never been trained as product sales people and are therefore more directly suited to a service-orientated business. For example, when (in the 1990s) the computer company Unisys first moved substantially into services, it recruited a partner from Andersen consulting (Victor Millar) to its board. He then created a job specification for a 'principal' consultant, to be recruited across the worldwide firm. The HR function set out behavioural requirements, background experience and skills required from these people. It then recruited candidates to take over the customer relationship. The logic was that clients were looking for more customised projects and that existing account managers could only sell tangible products, not intangible services. As a consultative approach was required, all account managers were to be replaced by consultants.

There are a number of advantages a specific, experienced 'service sales force' has over product sales people trying to sell services. For instance, it can be difficult for an existing product sales force to sell services that have previously been given away free. The temptation is to fall back on old habits rather than employ new techniques and approaches. Whereas service sales people, particularly experienced consultants, are conversant with selling intangible service concepts and adept at managing relationships with more senior people who develop strategy.

The services sales cycle is often more prolonged than product sales because the customer is buying the credibility of the sales person rather than a tangible product. Time is required to stimulate the necessary trust and credibility. The danger is that the existing sales people avoid services and concentrate on products because they deliver a quicker return. Experienced service sales people, on the other hand, know how to reduce the time lag through enhanced interpersonal skills and the ability to quickly establish rapport.

It is possible to train product sales people in new skills, but attitudinal shifts of this kind are difficult for some, impossible for others. Service sales people and consultants, on the other hand, are used to charging for their services and know how to handle specific obstacles to conceptual sales. They are not burdened by historical precedent. New people can easily be recruited matching an attitudinal profile, benefiting the company from the outset. They can sell services to senior people without any embarrassment or fear. Their skills are particularly important when a company is offering an attractive, high-margin service experience. Product sales people and, in fact, service engineers are likely to outline the point-by-point components of a service, detracting from its value. The right service sales people, by contrast, can excite the imagination of buyers, drawing them into an enticing experience.

Solutions Selling

The need to approach customers, particularly business customers, at the moment when their requirements are ill-defined has led to that most pervasive of sales strategies: 'solutions selling'. This is an attempt to sell a customised package of products and services rather than a pre-defined product. It first appeared as a concept in the computer industry during the 1980s. At that time suppliers' main propositions to customers were to buy a new product because it would be faster and cheaper than the one they already had; but what was often missing was any analysis of changing business needs. They were not, generally, listening to their customers or taking care to understand their real needs.

The 'solutions' approach was created to encourage sales people (who had become used to focusing on the upgrade of existing products or overhauling their 'installed base') to learn to focus on real customer needs. Because customers had a mish-mash of different technologies and products that had to be taken into account, a process evolved whereby the seller conducted a technical audit of customers' needs and proposed a 'solution', initially called a systems integration project, which customised new and existing technology to meet changing business requirements.

Solutions selling has since been used successfully by a wide range of different companies in a host of different industries, to act as a catalyst for a change in the behaviours of their sales force. It has been seen in: computing, telecommunications, document management, healthcare and heavy engineering, to name but a few. Companies in these industries have set out to capture additional revenue from associated services or, perhaps more importantly, by understanding the real needs of their buyers. They frequently edge competitors out of their customers' installations by offering packages that work across all technologies.

However, in the enthusiasm to adopt this approach, and use it as a catalyst of change, the difficulties it causes have been disguised, neglected or ignored. The first, and most important, is that it can cause an erosion of margin and a fall in price. When companies sell products there are established prices and controlled discounts which a sales person can apply. But when putting together a package of products and services, sales people will often use a discount to get the sale; and the discount is frequently applied to supplementary items outside the discount controls on products; services for instance. However, when a product company moves wholeheartedly into 'solution selling', their own products can be a much smaller percentage of most orders. If it has no cost information on the other elements of the sale or, perhaps more importantly, no discount controls in place for service items or competitor equipment, the move to solutions will lose money.

A 'solution', a customised offer, should be an attractive, high-margin service experience for which the buyer is pleased to pay more than the individual components. It should be a technology firm's equivalent of a Savile Row suit or custom car; something expensive, sexy and enjoyable. Self-evidently it should be more expensive than any pre-packaged offers available in the market and, glitteringly expensive though it is, it should tempt customers to buy.

Secondly, if a supplier is seeking to run or grow a large business on volume sales, then one of its keys to success is scalability. It needs to 'industrialise' its services, packaging its offer to customers as much as possible. Many large technology companies are based on this ethos. Their marketing, sales, distribution and delivery are all based on volume. Yet, this is in conflict with a sales approach that customises every single sale through a solutions selling philosophy. Unless the company genuinely uses some form of mass customisation process and technology, the conflict in this strategic emphasis will cause inefficiency, higher costs and poor quality.

The third problem is that the marketing of solutions doesn't tend to create or communicate a clear value proposition. Customers need a simple proposition if they are not clear about what they are looking for or if they are going to change their buying habits. An appealing description and attractive packaging are important ingredients in the marketing of service experiences. Solution selling has led, by contrast, to extravagant claims, odd-sounding marketing messages and some ridiculous product names. The satirical British magazine *Private Eye* regularly lists some of the sillier examples, such as:

- **'Chilled food solutions'** – ready meals.
- **'Carestream radiology solutions'** – X-rays.
- **'Integrated vegetable supply solutions'** – greengrocers.
- **'Client solutions team'** – complaints department.
- **'Intelligent solutions, a complete IT solution, including hardware, networking and e-commerce'** – consultants.
- **'Inbound solutions guarantee to take the pain away from all telephony and application-related aspects of your organisation'** – consultants.

The king really does have no clothes. This is poor English, bad marketing and really idiotic business strategy. Architects take an approach to their work that is similar to the 'solutions methodologies' suggested by those who have specialised in this field. They understand their client's needs, often helping them to craft a 'requirements specification'. They then analyse the site and design a 'solution' to their needs; before assembling a multi-disciplinary team to build the building. Yet at the end of the project they deliver a house, a garage or an office block; not an odd-sounding, meaningless package. Marketers need to get a grip on this and make up their mind what their market claim really is. Are they offering a bespoke, customised, expensive service or more of an industrialised, streamlined and packaged experience?

In fact, it is easy for technology firms to take share in markets dominated by companies who have taken a customised, solutions approach to every sale. If they package their offers, perhaps going as far as to offer products which capture many of the service requirements, they will be able to undercut the price of established suppliers and improve their own profit.

Selling through Business Partners or Other Channels

A comprehensive sales strategy considers all the existing and potential routes to market; and some of these may be indirect channels such as business partners (other companies wanting to sell the supplier's products and services). The set-up costs of working with partners need to be weighed against the potential revenue and, more importantly, net profits produced by any indirect channel over an agreed period of time. Another important aspect of channel strategy is designing 'synergy' and avoiding 'channel conflict'. Using a business partner makes sense where, together, the two organisations can do something that neither party could do separately. Channel conflict might occur, though, where one partner does not generate much business as a result of the partnership. It also occurs if the two companies are too similar or if there are no cost savings gained by working together, and no benefits to either of the suppliers or any customers. These obvious difficulties are frequently missed in the heat of exciting discussions about joint approaches.

Using a partner also makes sense where a large geographic region needs to be covered, and two (or more) service organisations can do this better than one. It also makes sense where a subcontractor can be used, with a much lower cost structure, to provide services that are either difficult for the supplier to provide (through lack of skills or resource), or which they chose not to provide (because the subcontractor can do them as well, or better, at lower cost). A computer vendor in Sweden, for example, needed to cover the majority of the country, but faced cutbacks in its own organisation. Their answer was to join forces with a service organisation that had more locations, lower costs and a different customer profile. There was sufficient overlap between the existing customers of the two organisations for services to be subcontracted both ways. Both sides benefited, as did the respective customers.

The Creation or Acquisition of Separate Service Businesses

Some large technology companies create sales capability through a separate business under an experienced consultancy leader. For example, when it first ventured into services, the French-based telecommunications company Alcatel created a new consultancy company to lead its venture. This was headed by a general manager, specifically recruited from one of the 'big four' accountancy firms to build the business. It had a different brand name, a separate reporting line and its own accounting system. The management of this business was therefore able to structure it as a completely separate entity with the practices appropriate for this type of operation. Similarly, the Swedish giant Ericsson started its venture into the more competitive sale of services, in the mid-1990s, with the acquisition of a strategy consultancy. Other sales capabilities were then added to this small unit as business grew.

GENERIC SALES STRUCTURES APPLICABLE TO SERVICES

There are several generic approaches to structuring sales teams, which need to be considered and adapted by service businesses.

New Business Sales

This is the most recognisable and, in some ways, straightforward type of selling. A new business sales person is normally briefed and trained on the details of a product or service in order to sell to an intended group of customers. They will usually have a defined 'territory' to cover, depending on the size of the total market, and sales targets to achieve each month or quarter. These sales people have to be managed and motivated through pay, training, administrative back-up and targets; some form of payment by results being the chief of these. It is then up to them to find and gain access to potential customers who will buy, with the help of 'leads' generated by the company's marketing programmes.

This approach has been widely used across a range of both product and service businesses (from cars and confectionery to banking and insurance) in a wide range of countries for over 300 years. In 1905, for example, Singer (even then a global firm, making sewing machines in eight factories around the world) sold 2.5 million machines a year through 61,444 sales people (see Edgerton, 2007).

Contrary to popular opinion, the best sales people are the best listeners. The caricature of the pushy sales person is actually bad practice and counter-productive. Even second-hand car sales people are taught to ask questions, get an idea of the potential buyers' needs and then begin to offer product that best suits those needs; to match the product 'benefits' to the 'prospect's requirements'. With more complex products, this approach has to become more structured and, at its most sophisticated (with, say, complex engineering networks) the listening aspect of the sale is more like consultancy. The sales people who lead these deals, often first-class scientists, are taught the sort of diagnostic and discovery approaches which are second nature to leading strategy consultants.

When selling technical services to consumers, such as a service contract for their central heating boiler, the sales team has the benefit of an 'installed base'. This often involves working closely with the marketing function to identify which customers of the product are most likely to buy such a contract, and approaching people with that profile among the population who don't have a contract. Often this type of sale is led by a direct marketing campaign and executed by telesales people, combined with an online sales mechanism.

Account Management

The concept of account management is based on the fact that certain buyers will give a stream of repeat business to a supplier whereas others will not. It is widely used by technology firms and, due to its significance, is covered in more detail in the next chapter. Three researchers who studied '2,500 businesses in sixty eight countries' (Zoltners et al., 2006) claim that companies which move from regional sales to the appointment of account management find that the change 'increases revenues, . . . boosts customer satisfaction and reduces the cost of selling'.

Regional Sales versus Industry Knowledge

Companies often take a 'field sales' approach to their sales organisation. This involves sales people being assigned to a geographic territory and handling the approach to all potential customers within that territory. They are often managed by an 'area manager' reporting to a 'regional manager' who, in turn, reports to the sales director. The ethos behind this approach is efficiency in terms of sales time. The sales person can plan what length of time to take to visit customers within a geographic patch and management can assess this efficiency by the number of 'sales calls per day' (or its equivalent).

By complete contrast, however, the consultancy industry tends to base its sales organisation on 'industry knowledge'; familiarity with the industrial sector in which the buyer works. As problem diagnosis is at the heart of the consultative approach, clients need to be reassured that the consultants with whom they work have a deep personal knowledge of their industry. Most consultancies therefore have a sector approach to organising their practices. Very often there will be a 'practice head' who leads a group of specialists dedicated to a business sector such as telecommunications, computing or banking.

This organisation is not really interested in geographic efficiency. Very often 'virtual' teams can be dedicated to an industry's specific problem from across the world, such as currently operating at Accenture. For them, the incremental revenue that can be gained from greater industry knowledge outweighs any consideration of the number of 'sales calls' that can be achieved in any one day.

Part of the value to buyers of outside advisors is the perspective they develop from handling the problems of several companies in the same industry sector. Business customers are often curious about how they stand when compared to peers and have an open ear for issues or trends that outsiders can spot from such deep engagement. They will even attend confidential discussion sessions and briefing programmes with competitors present if facilitated by a supplier they trust and if there are clear rules for the meeting. Such a powerful position as industry expert is a very rich source of projects and fees.

There is, however, a greater reason for deep knowledge of the sector in which customers operate. Buyers rarely know the full scope of the products or services offered by a supplier and often have latent needs which, they think, cannot be resolved easily. They may not even have formulated them into potential projects or a request for proposal. A consultant, who understands the priorities of firms working in a sector, is likely to spot these latent needs and identify those that their firm can meet. They can then create projects, as yet unimagined by the buyer, to solve problems. So, the primary skill needed to succeed as an adviser is industry and client knowledge, while the secondary skill is the knowledge of the firm's resources. This suggests that consultants recruited from an industry and taught a firm's products or skills will have an edge over internal recruits

Service firms might have both these structures in their organisation. An electricity or gas utility might, for example, have a regional sales force dealing with its business customers throughout a nation. Many also, however, have consultancies, some working on an international basis, advising on the use of energy as a means to gaining incremental revenue. These are likely to focus on a particular industry so that they can understand and apply concepts to their clients' needs more effectively.

Partnership Selling

Consultative selling in professional partnership is remarkably successful. The majority of product sales people are unable to whisper in the ear of senior people and walk away with the high-revenue, high-margin deals that many partners in professional practices do every day. 'Business generation', sales, in consultancies is led by their top practitioners who manage client relations. This is called 'partnership selling', a term from the professional services industry where it is often the role of partners in the firm to lead the dialogue with clients. These partners are often 'unconsciously competent' at breathtaking salesmanship. Their sales approach is often intuitive and un-codified but, nevertheless, imitated by the best technological services firms. It is explored in more depth in Chapter 8.

The Phenomenon of 'Rainmakers'

Rainmakers exist in all professional services (from architecture through to accountancy and law). They are also the reason for the success of numerous consultancies in the technology sector; from single 'one-man bands' to larger engineering consultancies. A true rainmaker brings in outstanding amounts of revenue. While most senior professionals generate income and revenue, these people tend to generate two or three times the industry average. So there is substantial effort to identify, recruit and develop rainmakers. In fact, the nurturing of this talent is seen by many in private professional practice as their prime method of growing their firm.

The skills and attributes of rainmakers generally are as follows.

- A driven individual. They need to achieve and put their prodigious energy, often caused by personal trauma, into business success.
- Market focus. They know the market they operate in very well and understand both developing issues and key individuals within it. They concentrate their team's skills and expertise on the market and apply their knowledge.
- Reputation management. They create 'thought leadership' to enhance reputation and understand the use of PR. They write articles and speak at conferences to maintain a high profile.

- Client targeting. They identify high-probability projects which yield high returns and seek out buyers with needs.
- Networking. They put enormous effort into building and energising networks of professional contacts. They have frequent contact with clients, building trust and relationships. They ask questions and listen to needs, identifying projects to meet those needs. They close deals and then sell other projects.
- Delivery. They become reliable by ensuring a good team works on their projects. They manage expectations and ensure that their team delivers high-quality work.
- Measurement. They ask for feedback, adapt and develop. Over time, they build a track record.

Yet really effective rainmakers are aberrations. They are individuals who have above-average capacity to generate business but by much more than just a sales person in a product company exceeding their targets. These are difficult, driven and erratic people who need the approval and comfort of huge social networks. Moreover, most of them do not understand how they succeed in such dramatic ways. It is neither possible nor sensible to build a sizeable business around their erratic behaviours.

The leadership of firms moving into advisory services should, of course, consider the strategies and processes by which they recruit and manage rainmakers, but they need to build the business in such a way as to make room for their erratic idiosyncrasies. For instance, rainmakers are often adverse to systems, processes and administration; so they may not be particularly good at management or leadership. The service firm should therefore have clear human resources policies to identify and retain them outside the normal management promotion process. They should be managed appropriately and given the correct support with increasing responsibility for either markets and revenue streams or major accounts as they progress through the firm. The culture of the firm ought to be friendly to such difficult human beings. It also needs to explain their importance and their behaviour to other members of the firm who have to tolerate their erratic behaviours.

Sales Support

Many firms employ specialists who assist with relationship management or sales. Some simply provide administrative support, helping sales people or consultants to spend as much time with customers as possible. Others arrange seminars or hospitality events aimed at providing opportunities to sell. They have one over-riding objective: to maintain focus on developing sales opportunities. They may assist with account planning (preparing, attending and participating in planning sessions) and they may manage the production of proposals. Specialist teams are often set up in large business-to-business service firms specifically to do bid management, forming part of a sales or account team. In some companies, sales support people also participate in account development, visiting contacts to create relationships and open doors.

THE PRACTICALITIES: MANAGING THE SALES PROCESS

There are a number of general principles (a distillation of theory, research and experience) that will need to be considered by service companies that are establishing an effective sales capability.

The Sales Cycle

The words 'sales cycle' suggest something that starts at the beginning and flows through to the end; to the sale. Whilst this is true, one of its key characteristics is that, at any one time, there must be a number of prospects, all of which will be at various stages of their sales cycle. Yet, only a proportion of these opportunities will convert into real sales. The skill in this aspect of selling lies in knowing how to 'qualify' accurately and being able to identify how long the sales cycle is likely to last.

Typically, a sales cycle might have the following components:

1. 'Targeting'; working out who the intended customers are and ways to approach them.
2. Pre-sales activity.
3. Formal offer, proposal or 'pitch'.
4. Closing.
5. Post-agreement negotiation.
6. Post-sales support.

The length of time this may take varies with the type of sale and will only be apparent with experience. Timescales tend to take longer where formal procurement processes and invitations to tender are involved. However, these do have the advantage that the timescales are accurately known and usually published. On balance, though, large service contracts tend to take longer to sell than a product due to the increased levels of decision-making, and complexity of the proposition. The important thing is to determine what is normal. If the sales cycle is taking much longer than the norm, the likelihood of success should be questioned. If, for instance, one sales person is taking much less time, leaders should find out why and apply appropriate techniques across the business.

Targeting

Like product companies, service businesses use the term 'targeting' for either the process of selecting potential customers or the act of approaching them. Methods vary enormously because they are often trying to communicate an idea or concept. Some consultancies, for example, tend to create ideas, which they proactively suggest to their intended customers. Others analyse their customers' businesses in anticipation of a general discussion. Still others look to develop a vague friendly relationship that they hope will develop into mutual work. Some create elaborate diagnostic tools or 'benchmarking data', which they use to stimulate interest. Others employ specialist telephone sales or marketing groups to 'open the door'.

By complete contrast, those service businesses who build 'demand pull' through a strong reputation and brand will have customers coming to them. Their 'targeting' is very different from product sales people who have to go out to get customers and tease out sales in a geography or industry sector. They are able to choose those that they wish to develop long-term profitable relationships with and reject those that they don't.

Pre-sales Activities

Technology companies employ a wide range of marketing techniques to create demand, stimulate leads and warm prospects for a direct discussion with a sales person. Sometimes, pre-sales work includes getting on to approved supplier lists, from which employees in an organisation can choose. Without this no substantial work can progress. For instance, many leading public companies maintain a list of consultants and other service specialists that their people must use. Getting on this list may, therefore, become an objective of an account or 'targeting' plan.

At the sale itself, people who sell services have to take as much care to glean initial impressions during early stages of discussion with a potential buyer as any other sales person. For instance, body language, which has been the subject of many research projects and learned publications, is a powerful indication of underlying thoughts. They must also look for other non-verbal signals such as the environment for the chosen meeting. (If the approach is by telephone, it may be that the issue is not seen as very critical; if in a restaurant or bar this may be a signal that the customer is at an early stage in their thinking process and is open to a range of ideas.)

At this stage it is important to adopt a consultative style; to listen and diagnose. This is much more than cursory attention to words. They must listen actively, demonstrating understanding and checking back with the potential buyer. With a serious problem it may be wise to ask for time to consider the issues and book a further meeting to suggest potential approaches.

First-rate service suppliers try, though, at this phase in the discussion, to avoid a formal proposal. This is more common than many who have to suffer formal tenders or 'beauty parades' assume. The ability to move straight to work is an indication that the supplier has developed deep levels of trust with the buyer. Companies whose marketing really entices a customer induces strong relationships, stimulates the respect of buyers and helps skilled service sales people. A common approach is to suggest different methods of solving the problem and ask which the customer most favours. At a further meeting to discuss the 'draft' programme the buyer, feeling in control of the process, adjusts the work plan to their own, unique environment. Very often the project proceeds from there and leaves those intent on the detail of pre-sales processes wondering how the deal was stolen from under their noses.

The Formal Offer, Proposal or Sales 'Pitch'

As in product sales, a response to a formal request for a proposal must be carefully managed. At the very least deadlines, formats and specified areas of information must be met accurately. This is particularly important in government or public sector proposals where non-compliance can mean that suppliers are automatically excluded. The proposal document, the pre-proposal process and the proposal presentation should all be treated as communications exercises. It is an opportunity to listen and respond, not to bore, pontificate or show off. If the organisation has specified meetings or people to consult before submission deadline, they should be used. In these, as much detail as possible should be asked about both technical issues, the problem, budget and those involved in the decision-making process. Some customers take a lack of questioning as a sign of lack of real interest. The team formed to manage the proposal process must plan the approach to warm up those who need to be influenced.

The proposal itself must first be written as a clear communications document. Something that starts with pages of description on the supplier's resources and history is unlikely to communicate effectively. It should outline the buyer's problem or need, suggest an approach to resolving it, articulate the benefits of the approach, say why the supplier is unique and give indicative pricing. The production of the document should be managed as a project, leaving enough time for proofreading and rehearsal of how it is presented. This last step is often missed, and an excellent proposal and document is let down by a boring or sloppy presentation to the prospective customer.

However, the team presenting the proposal needs to be chosen carefully. Not only must they have the right technical skills and experience, they must also include people who will actually work on the job. Buyers are irritated if a senior person presents the proposal and is never seen again. So some suppliers make a virtue of the fact that the team presenting will deliver the work.

'Closing' Service Sales

In all exchanges between sellers and buyers, there comes a moment when buyers need to make up their minds. Very often, success can be improved if that moment occurs with the seller present, so sales specialists have created tools to help focus on this moment. Called 'closing techniques', they include:

- Asking for the business. It is obvious, and therefore often forgotten, simply to ask buyers if they want to go ahead. A surprising number of sophisticated and experienced business people do not do this.
- Overcoming objections. Buyers are asked whether there are any reasons why they can't proceed and each objection is handled as they arise.
- Open-ended questions. This really is an extension of 'overcoming objections'. A series of questions are asked to get further and further into the buyer's need and to match the offer closely to those needs. As consultants frequently have to start working with a client by diagnosing need, the 'consultative approach' closely resembles this technique. This is probably one of the most practiced closing techniques in service industries.
- Exaggeration to the absurd. Here, the seller may take one of the buyer's objections and exaggerate it to the point of the ridiculous in order to overcome it as a barrier to purchase.
- The 'assumed close'. Here a buyer's body language signifies that they are happy with the suggestions and want to buy. The supplier moves to talking about next steps and assumes the sale is agreed. In fact, they may be concerned that, if the customer is asked for the business or asked whether they want to go ahead, barriers will be raised in their mind.
- The 'go-away'. In this context a seller is convinced that the offer matches the needs a buyer has. If the buyer tries to negotiate on price, or cut corners, the seller can suggest that they don't go ahead. This causes the buyer to re-commit to doing the work.

Yet the sales organisations of technology companies may need to rethink their use of closing techniques when they sell their services. As services are more intangible, conceptual offers than products, if the seller is too overwhelming and uses closing techniques too forcefully, once the buyer has a moment to think, he or she is likely to feel cheated and the deal might unravel. These approaches need to be handled with real care in a service context.

Post-agreement Negotiation

Once the sale of a service has been agreed in principle, contract negotiation takes place. This is often straightforward in consumer sales and, in a business sale, need not be as complex as it is frequently made out to be. Sometimes no negotiation is required; the contract is printed on the invoice and the customer has automatically accepted its terms by buying the services. For relatively small services this works well, and the law acts as a back-up in most cases. The larger and more customised the deal, the more the customer is likely to negotiate the contract in such a way as to maximise benefit to them and reduce risk. It is important to have a good legal or commercial department who know the business and are responsive to needs.

It is increasingly common for a service level to be agreed, sometimes with a penalty clause for non-performance and, less often, with a positive sharing of rewards for over-performance. With a new service, or service on a new product, this is risky because contracts have to be set up well before any operational data is available on reliability, mean time between failure, repair times and the like. Contract negotiation can also be used as a means of selling and discovering new opportunities; although if they are discovered at this stage it is likely that the existing sale will have to be completed. A new sales cycle will then start for the new business.

Post-sales Support

Post-sales delivery of service is primarily about excellent work, but it is also about reassurance. If something goes wrong, for instance, it must be resolved quickly. It is better to admit that the supplier is not perfect and remedy problems, than to ignore them. If the purchase of any item is emotionally challenging or if it is expensive, buyers experience 'post-purchase distress'. This is anxiety caused by the purchase and is allayed by the buyer admiring the purchased product or sales materials. If this anxiety is not managed, then problems occur. Yet, as services are intangible, there is nothing to offset it. As a result, methods of summarising the emotional relief of a well-executed project have evolved in some markets. Sales people should think through the moments of anxiety that their buyer might experience and create physical devices to allay this discomfort. If possible, find tangible expressions of progress such as weekly progress reports and 'contact' reports, summarising each meeting. A good service sales person anticipates and obviates this concern.

Pipeline Management

This is a management concept and sales discipline that builds the generation of business into the day-to-day life of a sales team, taking the sales cycle concept and ensuring there are enough opportunities at each stage to deliver enough sales. It is the foundation of good sales practice and the focus of sales management in many industries. Sometimes called the 'sales funnel' and sometimes 'the book of business', it is a discipline that translates easily into service businesses, particularly in project-based businesses.

Without good pipeline management, activity can be erratic. The business experiences peaks and troughs in workflow because there is no consistent focus on the generation of future sales. Pipeline management helps overcome these difficulties. Figure 7.4 illustrates the concept, which should be approached from a hard-headed, numerically driven perspective.

For example, starting from the right-hand side, if a consultant has a target of $2 million, and the average value of their projects is $500,000, they need to win four projects a year. The next section of the pipeline shows that, in order to win four jobs a year, they need to propose a number of projects to potential customers. If, in their marketplace, the conversion ratio is two to one, then they need to propose eight jobs to win four. In some service markets, though, there are no formal proposals and presentations. Sometimes a buyer will simply ask a supplier to start the work. Other times the project can be loosely defined. Nevertheless, there must be some form of discussion, presentation or scoping of projects that forms the basis of the agreement to go ahead. These all count as part of the 'pitches' section of the pipeline.

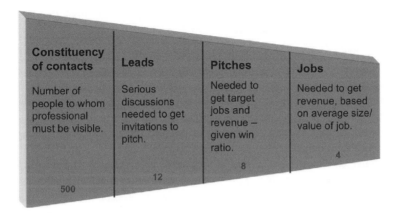

Figure 7.4   Pipeline management

The next category refers to 'leads' or 'serious expression of interest'. A sales person will be known to a number of potential customers but, at some stage, those buyers must express an interest in a product if there is to be a sale. This might be by telephone, in a meeting or in a social context. They discuss their needs and ask for more information. That may simply be an invitation to examine the need, or it might be an invitation to take part in a 'beauty parade' where the firm has to respond formally to a request for a proposal or presentation of their credentials, or both. If the conversion rate in a market is two to one, there will need to be 16 expressions of interest to get eight proposals to get four jobs. Already, then, the sales person in this example needs to receive at least one serious expression of interest a month if workflow is to remain healthy.

The final part of the sales pipeline, the wider end of the funnel, is the 'constituency of contacts' on which the sales person should focus. This will have several components. First, there will be those who are well-known business contacts; where there is a close relationship. They will be seen in a social context, and there will be a close relationship of trust, built on mutual value that has been given over the years. In service markets, the size of this constituency of contacts varies. It might be, for example, primarily in one major organisation; when, for example, a large customer has a dedicated account manager.

To use the funnel effectively, service sales people need to build these processes into their day-to-day lives. For example, they can make sure they maintain regular contact with their intimate professional relationships. They can schedule personal meetings, send information or articles they think will be of interest, and invite them to ad hoc hospitality or professional briefing events.

In the wider constituency, though they should work with their firm's marketing people to create a marketing plan. Famous marketing professor Phillip Kotler (Kotler *et al.*, 2006) pointed out that marketing and sales functions should be integrated in their ownership and cooperation in the sales funnel. He and his colleagues said:

> Sales and marketing are responsible for a sequence of activities and events (sometimes called a funnel) that leads customers toward purchases and, hopefully, ongoing relationships . . . Marketing is usually responsible for the first few steps – building customers' brand awareness and brand preference, creating a marketing plan, and generating leads for sales. Then sales executes the

marketing plan and follows up on leads. (Reprinted with permission from *Harvard Business Review*, Kotler, Rackham and Krishnaswamy, July/Aug 2006)

Their funnel had components in a sequence called: customer awareness, brand awareness, brand consideration, brand preference, purchase intent, purchase, loyalty and advocacy. Whatever the detailed components, however, the use of funnels will avoid the 'feast and famine' effect of erratic workflow if managed carefully between marketing and sales. The plan must be based on a knowledge of whom they are targeting, where these people go for professional development, and what they read or listen to. This information will guide the media chosen and the activities undertaken.

Pipeline management is also a powerful tool for sales managers. Many build the concept into IT systems so that the pipeline of the business can be seen easily by leaders. Some build in processes of performance review, to keep colleagues focused on the need to manage future business while conducting client projects.

SUMMARY

The sales discipline is a crucial part of any firm because it ensures that the company earns revenue. This chapter has explored the different aspects of selling services in technology industries. At one extreme are hierarchical geographic field sales structures and, at the other, the partner-led selling found in consultancies. Different sales principles are employed at different phases of the sales cycle and can be used with the relevant tools and techniques. A clear channel strategy needs to be developed which designs new service sales channels effectively and, as importantly, enhances the existing sales channels. Yet, real wealth is created and large contracts won when a service company's approach to market integrates marketing, sales and service operations. If an exciting, enticing service is positioned to tempt buyers, then the sale of service is relatively easy and price a mere consequence, rather than a barrier.

8

Marketing and Selling Services to Major Customers

INTRODUCTION

The reality of many technology firms and their service businesses is that they are dominated by important, repeat customers. In fact, in service businesses the phenomenon of word-of-mouth creating a competitive reputation tends to lead, more than in others, towards repeat buyers. So, for many marketers in service firms, their day-to-day job is focused on significant buyers. Yet, this perspective is barely covered in marketing courses and tends to be neglected in textbooks on the 'principles of marketing'. So, this chapter concentrates specifically on the need to analyse and service major accounts. Leading firms in the industry seem to no longer see this as just a sales strategy. They are creating services, value propositions, marketing strategies and communications programmes for one major customer at a time; and this is a significant contribution towards the development of attractive, high-margin branded business-to-business services. This chapter explores that phenomenon.

PRIORITISING CUSTOMER ACCOUNTS

Firms have to start by identifying and defining their most important customer accounts. This can be as crude as listing them by volume of business and ranking accordingly. Some focus on their 'top 100' (produced in this way) or some version of it. Others have tiered layers of prioritised accounts, each receiving different levels of attention according to the volume of business they generate. Unfortunately, though, these simple approaches won't meet all of the firm's objectives in its market, and it certainly won't reveal that it may only be receiving a small share of these companies' spend in its category. It may, for instance, want to identify customers with the highest potential and penetrate those further. Or it may have more generic strategies, such as wanting to penetrate a sector of the market or to take business from a competitor. So, leaders may set objectives to 'penetrate' certain types of premium customers in order to be recognised as a high-quality supplier. If these 'target customers' have low immediate spend, they may be categorised as a 'strategic account' and receive the same attention as important customers with a larger immediate business volume.

With these different issues to resolve, it is sensible to step back, and analyse both the existing major customers and the potential buyers before settling on any strategy or approach towards them. If, for example, customer accounts are mapped in terms of the size of their spend on services and in terms of the company's share of that spend (a derivative of the Boston matrix), there will normally be a spread of business, as shown in Figure 8.1.

Marketers can use this analysis to determine where best to allocate scarce resources for maximum return. This straightforward analysis suggests four strategies according to where

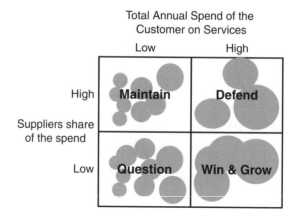

Figure 8.1   A representation of a business across its customer accounts

the accounts fall:

- **Maintain.** The intention here is to hold position with relatively little investment. Accounts in this 'maintenance' box should be managed with less resource by running the account through less expensive means than face-to-face account managers. Some, for example, set up telephone-based account management for those that merit less investment.
- **Defend.** These accounts are important and effort must be made to both keep the customer excited by the supplier's offer and competition out.
- **Question.** If there is one weakness that almost all executives have it's the reluctance to cut out unprofitable or low-potential customers. Those in this part of the analysis, though, detract from the resources and attention that should be showered on others. They ought to be culled.
- **Win and grow.** This group of customers is high potential and effort should be put into penetrating them further.

The size of the circles represents the size of the customers' businesses, so this gives an indication of which accounts are likely to be more lucrative and which to allocate resources to. If any of the accounts in the 'defend' box were lost, it would have a significant impact on the business revenues, while the future of the business may be in the 'win and grow' box.

The directional policy matrix (see the Tools and Techniques appendix) can also be used to help prioritise accounts and, at the same time, gain some consensus among the different people involved in account management activities. Involving business managers, account directors, sales and marketers in a project to score and rank accounts brings focus and alignment to account selection. Fujitsu's UK arm went through exactly this process, starting with a long list of over 300 companies and putting them through a screening process to identify non-starters. The remaining 100 or so were then subjected to the prioritisation process. The management team agreed what would make an account attractive to them, weighted each criterion in terms of its importance, and agreed how the scoring system would work. (A typical set of criteria is shown in Table 8.1.) They did the same for criteria that would indicate Fujitsu's relative strength in each account. Then the marketing team calculated the prioritisation of each account, prompting the company to agree different programmes for each one. This included: the personnel allocated to each account, long-range sales campaigns, marketing

Table 8.1   Weighting and scoring account attractiveness criteria

Account attractiveness criteria	Weighting	Scoring definitions		
		High (10 points)	Medium (5 points)	Low (0 points)
Company's turnover	0.2	>£500M	£250M–£500M	<£250M
KPI performance	0.3	Below industry par	Industry par	Above industry par
Propensity to outsource	0.2	High; have done so before	Medium; have outsourced other functions	Low; no outsourcing anywhere to date
Compelling event	0.3	Probable within 2 years	Possible within 2 years	Unlikely within 2 years
Total	1.0			

programmes and budgets. In fact, although the process had started out as an attempt to prioritise accounts for marketing and sales activities, it became a mainstay of mid-term business planning.

MAJOR CUSTOMER SALES STRATEGIES: ACCOUNT MANAGEMENT

Once the top-priority accounts are identified, account managers are assigned to develop them, taking them beyond the more 'opportunistic' approach usually adopted by sales teams. The concept of customer 'account management' is based on the fact that certain buyers will give repeat business to a supplier whereas others will not. It developed when product companies found that revenue improved if they took different approaches to existing buyers than those used with 'new business prospects'. They found that the skills of a sales person focused exclusively on getting new customers were different from those of a 'representative' dedicated to managing the orders from existing buyers. The latter is focused on creating longer-term relationships and gets involved in many issues other than direct sales (such as complaints, or administrative difficulties) which might threaten the business exchanges between the two sides.

Although used in retail, consumer goods and the car industry, it was when this concept moved into the young computer industry (with its then obsession on its installed base of products) that there was progressive codification of major account management as a discipline. The practices of global market leader IBM, where account managers had been chosen, trained and fêted for many years, were particularly influential. The concept has since moved into other industries and has been adopted, to a greater or lesser extent, by many different companies. It is seen throughout the service economy and even leading consultancies use relationship partners to focus on the needs of one important account.

Account managers are dedicated to one large customer. Their role is to get to know the client in depth. They must not only be the voice of their employer to the customer, but also the voice of that customer to their own organisation. They need to be capable of understanding and presenting the whole range of products, skills and services offered by their firm. At a minimum, excellent communications skills will be needed plus a recognised ability to generate revenue. Successful individuals in this role are also frequently creative, able to spot opportunities and

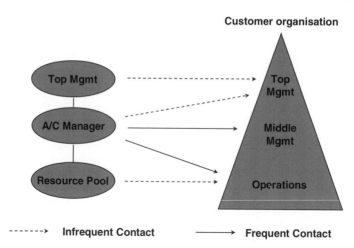

Figure 8.2 Representation of typical account management interactions

harness their own firm's abilities by forming teams to suggest ideas. They normally deal at middle or senior management level (as represented in Figure 8.2) but, on occasion, need to gain access to the heights of their customer's organisation (perhaps to discuss strategy) and the very bottom (perhaps to collect data).

Even at its best, though, there are limitations to the sales aspects of account management. Much modern practice has been based, for instance, on processes pioneered by IBM in the 1980s. It had client relationship directors and detailed account plans with sophisticated prospecting and sales support. Yet when, in the early 1990s, the company faced traumas due to dramatic changes in the computer market (like inter-operability), it nearly collapsed. This was, in part, due to the fact that its account managers had missed changing customer needs. Similar difficulties occurred at BT. After it was privatised, it invested heavily in account managers but, within five years, its share of its prime market, the city of London, had dropped from 100% to under 20%. Yet, if account management sales theory (which was designed when these markets had semi-monopolistic distortions) was correct, neither should have got into that situation because their account managers would be so close to their customers.

So, account or relationship management as a discipline has its flaws. Account managers might, for instance, hold onto information about their customers or miss more general trends because they are so focused on sales deals in one business. In fact, leading firms have found that account management needs to be supported by service and marketing strategies to major customers. Successful account management is more than just a sales strategy.

Xerox is one company that has an excellent track record for sales training and capability, but when faced with increasingly global customers that expected Xerox to deal with it on a similarly global basis, the company created an appropriate account management approach. It was supported by the executives of the company and has led to substantial increases in business among those global accounts.

XEROX'S EVOLUTION OF GLOBAL ACCOUNT MANAGEMENT

Xerox is the world's largest technology and services company that specialises in document management. With sales of $17.6 billion and 57,100 employees in 160 countries, the company helps businesses deploy smart document management strategies and find better ways to work. Global account management is not a recent phenomenon for the firm as it first appointed global account managers in 1988. But with increasing globalisation, customer requirements have changed and the need for more strategic account management has never been stronger.

Aware of the need to take their global account management to the next level, Xerox recently built a 'global accounts community', which concentrates on making it easier for customers to do business with the company on an international basis. This resulted in 27% revenue growth in two years which Xerox puts down to a stronger, more empowered, group of people.

Global Business

Global account management has emerged as one of the major strategic issues facing companies that need to manage their relationships with multinational customers that have a globally integrated approach. While there are some similarities with national account management programmes, the real value comes from understanding the customer from a global perspective. There is particular value in identifying opportunities that go beyond the traditional supply chain rationalisations and global price discounts. Yet it requires fundamental changes in the way such companies organise to do business.

As with many large corporations, Xerox was international in scope but it was not truly global. The company came to realise that although it had, over time, increased its number of global accounts to 100, it hadn't allowed for the considerable infrastructure or capabilities it needed to have in place to really support these accounts. It was also apparent that Xerox's approach to national and global accounts was 'cookie cutter', with no real customisation around the types of different accounts or geographical reach.

To improve its global account marketing, Xerox committed to pilot a new global programme with a refined focus on the very top-tier global accounts. Twenty five accounts were selected as being globally strategic for Xerox, and requiring an integrated and co-ordinated global approach through dedicated client managing directors and virtual teams across the world.

One of the big catalysts that drove Xerox in this direction was its move to being services focused. When providing services, the supplier is no longer talking to just procurement or strategic sourcing, but also the lines of business executives looking to enhance business processes. It's a very different dialogue than the 'cost per print' conversation typical in Xerox's market. With the move from a product to a partnership approach, and the capability to offer global services, Xerox has been able to step above a lot of its competition and create a very different offer.

Although integrated into the broader sales structure, this is about strategic account management, not just 'sales', and demands a very different set of skills and capabilities of the account leadership. It is a result of both external and internal pressure: to meet both the level of skill now required by the customer and Xerox's ambition. Global business is complicated, with a need to ensure good balance between country and global strategies. The elements Xerox attributes their progress to are: leadership, account selection, coverage, infrastructure and executive support. Removed from geographic boundaries and managed at the global level, the vision and strategy for these accounts are controlled by a much more empowered group of people today than they were when under the old regime of managing up through the organisation.

Leadership

Senior leadership is the critical factor for these global accounts, both at general management and account levels. At the outset, Xerox put the top tier of accounts under the vice president of global account operations, who reports directly to the chairman and CEO. Later, it moved to the president of Xerox Global Services. The credibility and leverage of someone who sits at the senior table, sharing progress on a consistent basis, is fundamental to success.

At an account level, the client managing directors (CMDs) who own the global strategy must build and maintain trusted relationships across different cultures, geographies and economies. This demands a high level of maturity and experience; articulate professionals who know the industry and how to engage at executive level. So, the CMD role has required a very different profile than the typical person Xerox would have hired ten years ago, and the company now recruits internally and externally to find these capabilities.

Account Selection

Like most companies, Xerox automatically gravitated towards the largest customers. Yet they found that it's important to identify accounts where the relationship is strategically important for both parties.

The main reason to limit the number of global accounts is to maintain clear customer focus. Xerox went from 100 to 25 tier-1 accounts, as it only wanted companies in the programme with a true appetite for global business. Today, three initial criteria determine whether a company can become a global account:

- The company has to be truly global, with operations in more than two regions and a substantial part of its business outside its home market.
- The size of the company, both in existing and potential business.
- The company needs to be organised in a global way, to have global procurement and be willing to collaborate with Xerox in a global manner.

Coverage

Xerox has found that a clear understanding of the scope of global accounts and determining how to add value delivers considerable return. Today, with a dedicated person to design global plans for each of the top-tier accounts, it has a clearer strategy for serving them. It has the ability to take a close look at the client, really study where they are as a business, take advantage of the emerging and growth markets they are pursuing, and set up a coverage scheme that matches the client's needs more effectively than in the past. As a result, Xerox is now realising much of its growth from new business with these customers in overseas markets.

Infrastructure

Xerox has found that setting up the right infrastructure for global operations improves performance and success; for example:

Global bid support – Xerox found one of the toughest jobs was trying to negotiate pricing from four operating companies and 160 countries around the world, particularly without a well-coordinated pricing or contracting programme. The importance of a coordinated bid increases considerably when moving into services, where its possible to build a more complex proposition. Today, all four operating companies have bid proposal teams and Xerox is able to pull together a much more effective and professional global account bid response. Whereas previously Xerox tended to bid for everything, it now first decides whether to even bid at all through careful qualification criteria. Now that everybody participates towards a winning bid overall, Xerox has the ability to smooth out inconsistencies that can occur on the global stage. Certainly, with this programme, Xerox has become more creative and differentiated, adding features rather than just competing on price.

Global agility – is a core capability that needs to be well resourced; global, regional, selective and responsive. Until recently, only the Xerox global general manager or CMD had the ability to step outside their country to help pursue a deal. Yet the pursuit team's cross-border capability is essential to go after these big accounts, particularly with real growth coming from the emerging market regions that don't have the skill-sets required to pursue those opportunities. Xerox is a lot smarter today in getting the people and teams who can really own the pursuit of opportunities into a variety of different countries.

Global compensation – Xerox introduced a new compensation plan for the top-tier team to help them drive global revenue. This comprises two key metrics: invoiced revenue and total contract value. Both are multi-year compensation plans to help people think longer term, which is a fundamental for global accounts. Ultimately, with senior people leading these accounts a level of P&L responsibility is needed and so Xerox is also working to measure bottom-line profitability.

Global planning – As global initatives can help to change a company's culture from a siloed, 'local geography' starting point, plans that were 100% country-based have now been changed to include global revenues. For example, Xerox's Japanese organisation now has its account reviews at North America headquarters.

Communication – is important for virtual teams and networks. A good reporting system with alignment throughout the organisation develops a more customer-centric rather than geographic focus. Xerox's own web-based 'DocuShare' has proved a valuable system for internal information sharing, and initiatives such as customised customer portals. From a sales perspective, Salesforce.com has enabled Xerox to manage global accounts more effectively and to provide information back to the countries as and when they need it.

Corporate Sponsorship

This global programme would not have thrived without the full support of senior management. Executive commitment is essential to help meet client expectations and drive the considerable organisation of changes needed. But it perhaps adds most value by its indirect support of the CMDs, which enables these people to have sufficient power to coordinate activities and influence priorities across traditional geographic and business unit boundaries. Xerox's executive education programme has now been revised to ensure that the top-10 executives in the organisation are absolutely aligned to each of these 25 top-tier accounts on a global basis.

In Summary

Xerox has built a strong global account community to meet the business needs of its most strategic global clients. In a short time, it has seen improved global bid success, a higher penetration of its services portfolio, greater levels of customer satisfaction and many more 'C-level' conversations in these priority accounts. It has garnered a 27% non-domestic revenue growth, which is five times above that achieved in the other global accounts as a direct result.

  The next step planned as part of this programme is growth through an industry-led approach if the top-tier accounts. According to Xerox, becoming a global company never ends, with the need to constantly evolve as strategic customers change; 'It's a race without a finish.'

Based on an interview conducted by the Strategic Account Management Association (SAMA) with Thomas J. Dolan, president of global accounts operations, Xerox Corporation, 1 August 2008.

PARTNERSHIP SELLING: MAJOR ACCOUNT MANAGEMENT IN CONSULTANCIES

The approach to major customer relationships in premier consultancies (and, indeed, in most high-end professional services) shows up one of the fundamental flaws in traditional sales account management: access to top-level, high-margin strategy projects. There are several unique elements to this, some almost the exact opposite of the sales equivalent. In these businesses, work is done by the leaders and owners of the business, the partners. These business leaders do not spend much time managing their own organisation. They are senior people, able to enter into a dialogue with the most senior leaders of even the largest businesses without a product to sell or an idea to push. Their role is not to administer their business, but to engage with clients and win work. They listen, identify needs and structure answers to problems, some of which have never occurred to their 'client' (as they call them) before they began the dialogue. Their normal relationships, access and dialogue is right at the top of their clients' businesses, as illustrated in Figure 8.3.

  Partners repeatedly say that success 'is all about relationships'. These range from the way individual clients are handled during an assignment, through the justification for spending on client hospitality and up to the sophisticated relationship programmes that huge global firms have with international businesses. All are designed to ensure that clients are treated well enough to engender respect and trust. Such a close, respected relationship means that work goes to tender less often and clients happily invest in a mutually profitable exchange. As a result, the ambition of most world-class consultants is to be a 'trusted adviser' to their clients.

  The need for trust begins at the very first encounter. When, for instance, a new client seeks out a professional for the very first time there comes a moment when they have to take a leap of trust. Even if they like the practitioner immediately and are convinced of their techniques,

Partnership selling

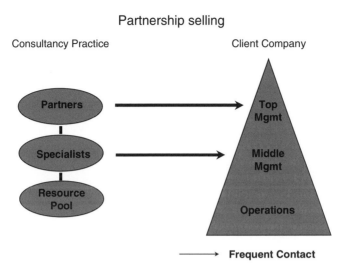

Figure 8.3   Major account management in a professional partnership

it takes a leap of faith to engage them. Called 'placed trust', this leap of faith is an essential step for anyone engaging a consultant or, in fact, any professional adviser for the first time. It is emotional, intuitive and frequently spontaneous (see Halliday, 2004). It ought to be the first focus of account plans and especially pitches to new clients. Those potential clients might look like the 'dragons' in the TV show: experienced, sophisticated or aggressive, but what emotions are in their head? What will induce the human heart to trust the practitioners' skills?

As clients return for different services, new levels of trust are explored. Is the organisation, for instance, 'trustworthy' in the sense that it always does what it says? There has not been too much good-quality research into the dynamics of consultancy services but, where it is done, one of the most common findings is the utmost importance of 'reliability' to potential clients. It's no good preparing really elaborate pitch documents if consultants do not return calls or if reports are delivered a day or two late. The very basic requirement of clients is that they must be able to trust the behaviour and response of the firm. Account plans should include initiatives to audit their firm's service quality, using something like the 'GAP' model (see the Tools and Techniques appendix), and create a serious improvement programme to deliver reliability as well as spending time chasing lots of individual deals.

As new clients turn into repeat clients, deeper levels of trust develop. They begin to like the practitioner they deal with and rely on their judgement. This induces a different kind of trust, more akin to that which develops between friends or relatives. It is nurtured by the client's respect for the practitioner's judgement and professional skill; but it flourishes if they feel that the person cares for them and thinks proactively about their needs. Yet even this familiarity is not what David Maister meant when he introduced the world to the term: 'trusted advisor' (Maister *et al.*, 2000). He demonstrated that the client/partner relationship can get to a point where there is deep trust which is very profitable. The client will understand the practitioner's skills and feel free to commission very lucrative work, on the basis of nothing more than a phone call. They will test ideas and ask for advice beyond the tight boundaries of the partner's professional skill. As a result, there is very little cost of sale and very high margin.

In fact, it is almost certainly true that a real test of client relationship success is a lack of pitches and tenders. The client does not feel the need for them because they know they get

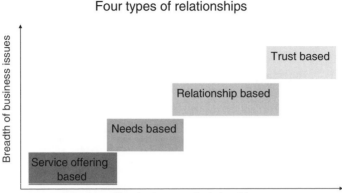

Four types of relationships

value. One advisory practice was very proud of the time they spent pitching for work. They had partners who were really good at spotting opportunities and getting onto lists. Once invited, they had a very slick process, which produced good documents and impressive, well-rehearsed presentations. They pitched for 90% of their work and were very proud of their combined pitches. Unfortunately, their marketing people obtained some very good intelligence about the behaviour of their competitors, prompted by the fact that 'profit per partner' and utilisation rates were below the industry average. Sure enough, there was a difference. Their competitors pitched for between 60% and 70% of their work. In other words, nearly 40% of that industry's work did not go to tender, it went straight to people whom the buyers trusted. This practice's relationship skills were so poor that no one trusted them to simply get on with the job.

Maister and his colleagues argue that this relationship can develop from a basic service offer to trust-based through various states (Figure 8.4). Their work is a sophisticated development of relationship marketing theory. They illustrate the differences and benefits of a trust-based engagement when compared with other forms. They also propose economic arguments to calculate the benefits arising from this approach.

CULTURAL VARIATIONS AND THEIR EFFECT ON ACCOUNT MANAGEMENT

Account and relationship management practice needs to be varied according to the culture of the organisation it serves. The most obvious example of this is variations in international cultures (explored more fully in Chapter 11). Some cultures emphasise one ubiquitous offer that is managed across the world. Many American companies work on the assumption that a similar management style, with similar policies and processes, is the way to deliver excellent 'global' performance. However, some other cultures (for instance, some Japanese firms) want their values and standards to be delivered through local subsidiaries that have freedom to express those values appropriately to local countries. Others are different again. The management of Swedish companies tends to reflect the consensus and debate (seeking fairness) which is at the heart of their culture. In Swedish multinationals, functions (like global HR and marketing) even have to persuade local operating companies to take on new ideas. They cannot just issue instructions. So, a supplier to IBM, Hitachi and Ericsson must vary its relationship management approach with each of these clients.

After their extensive study of different sales management systems (2500 sales people in 38 countries) two academics (Anderson and Onyemah, 2006) said:

> Firms need to bow to local cultural and legal norms. Outcome control systems are much better received in some cultures (such as the United Sates, Canada, Argentina, Italy, Southern Nigeria or parts of India) than in others (such as Sweden, Japan or Korea). A firm with global reach, therefore, should have multiple control systems for its various sales forces. (Reprinted with permission from *Harvard Business Review*, Anderson and Onyemah, July/Aug 2006)

There are, though, other cultural variations in organisations. Some, for instance, are 'innovators' whereas others are more cautious about change. So far a long time, Microsoft segmented its technical services business customers on their attitude to innovation and risk. The way that relationships were managed with these customers, and the design of the technical service offered, varied by segment. Many companies offer 'innovation workshops' to their major customers as a way of deepening relationships and identifying new ways of building value for both companies. The way these workshops are positioned, the style of their delivery and the degree to which the content discussed is blue-sky thinking (as opposed to good ideas already in practice elsewhere) will depend on the customer's attitude to innovation in general.

ACCOUNT PLANNING

The best account management normally involves an 'account planning process' of some kind, frequently undertaken on a six-monthly basis, which adds discipline to relationship management. It normally involves an internal meeting of employees who have an interest in the account, led by the account manager. At this meeting, the team discusses a number of issues, including:

- Objectives for the account. These might be financial, relationship-building or strategic.
- Environmental awareness. A review of the customer's market and challenges, used to understand issues and identify potential opportunities.
- Business profile and performance review.
- Creation of potential projects and sales.
- Sales and marketing programmes specifically for that account.
- Proposed annual investment in the account.

The account team needs to understand and discuss its business goals for the account and agree objectives because they are the basis of the plan. As with all business initiatives, these will fall into a hierarchy, starting with the overall business goals (for example to defend it or to grow it to a particular amount per annum account). The next step is to map the value the supplier can add to the account as it addresses its business issues, and identify one or more opportunities to pursue. The key here is to remain focused on what the customer needs to achieve, always starting with their issues rather than the list of services the company has to offer. A specific working session where issues are matched against the company's portfolio of services normally identifies a number of possible opportunities.

From this analysis of opportunities, the team should develop specific sales objectives, such as to win one or more of the opportunities identified. Yet it should also create supporting marketing objectives; such as to reposition the company as a consultancy partner to the account, or to build awareness and preference among all key executives for their company.

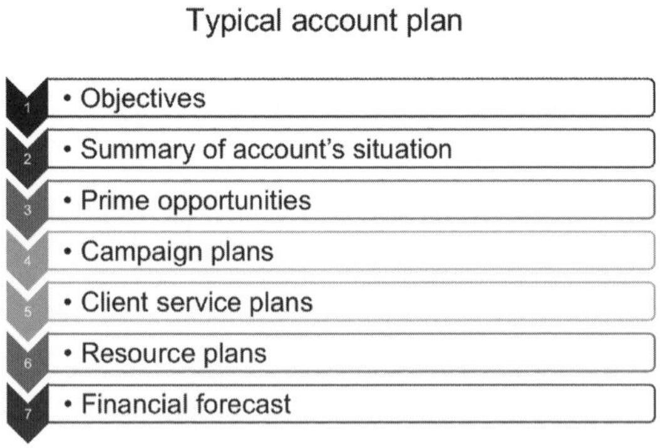

Typical account plan

Figure 8.5   Typical account or relationship plan

These objectives must be specific and measurable, allowing the team to track progress. As important as the objectives themselves is the conversation. The account team has to agree how often progress against the objectives will be measured and how it will be reported to senior management.

Sometimes, account leaders will involve members of the customer's organisation in the planning session to add perspective and depth to the debate. The output, often in the form of an account plan, is shared with relevant people in the firm. A typical format is shown in Figure 8.5.

Account plans come in all shapes and sizes, often PowerPoint presentations to aid the account governance and review process, and ranging from six slides to 60. Global account plans are often bigger, with an appendix showing each individual country plan inside a multinational company. The draft plan should be signed off by a number of stakeholders usually, including the account director, marketing director and the sales director. Once approved, it should form part of the regular account reviews and account governance procedures.

Developing Insight into Major Accounts

Many large businesses are so vast and complex that they can be regarded as markets in their own right. Some multinationals, for example, have greater revenues than the GDP of entire nations and others are larger than industrial sectors in some of the leading economies. They justify as much investment in research and analysis as more generic markets, and this is one area where marketing can contribute insight and direction in the account planning process.

Data is gathered from multiple sources to understand specific customer needs, how services can be used to address those needs, and also how best to communicate the value of those services to the different buyers and influencers in the organisation. Some marketers in leading service organisations have come to the conclusion that, if large business customers are a market in their own right, then this analysis can be conducted in a similar way to a full 'market audit' (see the Tools and Techniques appendix). For instance, to build an understanding of the account's situation and plans, information can be collected and used in a similar way to which macro-environmental factors are gathered for generic markets. This environmental analysis works through the external forces, like economic and legislative changes, which buffet the customer's business. Many use the 'PESTEL' mnemonic as a structure for this thinking (see the Tools and Techniques appendix).

Some go one step further and analyse how their customer's market is evolving. They might cover: competitive intensity from changes in the balance of power in the market between suppliers, buyers, new entrants and possible substitutes for their services (Michael Porter's 'five forces'). They might also examine the way in which their customer's customers segment themselves and how they are buying. A clear view of their customer's competitors is also collected. They analyse the way they group and behave, and their performance relative to the account on key indicators for the industry (such as average revenue per user in telecoms or number of products per customer in retail banking). One of the most powerful aspects of this approach, though, is 'value chain analysis'. The aim of this perspective, used routinely in first-rate consultancies, is to focus on what the customer provides for its customers. It examines the contribution of all functions within the organisation and what exactly the business does for its buyers. It will enable the account planning team to match its services to the customer's needs and to be proactive with proposed projects. It can unleash powerful, innovative propositions, which revolutionise a customer's business and approaches to its work.

Publicly available information sources open to the marketer include: the account's website (including investor relations information if it is a public company); press coverage (Google alerts can provide daily media coverage of the account, for example); social media (blogs, tweets and linked-in profiles of executives); and conference presentations and articles given or written by its employees. Sources of information available for the marketer to buy (and perhaps already subscribed to by their company) include: information aggregators (such as Factiva, One Source or Boardex); investment analyst reports, industry analyst reports and academic case studies of the organisation.

A profile is needed of each customer, containing information on its strategy, key initiatives, performance and associated issues, organisational structure, culture, buying priorities and behaviour towards technical services. In addition, profiles of the key executives within the business are important. What are their roles, their background, their objectives and their perceptions of service suppliers? Their likely buying behaviours can be understood by creating a history of their previous decisions.

Finally, an overview of the competitive landscape within the account is crucial. How much does it spend on services in the category and what share do the various suppliers have? What are the perceptions of those suppliers in that category? What is their perceived quality of service? How much do they earn from the account and with whom do they have key relationships? It is important to understand the competitors' personnel and their networks in the account. Every day executives in the customer organisation see the supplier's employees and those of competitors. Sometimes they proffer ideas and sales bids on the same issue or needs. What do the executives in the account see and hear? How does it seem to them? What are the strengths and weaknesses of the various sales people offering them advice?

A 'map' of the key decision-makers and influencers is a useful tool. It should reflect their relationships with each other and indicate the strength of influence of each person on the buying decision for a specific category or opportunity, together with their perception of the supplier. It is important to be honest here: sales people have been known to exaggerate a relationship with influential executives who are positive towards the company and can assist the sale. They might feel obliged to create the impression that they know the customer's organisation better than they actually do. 'Positive support' for a proposition might turn out to

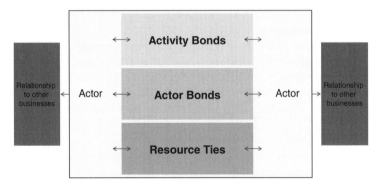

Figure 8.6   The ARR tool, a method of understanding depth of relationships
Reprinted with permission by Thomson Learning, from *Understanding Business Marketing and Purchasing* by Ford, 1992.

be a passing meeting with an executive in the corridor following an introduction by another executive in the account. It can be difficult to admit that, after working on an account for several months, the extent of an account manager's relationships are no further advanced than the original people known at the beginning. Marketing can add value here, by identifying and profiling key executives; and finding ways to open doors for the account team to meet them.

When mapping networks and the people that influence buying decisions, external advisors and contributors should be included. These are likely to be: industry analysts, procurement advisors, trade or professional body representatives, journalists or editors, and peers in similar roles in other companies. Many of these are sought out for opinions on proposals, or as sources of potential suppliers.

The actors/activities/resources model (ARR, see the Tools and Techniques appendix) can also be used as part of the analysis of the relationship between the account and its supplier. This model, shown in Figure 8.6, divides business relationships into three levels: activity links, actor bonds and resource ties. It is a reasonably well-developed tool to plot, rate and understand the relationships a service firm has with its customers. Some customer interactions will be based primarily on work. The supplier simply undertakes paid activities for its customer. Clearly, this stream of business is vulnerable to a competitor who develops more powerful and positive relationships with the customer. To develop this type of account, suppliers should create the next level: 'actor bonds'.

Actor bonds are relationships that develop over time between individuals in the supplier's organisation and those employed by the customer. Clearly, these will be of varying depth and nature. They should be mapped out objectively and plainly as part of this analysis. No one person, including the most energetic and ambitious account managers, can get on with everyone. If chemistry does not work, for instance, and some people are disliked, then others need to be introduced. Some supplement account relationships with direct links between administrative staff. Others ensure that there are senior relationships between top managers. Some service suppliers, for example, assign their most important accounts to each of their top directors. Their role is to assist the account manager in deepening initiatives and relationships. Others create social and sporting occasions to foster deeper appreciation between the two sides.

'Resource bonds' reinforce the vulnerabilities of the previous two relationship types. This describes the organisational links put in place between the two businesses; and it is frequently the marketing people that have to initiate them. They are needed because mere informal relationships between individuals are not sufficient safeguard for the interplay with a firm's most important customers. It may be, for instance, that an important player on one side or the other retires or moves job. In which case, large recurring revenue streams will be at risk. This level of communication between the two suppliers occurs instinctively after a while but can be more effective when planned and structured. They give to each other. The customer gives time, information, briefing on strategy, quality feedback and opportunities. The supplier gives information, training and secondments. Both gain from this institutionalised interaction. It ought to be a significant part of the account plan.

There is also the option of using research to gain insight into an account. This is particularly useful in understanding the priorities of influential executives within a business and their perceptions of various service suppliers. HP and Fujitsu have both used this approach to good effect. They conduct in-depth, qualitative interviews with approximately 10 or more executives working for each of their important clients and then compile a summary of their responses; capturing the insights they glean. Accenture has gone further, conducting, in one case, quantitative brand and perception research among more than 100 company executives around the world and then using the responses to form the subsequent marketing plan to the account.

Other, more simple, approaches to collecting primary research include meeting with executives in each organisation and discussing their issues with them. If this route isn't available (perhaps the organisation isn't a current customer) then inviting a representative to join a 'board of advisors' and exploring their views in a group discussion may be more appropriate. Also, when executives leave senior positions in large organisations, they are often available, on a consultancy basis, to share their knowledge and experience of the account, and can be retained to help build account insight. They might also advise and coach on a specific opportunity while their knowledge is still current.

At this stage it can be helpful to use techniques (such as SWOT, see the Tools and Techniques appendix) to capture opportunities and potential issues that the customer must address. A succinct overview of the situation and specific opportunity analysis should be part of the account plan if the service is going to be proactive, lively and challenging. In many cases, the accounts selected for account management attention reflect their current importance to the business rather than their future potential. Good opportunity analysis can get people to agree where the best future potential lies.

This analysis, although detailed, time-consuming and thorough, is important. The very best service firms lavish time on it because wider perspectives and insights are not always seen by individual account directors and sales teams. They are too close to the individual buyers. In an ideal world, most of this would be contained within the account plan (which the account director and his team maintain on a regular basis) and routinely reviewed through account or sales governance processes. Unfortunately, this is, at the time of writing, rarely the case except in the most prominent businesses.

Creating Account-specific Value Propositions

One of the most recent trends in business-to-business service marketing is the creation of unique and specific 'value propositions' for individual important customers. Marketers in leading firms set out to specifically create a service that resonates with and excites their customer. It is a deliberate attempt to move away from vague, meaningless and commoditised 'solutions'. For each opportunity identified in the account plan, a value proposition is created that demonstrates an understanding of the buyers' issues, makes the benefits of the offer clear in their language, highlights how the value of the offer outweighs the price, and reinforces the unique selling points or differentiators of the company. Marketers then try to create a snappy 'elevator pitch' for use by the account team when meeting buyers and influencers.

When creating value propositions, it's useful to review competitive materials to understand the messages they are using and the differentiators they claim. Unfortunately, most value propositions in technical services are woefully inadequate, focusing more on the supplier and its technology than on the buyer's issues and the benefits the proposition can bring them. They also suffer from overly complex language.

Working through a value proposition template helps to define the content that will be needed for an integrated sales and marketing campaign into the account. For example, the issues identified in a Xerox workshop about a leading car company included the need to personalise as much documentation as possible for the car buyer, such as that included in the glove box. This gave rise to the concept of a thought leadership campaign around personalisation, which would be of interest to the intended buyers and position Xerox as having something relevant to say on an issue close to the account's heart.

Similarly, the benefits presented in the value proposition provide clues as to the customer references that would be needed to build confidence that the company has a track record of delivering results. In another workshop, focused on the document management of a leading bank, the importance of running a smooth outsourcing process was raised as an issue. So Xerox created a documentary-style video featuring other customers who had outsourced, sharing the lessons they had learned through the process. When it was sent to the buyers in the bank, they liked it so much that they telephoned Xerox to ask for more copies.

Designing Bespoke Marketing Plans for Key Accounts

In 2003, the Boston-based IT Services Marketing Association (ITSMA) noticed an emerging trend among its members: they were applying marketing to important, individual customers and not just generic markets or sectors. This was not the usual sales or bid processes dressed up in marketing jargon. Accenture and Unisys, for example, were using this technique to change perceptions among their important customers. They were concentrating scarce marketing resources on a handful of companies. And when these companies were of the scale and scope of, say, Vodafone, BP and HSBC, sitting on huge lifetime value for IT services businesses, it's not hard to understand the logic behind their approach.

This 'account-based marketing' (ABM) aims to help companies broaden and strengthen their account relationships, increase awareness and demand for their services, and ultimately improve their financial results. The approach has been given various titles by the marketers involved (including: client-centric, key account and one-to-one marketing). After some initial research, the ITSMA defined it as 'treating individual accounts as a market in their own right'.

It seems that many companies across the IT and telecommunications sectors are adopting ABM to build better relationships, boost profitability and achieve deeper working relationships with their key accounts. It has helped them to:

- Increase sales effectiveness based on a better understanding of the client's business.
- Apply innovative thinking to client issues, sparking more strategic conversations and relationships with influential players.

- Increase awareness and demand as a result of more targeted and more relevant marketing and sales campaigns.
- Improve perception of their company as a preferred provider who understands their business.

HP, for instance, reported that it had achieved a 200% increase in its 'sales funnel', a 16% increase in account revenue and up to 13 times the normal campaign response rates from adopting a bespoke marketing approach. While just the pilot schemes at Xerox saw a 200% increase in new services opportunities, a 400% increase in new meetings with executives and a 200% increase in bid shortlist attainment. Dr Charles Doyle, who initiated the original adoption of this technique at Accenture, said:

> Client-centric marketing is about managing client perceptions of your services and capabilities . . . to move client perception in a positive direction over the long term. The perception should be able to endure beyond changes in client personnel and changes in the company. Generalist approaches are no longer enough in today's competitive market. No more going to big industry shows and trying to get clients to attend. This time, you customise the show, the roundtable and the thought leadership. You still have to do the big shows to create the awareness, but it's no longer the basis of business to business services marketing – client or account centric marketing will be the model of the future. (Reproduced by permission of ITSMA)

At time of writing, Accenture reportedly has 80% of its industry marketers focused on ABM; over 200 people. Fujitsu, HP, IBM, Unisys and Xerox have piloted the approach and have used it in different geographies, while a whole raft of other companies are at the stage of piloting a programme. If customers are getting this level of attention from some of their suppliers, it will affect their expectations of all their suppliers across service categories and industries, creating a new benchmark.

ABM can ensure good analysis of accounts and prioritisation of remarkable opportunities. It can bring discipline to the mapping of buyers as part of the account planning process and can create compelling, competitive value propositions. Once this is done, an integrated sales and marketing plan can be created, in the same way as is done for a generic market or segment. Usually there are two aspects to this: building credibility for the new value proposition inside the account; and getting the account team access to the right buyers and influencers for that proposition. Some of the tactics used at this stage include: customised thought leadership (on issues uncovered in the account analysis and summarised in the value proposition), dedicated portals or extranets, high-value direct marketing, private seminars and workshops, content-rich hospitality and customised references or case studies. The channels most suitable for the messages and the best sequence of activity should be planned in as much detail as that for a more generic market. It is important to develop an integrated approach that incorporates multiple channels that the buyers and influencers use and trust, as demonstrated in the Northrup Grumann case study.

NORTHRUP GRUMMAN USES ABM TO WIN A $2 BILLION DEAL

Northrup Grumman's information technology sector team knew a good opportunity when it saw one. In 2003, the Commonwealth of Virginia created the Virginia Information Technologies Agency (VITA) to consolidate all state IT services under one single agency. The newly formed VITA began looking for a partner that could help the state improve its

IT infrastructure and deliver managed services to more than 90 executive branch agencies. The $2 billion IT Infrastructure Partnership, the largest IT award in state government, would span 10 years and support economic development in multiple regions of Virginia. Northrup Grumman knew it was well qualified to do the work, but it had a perception problem to overcome: in Virginia, the company was viewed largely as a builder of ships and submarines. Its depth of experience in IT and at state levels was not well recognised or understood.

Understanding the Customer, Collaborating with Sales

To better position itself to win the contract, Northrup Grumman created a variation on the ABM approach. Early in the process, the company devoted significant effort to networking with key players to understand VITA's issues, concerns and objectives as well as to gain insight into how those influencers perceived Northrup Grumman.

It quickly became clear that VITA was looking for a partner with:

- Best-in-class IT expertise.
- A significant presence in Virginia.
- A commitment to economic development in the most depressed areas of the state.
- A focus on delivering the best possible options and offers to the state employees potentially impacted by the programme.

According to Liz Schwatka, director of marketing and communications for Northrup Grumman's Commercial, State and Local Group:

> Building a deep understanding of VITA's goals and objectives was a critical first step in winning this contract, and our business development team played an invaluable role in defining key themes and identifying target audiences. Marketing and business development were "joined at the hip" on this positioning and branding initiative for three years. We worked together to figure out how to reach the right people with the right messages at the right times and places. It was a very collaborative effort, and we couldn't have implemented the campaign without them.

Planning and Executing the Marketing Campaign

Building on its knowledge of VITA's goals and objectives, Northrup Grumman's communications and marketing team designed an integrated marketing campaign to raise awareness of the company's technology expertise and solid Virginia roots. Consistent messaging across all communication channels was key to the campaign:

- Advertising focused on Northrup Grumman as a vital, long-time member of the Commonwealth's business community as well as on the company's people-focused culture.
- Sponsorships and speaking opportunities were selected based on the technology content, and subject matter experts were chosen based on their technology know-how.
- The company's grassroots efforts worked across the state, particularly in southwest Virginia, where company executives spoke at local community events and higher-education venues and submitted opinion pieces to local newspapers.
- The company also developed a video featuring Northrup Grumman employees talking about their individual experiences working for the company and working in Virginia.

As a result of its branding campaign, Northrup Grumman received more than 200 media hits in local, state and national publications regarding the VITA contract. Anecdotal feedback also suggested that state executives were taking notice of Northrup Grumman, its presence in Virginia and its technology expertise.

Communicating Via the Customer Experience

Schwatka was quick to point out that although the branding campaign certainly had a big impact on the success of Northrup Grumman's bid, it was really the customer experience that clinched the deal:

> At every single touchpoint with the client, Northrup Grumman wanted to reinforce that it had the expertise and Virginia roots that VITA was looking for. Our employees—from the most technically oriented delivery people right on up to our Chairman and CEO—went to great lengths to make sure that VITA felt confident in Northrup Grumman's abilities and comfortable with us every step of the way.

She stressed that the participation of senior-level executives was particularly important because their participation in media interviews, speaking engagements and community outreach demonstrated to VITA just how committed Northrup Grumman was to the state's success.

Results

Northrup Grumman did a number of things right to win the 10-year, $2 billion contract with VITA. First, it made sure that the marketing and business development teams worked closely with each other to gain a deep understanding of VITA's issues, priorities and needs. Second, it took that insight and translated it into a highly focused branding campaign that was designed to widen the perceptions of key decision-makers within the state to include knowledge of Northrup Grumman's IT expertise and its deep Virginia roots. Third, it ensured that every interaction the customer had with the company would reinforce this new perception.

Today, the company is focused on continuing to demonstrate its commitment to the Commonwealth of Virginia and is looking to leverage its success in other states and municipalities across the country.

Source: Reprinted by kind permission of ITSMA.

Relationships develop in stages, evolving from a point at which the buyer may not have heard of the supplier, through phases of increasing familiarity (when they learn more about what the company does), through to the stage at which the buyer seriously considers buying and begins to prefer the company to other suppliers. After purchase, once the buyer has enough confidence to sign a contract, the relationship continues as the service is provided and perhaps other services are bought. Different types of communications have relevance at different stages in this development with major customers (see Figure 8.7). More traditional

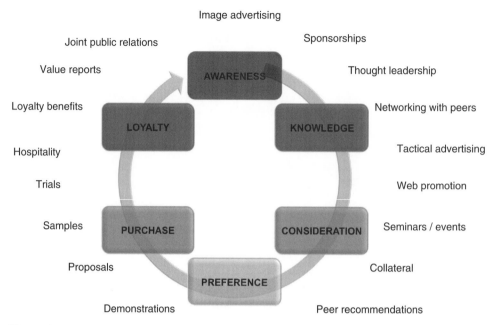

Figure 8.7 Communications across stages in relationship development

broadcast media is appropriate at the early stages to build awareness and familiarity. More content-rich and targeted media helps to create understanding and preference as clients get near to purchase. Increasingly personal and interactive media establishes confidence, drives purchases and develops an ongoing business relationship that leads to repeat purchase and loyalty.

As with any marketing programme, the success of ABM relies on good programme management. This includes: agreeing the campaign timeline, creating a critical path of activities and milestones with the account team and agreeing the governance process. It is useful, for instance, to agree escalation procedures because the bespoke marketing plan relies on significant time and effort from various members of the account team and company executives, who may all be subject to other priorities and might miss the deadlines set for them. Where this puts the plan at risk, it must be escalated according to the process agreed. Reporting progress against objectives and plans is important, particularly if this is part of a carefully monitored pilot project.

An example of typical success measures used in ABM is shown in Figure 8.8. Accenture, for example, put great emphasis on measuring the impact of ABM when under the direction of Dr Doyle. They started with a perception study before any ABM took place. They asked: 'Did this client's perception of what we do in outsourcing go up 10% over the last year?' This was more powerful than monitoring generic measures like share of voice. Share of a client's mind is much more important and precise. Other companies track ABM impact on three levels. The first is at a 'campaign level' within each client, the second is at a client account level and the third at an ABM initiative level (compared with other sales and marketing initiatives).

A sample ABM scorecard

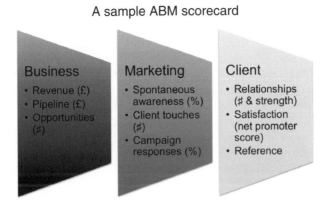

Figure 8.8 Example metrics scorecard

Setting up an ABM Programme

There are a number of important factors to consider in setting up a bespoke marketing programme for key clients. The first is senior-level sponsorship. Some programmes have failed because they started as a good idea within marketing and did not have the support of the business. Leaders need to give the programme visibility and energy, clearly explaining its importance among the range of competing initiatives that sales and account teams will be grappling with, and updating the business on progress regularly to maintain momentum. Ideally, it will become part of the standard governance and reporting procedures of the business, with its own agenda item to report progress at management meetings.

There is often some confusion about budgeting for ABM. Given that it is so focused on an individual account, marketing tends to pay for some elements while the account team will also stump up funds from its own P&L. Where there are templates that can be re-used, for example, marketing will often fund these. The function is also likely to fund a programme office to oversee the overall initiative and a training schedule to build the competence of its personnel. If a large deal is involved, further money can be found in a supplier's bid budgets. Both bid budgets and the account P&Ls can dwarf the typical marketing budget. So an early discussion about how ABM will be funded is crucial. Absolute spend varies by company, but can be anything from a small, basic budget (say for a 'defend' account) through to more extensive funds for a 'win' account.

Another important issue is the allocation of marketers to do ABM work. Some companies make ABM just another part of the marketer's job, while others allocate one marketer to individual accounts to become part of the account management team. Since ABM involves end-to-end marketing, it is most successful when the resources allocated are well-rounded marketing managers and not communications specialists. Broadly speaking, account-based marketers need to be competent across four areas of marketing: marketing planning, value proposition development, marketing communications and marketing programme management, as shown in Figure 8.9. This is a wide range of competencies usually found in general marketing managers rather than in communications specialists, which explains why most account-based marketers are drawn from industry marketing teams rather than from communications or campaign managers. Account-based marketers within account teams need to apply a combination of marketing competencies, commercial acumen and personal impact. The role requires

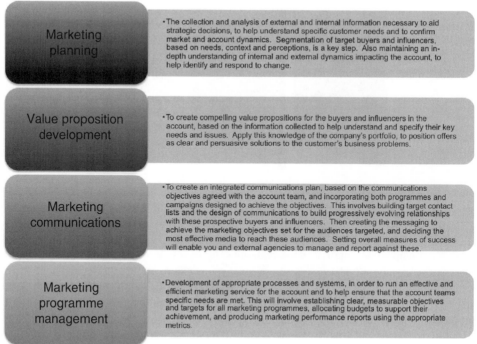

Figure 8.9 Marketing competencies required for account-based marketing

marketers to work closely with account directors and sales teams and they need to be confident in their ability to facilitate and challenge senior sales people. They must provide the professional advice needed to achieve results for the account.

SERVICE QUALITY FOR MAJOR ACCOUNTS

Another aspect of major account management is the design and delivery of specific after-care or service quality initiatives for individual major customers. Businesses will often craft individual, customised service support for large customers. This is most clearly seen in organisations that have been through dramatic change. Before BT was privatised, for example, it provided one ubiquitous service for all customers. After its market changed it set up 'major customer service centres' which gave dedicated service to individual major customers. Orange Business Services, on the other hand, has evolved its customer satisfaction survey to focus almost exclusively on major accounts, with independent research conducted every six months among a group of interviewees agreed up-front with the client representatives in the account. Results are reported at an aggregate level to help Orange Business Services improve across its organisation, but are also reported at an individual account level, forming part of the input to an account-planning session that creates plans to deal with the specific issues raised.

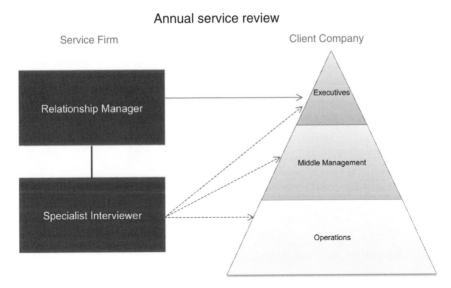

Figure 8.10   A typical account review process

Most businesses have a customised process (represented in Figure 8.10) to collect measures of service quality from their major customers. The process, conducted regularly with each account, aims to collect the following data:

• A perspective on the health of the firm's relationship with the account.
• A profile of the views of different people in the account.
• The client's key service requirements in order of priority.
• Any gaps between their ideal and the industry supply which can be exploited.
• Comparison of the supplier's performance against the customer's ideal and the industry average.
• Detailed comparison against at least one competitor's performance.

An individual who is independent of the day-to-day relationship (from the firm's leadership or quality assurance team) meets with the client and agrees a profile of people to be interviewed. Interviews are then set up and conducted about past performance and future need. When the interviews are completed they are summarised into a report, which is agreed with the client and then fed back to the account team. This process will be built into the client contract. Businesses that use this approach generally have a revolving programme. So, if 36 major accounts were to be tracked, there will be say, three projects a month.

SUMMARY

It is quite normal for sales and marketing to work together to maximise value from and for major accounts. This process starts with clarity over which accounts matter most, and agreement over the objectives for each priority account, and how it will be resourced. The recent trend to design bespoke marketing plans for major accounts increases the power of account management programmes and accelerates results. This seems to have occurred in service firms wanting to expand the effectiveness of their marketing to major accounts because of the greater propensity of service businesses to concentrate on repeat business, reputation, trust and relationships.

9
Communicating with Service Markets

INTRODUCTION

It is not an efficient use of resources to rely on sales teams and major account managers alone to generate revenue. Nor is short-term, tactical sales support enough. Numerous businesses in many different markets have demonstrated that the use of carefully designed broadcast communication techniques paves the way for sales. They can be used to attract initial attention, to stimulate interest and to induce customers to buy; sometimes without the intervention of a sales person at all. Even in markets that are heavily reliant on a sales person or a consultant to close and deliver sales, the use of marketing communications can, cost-effectively, attract customers to the point of sale and reassure them afterwards that they did the right thing. Service firms should take advantage of the complete range of modern communications tools available to build their businesses. While many of the most commonly used channels (advertising, public relations, direct marketing, social media, etc.) can be used to market services, marketers should also employ concepts that are specifically relevant to services, such as reputation enhancement, relationship marketing, viral marketing and thought leadership. However, any communications programme must be built around a clear communications strategy. That strategy should give direction to a number of marketing communications issues, such as understanding the target market and designing a meaningful message. This chapter sets out to address these issues, giving an overview of effective marketing communications; from strategy to implementation.

EFFECTIVE COMMUNICATIONS STRATEGY

Communication is much more than a 'p' in any marketing mix (see the Tools and Techniques appendix). In fact, the way it was described in the early days of marketing theory ('Promotion') is completely inadequate because it implies that any activities are one-way; directed from the supplier to the customer. It is true that suppliers need to put clear, simple propositions to groups of customers. It is also true that they need to be put in attractive, distinctive and meaningful ways that stimulate their desires and appetites. If suppliers do not communicate effectively, they cannot overcome the noise of other claims on their intended customer's attention. Buying simply cannot begin if potential customers do not know what the company is or what it stands for. As the adverts for McGraw-Hill once ran: 'I don't know who you are, I don't know where you come from, I don't know how long you have been in business, I don't know who your customers are ... now, what are you trying to sell me?' Yet, although marketers want to capture the imagination of modern buyers, they must also communicate with them by getting a response and adjusting their message in the light of that response. They must listen and respond as much as they plan and pronounce. Effective communication is a two-way process.

Good communication is also carefully planned and sustained. It is simply not effective to put out erratic, short-term tactical programmes hoping they might elicit responses that lead to sales. Neither is effective communication a collection of ad hoc tactical under-funded programmes such as 'email blasts' or 'webinars' which change month by month. These techniques work well in the context of a carefully designed, integrated communications programme that has cumulative impact on a well-understood group of customers. On their own and out of context, though, they are almost a complete waste of money.

Effective communication is based on a clear strategy and is carefully planned. Marketers and marketing organisations must learn to craft carefully targeted communications that are digestible and easy to understand, while at the same time cutting through the myriad of other messages to reach their intended audience. This is a precise and carefully orchestrated, long-term operation, not the whimsical and erratic activities of people trying to impress their bosses by grabbing after the latest digital media fad. In short, marketers in firms of all sizes and sectors need an effective, clear marketing communications strategy.

This also applies, for the sake of doubt, in business-to-business markets. Historically, academics and marketing writers have tended to play down the role of the full marketing communications mix in business sectors. They have argued that, because of 'industrial buying processes' and the use of specialists (like purchasing managers), the ability of communications to influence decision-makers is limited. They have also argued that the sales process has been managed by sophisticated sales people and therefore the supplementary effect of other communications material is negligible. Yet, a combination of the recent experience of leading business suppliers, more recent research and a deeper understanding of human behaviour has challenged this view.

There is a considerable amount of emotion within the decision-making unit of businesses. For many important purchases, such as outsourcing services, the various people in the decision-making unit are influenced as much by emotion as consumers. In fact, some modern behaviourists argue that almost all decisions have an emotional element. Business buyers might, for instance, be worried about the perception of the effectiveness of their work, or their political position, or looking to use the purchase decision as a means of gaining influence. If decisions are emotional as well as rational, then decision-makers can be influenced by the appropriate communication.

The process of buying is as different as the different organisational structures and cultures of each business. Some might use large, process-based bid techniques while others in exactly the same sector might turn on the whim of a forceful leader. Sales people need the help of marketing communications activities that allow for this variety and warm up different buyers to the point of sale. Good communications are absolutely essential to enable sales teams to function effectively.

So, marketers must think before they do. To communicate effectively with a group of buyers and to influence their purchasing behaviour, they must resolve a number of important and sometimes complex strategic issues. These include, for example, a deep understanding of the intended market, the crafting of a message that is meaningful, and the choice of media through which to communicate. These need debate and resolution at senior levels in the firm before planning and implementing the detail of any campaign because they have financial implications. Marketers must develop clear communications strategies which give precision, clarity, narrative and impact to any practical marketing communications.

Any firm attempting to communicate with a market should also be very practical in its approach. They need to put in place the skills and management processes that ensure these issues are properly handled. All too often management teams will launch a marketing or brand communications piece without it being properly worked through, using phrases such as a 'soft launch' to hide the fact that insufficient brain power, time or funds have been made available. Firms that do this are often inexperienced at marketing communications, and have been able to behave like this because of benign market conditions. However, most of the leaders in the service sectors of technology industries (such as IBM Global Services, Apple and Accenture) are now leaving such amateurishness far behind. Like FMCG companies, they have set up expert communications units that learn from their previous experiences and bring first-rate communications disciplines to their firm-wide marketing communications. This will eventually give them a competitive edge which others will have to mimic if they are not to fall behind. Marketing communications is not an expensive luxury. Companies get the return that they pay for. Careful systematic investment in good communication can be competitively priced and will save sales costs.

The essence of good communication, then, is: good strategy, clear objectives, good targeting, a poignant message, imaginative creative (particularly writing), responsiveness and effective organisational capabilities.

ASPECTS OF COMMUNICATIONS STRATEGY

Behind virtually all the best communications campaigns has been a clear, effective communications strategy. As with business strategy, this might be produced in different ways and be very different in format but it gives clear shape and direction to all communications aimed at different audiences. The strategic communication issues that should be considered include the following.

The Unique Dynamics of Service Marketing

Chapter 2 outlined the unique characteristics of services. They are, amongst other things, intangible processes, which mean that ownership does not pass to customers. Users experience 'a journey' that encompasses technology, process, people and environment to varying degrees. This means that service marketers must build a number of these unique characteristics into their marketing communications strategy. They might, for instance, make the intangible seem tangible by developing collateral such as: case studies, brochures and other forms of packaging. This makes the service easy to access and buy. Any physical or virtual points of contact with customers (such as websites and premises) should be carefully designed to reflect the nature of both the claim to market and the desired service experience. Good, clear collateral must be deployed in post-purchase communication as much as in pre-purchase; to reassure the audience and to lessen the effect of post-purchase distress.

They should also set up mechanisms whereby would-be buyers can test a service as much as possible before purchase. They might, for instance, offer free trails or give away service experiences as prizes. The more people experience a new service, the more word-of-mouth and reputation will grow. They might also use celebrity as a device to kick start word-of-mouth about the service. This works as much in business-to-business as consumer markets, and may involve giving a famous figure free use of a service for a limited period, on condition that they talk about using it in publicity. If this celebrity is well known, they are likely to reach a wide audience, kick starting the firm's reputation.

Unfortunately, the nature of services demands that marketers become involved in quality issues. They cannot, in a service company, restrict their role to the more traditional functions of the marketing department. Their customer's experiences of the service can cause them to use it again or to turn away from it, as much as (some would say more than) any marketing initiative. So, marketers must give direction to the operations of a service firm to ensure that the service experience matches the market claim. Marketing communications programmes must influence the expectations of the intended customers, creating an idea of the service they will experience. Any communication claims that stimulate them to use a service will have implicit or explicit promises. It is ridiculous, for instance, to claim that a company's quality is the 'best' or 'world class' if it clearly isn't or is not known for being outstanding in any particular way. Even if this is well-intentioned it will make customers cynical and cause them to doubt the integrity of other messages.

It is worse, of course, to lie to customers about service experiences. Yet, many do; albeit unintentionally. At the time of writing, for instance, many Internet suppliers keep the costs of their telephone service to a minimum. Although that might make economic sense, the standard message used routinely by service managers ('we are experiencing unusually high demand at the moment') is, in fact, untrue. Any supplier who has run a call reception service for a year or two will have accumulated call patterns which will enable them to calculate quite precisely the number of agents to employ. There are also routine mechanisms to cope with peaks in demand that can ensure customers do not have to wait too long. The customers themselves sense that this is unnecessary and become frustrated. These lies undermine any marketing claims about quality and intended service excellence. Ironically, marketers working for the firm can be completely unaware that this is routinely experienced by their customers and dramatically undermines their work. They must step out of their functional silo and get a grip on this type of operational communication, to limit the damage it causes.

The need to involve marketers in the implications of expectations raised by service brands has prompted an interesting new concept called 'customer experience management' (CEM) to emerge. Similar to a planning approach in the financial services industry called the 'customer journey', it charts the course through a service process from the customer's perspective. This ensures that the expectations raised by brands are experienced by each group of customers at each point of interface with the company; and that marketing communications are appropriate at each of those points of interface. It involves communication specialists in operational documentation (invoices, timetables, notices, etc.), signage (into and within the premises) and the design of the service environment.

Understanding the Intended Segment of Buyers

One of the most important components of effective communication is a deep understanding of the human beings for whom it is intended. It is in this field more than any other that a behavioural view of markets and economies is important. Marketers must understand their intended group of customers in as much depth as possible. How many are in the group? What do they read? What do they listen to? Which conferences do they attend? Which voices of authority (individuals or organisations) do they respect? What professional associations do they belong to? What do they think, feel and believe? What are their aspirations, dreams and ambitions?

Marketers must decide which buyer groups to focus on and concentrate on them to the exclusion of others. They must then invest in understanding both the rational and emotional needs of this group of buyers. If there is one difference between marketing-literate companies and others, it is their concentration on customer knowledge and their willingness to invest in research to get it. The leadership of their companies know, through bitter experience, that their sales success hangs upon the intimacy of their customers' responses to their communication.

They invest in regular, deep research to understand this dynamic and to adjust to it. Moreover, they invest in recruiting and training experts whose job it is to understand this relationship.

Media planning specialists in advertising companies can determine which media has the most impact upon an individual customer group, whereas good-quality research will give insight into their attitudes and needs. By understanding this level of detail, marketers will create communications that will be effective and appealing; the investment will be well directed.

Creating and Managing the Message

Experienced marketers have learned that effective communications often depend on the use of one simple, relevant message. It has to be relevant to the target group to ensure impact and responses and simple enough to be taken in by people in a brief moment of time. It also needs to be capable of sufficient repetition to become familiar to the intended audience without becoming irritating or boring. The message needs to contain a definite proposition which should be unique and strong enough to persuade people to act. It is likely to contain both rational elements (giving the reasons why the service should be bought) and emotional appeal. It might be testimonial-based (using examples of others who have bought) and 'dramatic' (using a narrative to convey meaning). Some are humorous, using laughter to appeal to humanity and create empathy, whereas others focus on fear, raising concern about consequences ('nobody ever got fired for buying IBM').

Yet, it takes enormous effort to achieve the elegant simplicity of an effective message. The firm needs to work with communications specialists in order to frame what it says to the marketplace appropriately. The credit card company MasterCard created, for example, an enduring and simple message, which has earned it awards, impact, revenues and distinction. At the time, American Express and VISA were their main competitors. Both had strong brand values and personalities built in the 1980s era of materialistic, outer-directed, success-focused values. MasterCard, by contrast, lacked emotional currency or aspiration. It was seen as ordinary, unassuming, unpretentious and practical. However, the consumer climate was changing. Economic pressure and growing debt began to take its toll, so that the materialistic, outer-directed consumer culture was giving way to a growing search for quality of life, relationships and greater balance between work, family and leisure. This, in turn, was leading to a more enlightened, if wary, view of credit cards. MasterCard called the consumers who adopted this mindset 'Good Revolvers', because they used credit cards to acquire the things that mattered to them, things that enriched everyday lives. They summed up their selling idea as 'The best way to pay for everything that matters'. Their communication idea, based on a detailed understanding of the people they were aiming at, became: 'Priceless moments'; and this, in turn, translated into a message:

> There are some things money can't buy. For everything else there's MasterCard.

Accenture has also used this approach to good effect in business-to-business marketing, an industry that often communicates features and benefits rather than short, powerful messages. Its message, 'We know what it takes to be a tiger', was built on its compelling brand promise ('high performance delivered') and appealed to the emotional and personal aspirations of its intended buyers, who all wanted to succeed and stand out in their business lives.

The timing of the message also needs to be carefully considered. At the lowest level this involves timing in relation to customer needs or in terms of other messages. In Europe, for instance, the culture of many countries is to take long holidays in August and many business people return in September ready to start new initiatives; a good time to start communications campaigns. However, suppliers should also think about the timing in terms of the customers' phases of thinking and progress along the buying process. Messages, which introduce a new concept, for instance, can be tailored to customers who are at the early phase in their thinking. Whereas messages aimed at those near purchase will emphasise benefit and include a call to action. Messages aimed at those who have purchased, however, may be about service, process and post-purchase reassurance.

Allowing for the Maturity of the Market

As discussed in Chapter 3, markets tend to follow a pattern as the ideas on which products or services are based diffuse across society. This, in turn, creates a phenomenon called industry or market maturity (see the Tools and Techniques appendix). So, marketing communications strategy needs to be different at different stages in this diffusion process.

For instance, when an idea is completely new, a true innovation (as Blu-ray was in the 2000s, mobile phones were in the 1990s and video recorders in the 1980s), suppliers need to make people comfortable with the idea before they can sell individual products or services. They will, for example, use PR to educate. Articles will appear in magazines and newspapers that discuss the innovation and how it has occurred. Exhibition stands and news items will be dedicated to it. The IT industry has been particularly good at stimulating this form of PR because of its genuine excitement about new technology and the links between places like Silicon Valley and venture capitalists (which prompt news stories about new technology deals). In parallel with this, sophisticated marketers will focus on individuals who buy new concepts. As everyone who has studied elementary marketing understands, these are the 'early adopters' in markets. They are the geeks who enthuse about new ideas and love to buy new technology; called by some 'Freds in sheds'. In recent years, firms like Apple have had considerable success by identifying 'mavens' (see the Tools and Techniques appendix) from among these people. By creating a group of these 'super users', they involve people in their brand and stimulate word-of-mouth about their innovations.

Marketing communications strategy at another phase of the industry's maturity is likely to be completely different, however. In a mature market (like computer maintenance, for instance), customers are very familiar with the concept. In these market conditions, suppliers are after profit by teasing out defined segments and creating fascinating forms of differentiation.

Brand Building

As outlined in Chapter 5, a firm's brand is one of its most precious intangible assets, affecting both the price and quality of every job it does and every service it sells. Part of the strategy to enhance its value is likely to involve programmes that communicate the brand specifically to the general marketplace in order to make it famous.

One of the inhibitions that inexperienced firms often have about brand building and brand management is the wrong perception that it is based on expensive advertising alone. Yet several well-known, successful brands (like Virgin) have been built without advertising. See, for example, the research by the academics Joachimsthaler and Aaker (Joachimsthaler and Aaker, 1997), which demonstrates that there are several ways to communicate a brand promise to a market, particularly for service businesses. The first is through the experience of the service. If a firm has a particular style and approach in its work, it will cause a reputation to evolve through word-of-mouth which will, in turn, create its brand position. Even in the early days of their business, Starbucks, for example, launched new coffee shops by handing out free coffee and used no associated advertising at all. The second is by investing in and managing internal communication so that employees are clear on the firm's position in the market and speak with a common voice to customers.

Another tool is PR. There are good examples of brands built through PR and publicity alone because they have gained fame or a strong reputation through sustained public profile. Several companies which started with a modest budget have, through targeted reputation management, turned their brand into a substantial intangible asset. Their techniques include media relations (proactive work with journalists and editors), sponsorship and publicity. Yet this must be realistically based on the actual reputation that the company or its leaders has. The firm must first identify and measure its reputation and only then use sustained PR to amplify it.

However, the main tool is, of course, broadcast marketing communications: advertising consistent marketing messages. Whatever the chosen method, the firm should ensure that regular funds are set aside to enhance its brand through building fame by consistent communication to the chosen market. Brand needs to be a clear and well-understood component of all marketing communication. It needs to be a dominant theme of communication strategy.

Competition, Share of Voice and Other Influences

Suppliers need to consider the impact of their communications and the investment needed in the light of other messages that the intended customers will receive. These might come from direct competition or competing concepts. Both can influence the clients' attention to the message. Yet the overall demands on people's attention also need to be taken into account. In a busy modern lifestyle they are assailed by many different messages through many different media. Too little communication will mean that the message will not reach the threshold necessary to command attention, in the light of other demands. On the other hand, too much will lead to saturation, undermining the impact as buyers become bored.

International and Inter-regional Communications

It took one of the world's leading novelists, Graham Greene, to capture the essence of the difficulties that business people have with international marketing. The anti-hero of his 1958 comic novel, *Our Man in Havana*, was James Wormold, a British vacuum cleaner franchisee based in pre-revolutionary Cuba. Early in the book he says to his cynical friend:

> ... And that's another thing. You know what the firm has done now? They've sent me an Atomic pile cleaner ... of course there's nothing atomic about it – it's only a name. Last year there was the Turbo Jet; this year it's atomic. It works off the light-plug just the same as the other. They don't realise that sort of name may go down in the States, but not here, where the clergy are preaching all the time about the misuse of science ... Father Mendez spent half an hour describing the effect of a hydrogen bomb. Those who believe in heaven on earth, he said, are creating hell. How do you think I liked it on Monday morning when I had to make a window display of the new Atomic Pile Suction Cleaner? It wouldn't have surprised me if one of the wild boys around here had broken the window.
>
> They won't let me (change the name). They are proud of it. They think it's the best phrase anyone has thought up since 'It beats as it sweeps as it cleans' ... a woman came in and looked at the Atomic Pile and she asked whether a pad that size could really absorb all the radioactivity. And what about the Strontium? she asked.

This is not an irrelevant piece of fictitious writing. Businesses have been conducting international marketing campaigns for at least 300 years; and making fundamental mistakes for at least as long. What might look sensible in headquarters can look ridiculous in remote locations. One European vice president of marketing resigned soon after the start of the new century from a leading global technology company because of growing exasperation with this phenomenon. The American chief marketing officer ran worldwide marketing with a very tight 'dashboard' approach. Events, mailings and campaigns were expected to run at set times in different parts of the world and local marketing teams were expected to concentrate entirely on execution and administration of centrally designed programmes. This VP decided to leave when asked to run a seminar in Paris on 14th July (Bastille day and a national holiday as important to the French as 4th July is to the Americans).

International marketing is examined more extensively in Chapter 11. It is essential, though, that the firm's marketing communication strategy covers this subject explicitly. What degree of latitude is necessary to make the message and the emphasis understood in different cultures? Which communication channels will be most effective? How much must the creative execution be adjusted to the taste and prejudice of local audiences? Where is the balance between pragmatic acceptance of customisation and the incremental cost of this work? In which countries is any transliteration of communication into local cultural taste uneconomic?

Internal Communications

Internal communication with employees of the firm, particularly with intelligent and self-motivated people such as those found in consultancy firms, is an important aspect of a service firm's success. It helps to keep employees informed of important issues, contributes to a common sense of purpose and helps develop positive motivation. However, internal communication that is ill-conceived, poorly planned or erratic can have a detrimental effect. If the leadership of the firm doesn't ensure that there is good internal communications management, the leaders and people responsible for various activities will copy emails, send out erratic messages and pass on ad hoc technical changes, causing confusion. If there is no communication discipline, there will be a multitude of conflicting, disparate messages, which will become a deluge that employees are unable to comprehend. The firm will suffer from over-communication and employees will be overwhelmed by a discordant cascade, which conflicts and confuses.

Service marketers need to construct an effective internal marketing programme with similar characteristics to an external programme. Just as with external communications, the firm needs an internal communications strategy. What is it communicating, to whom and why? What are the objectives of the leadership in communicating with its own people? What are the core messages and what are the outputs and behavioural changes that they want to achieve through the investment in communication?

These different elements of communication need to be thought through and included in the marketing communications strategy. That strategy, as with general marketing strategy, ought to be produced to suit the style and culture of the firm. For some, it will be a detailed document produced after months of detailed analysis. For others it will be a lengthy PowerPoint presentation prepared over several collaborative meetings; while for others, it will be a short, pithy statement of intent. Whatever its format, this strategy is likely to encompass: the message, the communication objectives, the aspirations of the human beings it is aimed at, the budget and the intended outcome. It must, though, give a clear direction to all campaigns and programmes.

COMMUNICATIONS TECHNIQUES PARTICULARLY IMPORTANT TO SERVICE COMPANIES

There are a number of communications techniques which, while used by many industries, are particularly relevant to service markets. They include the following.

Employee Interaction and Networks

As discussed earlier, one of the main communication channels to customers is the employees of the firm. If a service firm has 10,000 employees talking to three customers a day on average, there are 30,000 opportunities to communicate with customers every day. Marketers should therefore take into account the views of employees when preparing any external communications. At the very least, people should be told about any campaign before it is released to customers. However, effectiveness is dramatically improved if employees are drawn into the campaign and enthused to talk to customers about it.

Internal communications will also need clear message management. This involves crafting the message to employees and prioritising against the many other internal business communications. Since putting out multiple messages can give rise to conflicting communication, it is not only important to make sure these messages are simple and straightforward, but also that they do not conflict with other internal messages. As with external communication, they need to be sustained, simple and relevant. Yet, with internal communications, one of the most important considerations is credibility. Employees must be able to give some credence to the thrust of senior management claims. The ideas and actions of leaders need to be properly aligned so that, either overtly or subtly, they don't undermine the professed direction of the firm. If they do that, there is a danger of lowering morale and that, in turn, can have a detrimental effect on customer service. As with external messaging, it takes time for internal audiences to accept and internalise messages. Messages about the firm's values and priorities should be unchanging or, at least, slow to evolve.

Marketers need to be just as careful in planning the use of internal media as external. The prime and best media through which to communicate with employees is their own line management. The second best is gossip. Both need to be valued and influenced by communications planners. Others include: 'town hall' meetings, internal email, company intranets, blogs and wikis and staff magazines. Some service firms even design their external advertising with an eye to motivating their own people. This was a consideration, for example, in Accenture's choice of advertising in airline departure halls. Many of their staff fly regularly and gain encouragement from being part of a large firm, able to communicate on this scale.

Modern relationship marketing techniques can be applied to internal communications. The marketer sets out to communicate through internal networks, particularly in a large firm which will be dominated by interconnected relationships of people. These might be people who have been in the same firm for a decade or more, who have learnt to trust each other by working on projects or in managerial teams. They are circles within circles and networks of trust. There might be, for example, a network of specialists around the world concentrating on one business sector. The people involved have a common interest and rely on their internal network to receive information. Their leaders and the immediate group of contacts with whom they work daily are much more relevant to them than other messages. So gossip becomes more effective than the deluge of emails, magazines and other publications. Marketing specialists can map these networks and target communications through them. This is a form of viral marketing, where word-of-mouth is used to influence the behaviour and motivation of a company's people. It is a powerful internal marketing mechanism if used effectively. Moreover, all of these networks will interconnect with customer networks, and so good, credible communications from the leadership through these internal networks will also influence customer thinking.

So, making a member of the internal communications team responsible for viral and relationship marketing techniques makes sense. They need to understand the internal networks, how those networks communicate, what the word-of-mouth is and how to influence it. As with external approaches to relationship marketing, the internal approach enables the identification of network nodes, or points at which the networks interconnect. These might be conferences of different consultants who focus on industry marketing or an informal meeting of all technical staff interested in one professional area. All are opportunities for the firm's leaders to target communications more effectively.

As with external marketing, internal marketing is improved by the use of communications campaigns. A campaign is a group of activities coordinated together through the available media over a period of time in order to reinforce one message with the internal audience. As with external communications, an internal campaign needs to be planned carefully. The communications manager needs to have a clear idea of who the target audience is, what the key objectives are and what the available budget is. They need to refine and agree the message and concentrate on the media to be used over whatever period of time. A firm adopting a campaign approach to internal marketing will ensure that its key campaigns are known to the communications manager and there is no conflict when received by the internal audiences.

There also need to be open and reliable methods through which the response to messages from employees can be gathered. Many firms carry out internal staff surveys to get objective measures of reactions to messages. These frequently have to be anonymous, in order to encourage honest feedback. But this is much more than merely measuring whether they have received and understood the leader's messages. The best service companies set out to understand the views, concerns and motivations of their employees. Just like external communications, effective internal communication is two-way; a dialogue more than a monologue.

Viral Marketing or 'Word-of-mouth' Communication

Academics have demonstrated that 'word-of-mouth' plays an important role in building services revenue. At the same time, techniques to enhance word-of-mouth have come to the fore with the Internet and have been called 'viral marketing' (viral because the aim is to spread an idea among a target community in much the same way as a virus spreads in a body). This technique exploits social communications and gossip. Internet users have demonstrated the propensity of human beings to chat and gossip. They copy emails and jokes to each other. There have, for example, been several famous instances of sexual indiscretions being discussed in emails with friends and then being copied to so many people that, within days, they are reported in national newspapers. Some companies have exploited this by creating short, humorous video clips that are for the Internet alone, tapping into the power of YouTube. As the technique involves the transportation of an idea from person to person, it is a powerful way of judging popularity.

However, the Internet version of viral marketing has distracted attention from the fact that it is a powerful, long-standing marketing technique neglected by marketing academics. One of the earliest examples of it was the launch of a soap by a London hairdresser called John Pears in the 1780s. He paid school children to run around the city with catchphrases ('What's up?

Pear's soap!') and started a successful 300-year-old brand. Another example is the use of the technique exploited by consultants and IT companies to market thought leadership output: the marketing of business ideas. A new idea will emerge from a business thinker or a consultancy in the form of a 'white paper' or book. It will be presented at a few leading conferences. (It is most powerful if it gets a hearing at a chief executive-level conference such as the World Economic Forum held each year at Davos.) Product managers in conference companies then spot the idea, allowing time on agendas for it to be presented. Soon whole conferences are dedicated to it, articles published and books authored. Leading consultancies then dedicate partners and staff to practice in the area. They adjust case studies from past projects and present briefs to their clients. At this stage, academics get funding for reputable research into the idea and add to its visibility. So, viral marketing has been the primary tool for communicating thought leadership pieces to business buyers. The new idea is 'viral', carried and reinforced by debate in the business community, which can enhance the reputation and position of the firm if properly resourced and managed.

Reputation Enhancement

As Chapter 2 showed, an important part of the momentum for the growth of revenues in some service firms, particularly professional practices like consultancies, is the creation of 'demand pull' based on a reputation for excellent work. Marketers should set up mechanisms to enhance this natural reputation. This will increase repeat business and referrals. The issue here is the competitive reputation in the minds of customers. When buying services, how does the firm rate in their buying criteria? Everyone takes pride in the quality of their work. None sets out to do a bad job. However, customers may use different criteria to assess a firm's performance, based on more than technical expertise. Any marketer beginning to tackle the communication of a service business ought to start with an audit of competitive reputation. Having understood that, the second step is to put in place mechanisms to amplify that natural reputation. The sort of mechanisms likely to do this are reference or customer evidence initiatives that make visible the expertise and accomplishments of both the firm and individuals. Microsoft uses just such an initiative, to amplify reputation and provide credibility and reassurance at key stages of the customer's buying decision.

CAPTURING AND COMMUNICATING CUSTOMER EVIDENCE AT MICROSOFT

For the past 3 years Barbara Jenkins has been running the customer evidence programme in the business and marketing organisation of Microsoft UK. Bridging the needs for customer success stories across both sales and marketing campaigns, the programme also links into Microsoft's global evidence programme, particularly where large deals are in the pipeline and evidence is needed from the world's biggest organisations in different industries around the globe.

Reasons for the Customer Evidence Programme

The drivers for setting up the customer evidence programme initially came from both the external environment and internal demands.

Externally, in the business-to-business arena, references have always been important, but two things happened to make them more important than ever. First, IT procurement became more sophisticated. Prompted by the public sector, where there was increasing demand to demonstrate value for money from public funds, the burden of demonstrating that value fell onto the suppliers of IT. They have used past examples as well as ROI modelling tools. Although Microsoft had a solid track record in technology, it was now charged with demonstrating business value.

Second, the rise of social media and consumer recommendations, driven by suppliers such as Amazon, also influenced IT decision-makers. This meant that the procurement process for IT services needed references much earlier on in the buying process and sales cycle.

Internally, in response to the external drivers and as marketing became more sophisticated in Microsoft, the ad hoc process previously used to capture and share evidence of successful projects needed to become more effective and efficient. This ad hoc process meant that sales people often relied only on the references that they themselves were involved in or knew of from their personal networks in the sales team. But as sales grew in size and scope, and references increased in importance, this became an inefficient way of working. A new approach was needed to deliver better knowledge management, and linked to that, better corporate memory (as people moved jobs or left the company).

At the same time, Microsoft stepped up its focus on customer-centricity and satisfaction. The previous ad hoc approach meant that often the same customers were asked to act as a reference, and were over-used. It became obvious that, at least in English-speaking markets, the top tier of customers were experiencing this; they needed to be ring-fenced and managed more sensitively.

So the customer evidence programme was established with the objective of assuring sales. The programme allowed the production of authentic, credible materials more cost-effectively than many other approaches to messaging and content. Longer term, the programme was expected to increase customer satisfaction, producing an army of advocates. So, it was critical to look at the customer's objectives in buying Microsoft solutions and evaluate how well these objectives had been met. A constant feedback process was needed, to make sure that value was delivered and customer satisfaction achieved.

A Systematic, Data-driven Approach

The customer evidence programme was built from the ground up, starting with a process of analysis to determine what the business needed. The first step was to produce a matrix of market sectors against Microsoft solutions, in order to prioritise where the best opportunities lay and evidence needed. Two further filters were overlaid onto the matrix: the first showed where gaps in sales support materials existed in each cell; and the second showed how tough the competitive situation was in each cell. The completed matrix provided a gap analysis that showed priorities for investment for the customer evidence programme.

The second step of the analysis was more granular, looking at the specific nature and needs of each priority cell. For example, local and central government cells might need a different approach to the hi-tech communications cells. In the latter, for instance, the medium of the reference is as important as the message. Buyers expect customer evidence to be 'all singing and dancing'; using multimedia and developed in an engaging way that shows Microsoft understands the sector and is at the cutting edge of media.

From this analysis, the plan for the customer evidence programme was crafted, built on an annual basis with specified quarterly deliverables. The pipeline of evidence stories was represented, showing the timelines and critical milestones in their development. Milestones were created by working back through the sales cycle of large, future deals, to highlight when specific evidence would be needed. The pipeline of customer evidence was filled by the sales teams. They were quickly bought in to the new approach once they realised the value of the evidence programme to their own sales, and their own role in helping to populate the pipeline.

A fully automated workflow process was set up to enable the production of case studies, and to ensure efficiency. To start the production process, an outline form populates a workflow system. This triggers a series of emails to allow a potential story to be evaluated for its relevancy and potential impact. Then, once funding is approved, the production process is triggered. An external case study writer contacts the customer, and the workflow system is updated in real time as the story is drafted, edited and approved by both Microsoft and the customer. Delays trigger automatic alerts and warnings, so that issues can be ironed out as soon as possible.

Stories are produced in a number of formats, depending on the business need and customer preference. They range from a simple quote through to a written case study, but can involve an audio or video piece, presentations and site visits to the customer premises.

The final milestone is complete once the story is posted to Microsoft's global database, where it triggers emails to thank everyone involved and to alert Microsoft people to its availability. But while this is the final milestone in the evidence production, it is only the start of the process of getting value out of the story. Each story is tracked in terms of how it is used, both internally and externally, and a feedback loop is being established to identify the return on investment for the story. This is no mean feat, as the stories are used as part of integrated marketing campaigns as well as in specific sales situations. For example, while it can be easy to monetise the value of each reference used to support a successful sale or reflected in a piece of press coverage, it is more challenging to estimate the return on using a customer quote in direct marketing material or a major presentation.

Benefits and Lessons from a Best Practice Approach

The results of the programme, which achieved its objectives, were impressive. The UK programme demonstrated best practice in a number of areas, winning awards for its use of evidence as part of new product launches and for the quality filters that were applied to evidence creation. The latter was a scoring system examining those stories that demonstrated value for money, or other key metrics. These filters have meant that Microsoft rarely produces a poor case study; poor-quality references are weeded out early in the process.

Reflecting on the way the programme has developed over the past few years, Jenkins shares a few lessons learned. First is the size of the task of changing sales behaviour, so that they use a system rather than their own network and ad hoc approaches. Time and attention is needed, together with a clear demonstration of the link between the value sales people get from using references and their role in developing them.

The second lesson is that top management sponsorship is vital to its success. They simply must understand the programme's value as a business asset and look regularly at high-level reports of the evidence pipeline. They must act as an escalation point for issues and also as recruiters of customers to feature in the programme.

The final lesson is the importance of the quality filters in a programme of this kind. Without these, and despite the systematic approach to evidence production, it would still be possible to waste money on mediocre stories. Jenkins says:

> this programme will only become more important to Microsoft as the economy goes through and comes out of the current recession. In the past, references have been more about giving prospective customers comfort in their decisions, but now references have to demonstrate that they are de-risking those decisions. This recession is re-setting expectations and business practice, to the extent that IT procurement will never be the same again.

Thought Leadership

Thought leadership is a term used commonly within the technology industries for the publication and dissemination of ideas for commercial advantage. It was discussed as an innovation tool in Chapter 6. The communication programmes that surround it can range from an article in a magazine to a major, sponsored PR programme or a complete book. Many management concepts, both credible and not-so-credible, (such as process re-engineering, CRM, Total Quality Management, globalisation and shareholder value) have been promoted in this way. Some of these management fads then fail. Others become established concepts, validated by academic research. (The 'Boston matrix' as a strategic planning tool, process management and quality management are but a few.)

Service firms can exploit this phenomenon to gain revenue and competitive advantage. They can pioneer ideas (the most costly approach), validate new ideas (an approach open to leading brands), join the wave as a sales tool (most powerful when professional services, IT and City firms all join in) or play the sceptic. They can aim communications programmes about specific ideas at different customer groups. For instance, certain, generally large, corporations are nearly always the same 'early adopters' of a new idea and are ideal targets for innovative ideas.

Thought leadership can be expensive in terms of both cash and time costs. So, it is sensible to create a discipline to manage the investment. This might include:

- Cost management. Cash costs include research, printing, promotion and publication. Time costs include the opportunity cost of staff taken away from chargeable work and support staff.
- Criteria for approval. These might involve: payback, commercial opportunity, quality, reputation enhancement.
- Anticipated revenue or leads stimulated.
- Timetable of all projects.

MANAGING THE CREATIVE EXECUTION

One of the aspects of marketing communication which can be the most powerful in convincing customers to respond or buy is the creative components of the campaign. This can range from the text used in emails through to the music and film which are part of a full broadcast advert. It is beyond doubt that the quality and appeal of the creative representation of a message can have a dramatic effect on an audience, affecting buying behaviour and recall. It needs to be taken seriously, even by business-to-business marketers.

Some service firms invest heavily in the creative execution of their communications needs. They use graphics designers for client proposals or brochures, Web designers for Internet-based work, video teams for both staff and client communications and advertising agencies to design advertisements. The variety and range of work is as wide as the methods, or lack of methods, used to manage this resource. For instance, some firms have used elaborate design techniques for individual client proposals. Many have large in-house design teams, often supplemented by ad hoc design consultancies.

In an ideal world, creative work ought to be selected using competitive processes and an agency brief. While many firms work well over many years with small agencies that learn their culture, a new, complex or significant project should be subject to competitive tender. The process of agency selection is shown in Figure 9.1.

Step 1: Write agency brief
This document is to be sent to agencies as a basis for their proposals. However, it is also useful as a mechanism to summarise the firm's view of the project. It should contain the issues listed in Figure 9.2.
Step 2: Create a shortlist of agencies
Create a list of potential agencies from recommendations, contacts and directories.
Step 3: Contact agencies to see if they will compete
Make a telephone call or write an email to the new business director. Not all agencies will be able to pitch because some will have a potential conflict of interest.
Step 4: Send brief to agencies
Step 5: Create selection criteria
Define the pitch process by creating a set of criteria on which to rate agencies. This will help ensure objectivity on the day. Criteria might include: previous experience, understanding of the brief, etc.
Step 6: Agencies present to team
Invite agencies to a day of presentations to a team of leaders.
Step 7: Select final agency and engage
A contract should be negotiated using appropriate functions or skills.

Figure 9.1   Agency selection process

The first step, the agency brief, should include the elements outlined in Figure 9.2.

Step 1: Communications objectives
Step 2: Other enterprise or business objectives
Step 3: Description of target market
Step 4: Message and desired take-out
Step 5: Agency dynamics
Step 6: Constraints
Step 7: Time scale
Step 8: Budget
Step 9: Internal clients and resources

Figure 9.2   Agency briefing process

However, the best agencies have their own briefing formats, and some even refuse to engage with a client until the brief is completed to their own satisfaction. This is their way of avoiding risk by educating customers so that the creative execution, and the campaign overall, has the best chances of success.

MEDIA PLANNING

As important as the creative elements of a communications campaign is the precise and careful planning of which media to use. There is a variety of media that influences the intended market group; including, for instance, broadcast communications, online activities, the firm's sales people, service employees and referrals from people that customers respect. All need to be understood and used carefully. The frequency, cumulative impact and style of the message being communicated through that media needs to be planned very carefully. In an ideal world, an intended customer will receive the same message, whether it is from the professional association to which they belong, an article they read, an informal discussion with a colleague or a presentation from the supplier. Well-crafted messages will get through to the marketplace, with a subtlety that hides the calculated planning of appearance in relevant media.

The media available to service marketers includes the following.

Advertising

Advertising is the practice of buying space in various public media in order to communicate a message. The media used includes broadcast media (television, radio and cinema), print media (newspapers and magazines), online media (websites and online publications) and outdoor media (like posters). Marketers need to plan carefully the use of this media to influence the intended audience. In fact, advertising campaigns are calculated on the basis of the likelihood of their audience seeing or hearing the message a given number of times over a sustained period.

Advertising (known as 'one-to-many') is attractive in that it communicates a message to a large audience very effectively. It is particularly good at communicating messages that require impact, like the launch of a new business. It is used extensively by both consumer and business-to-business technology services, but their commitment to it can be erratic and ineffective.

It does, however, have a number of drawbacks. Firstly, it can be expensive, requiring large investment to reach audiences through television or radio networks. Yet, the ability to target messages very finely has improved considerably over recent decades as a proliferation of media has occurred, and this has also reduced costs. A business-to-business campaign, for example, if it uses radio, business or news-orientated TV channels and specialist publications (such as in-flight magazines and trade papers), can be cost-effective. Called 'narrow casting', this can sometimes be more effective than the expensive, generic broadcast advertising of the past. A buyer listening to a jazz station in the car on the way home or viewing a favourite TV channel on a specialised station may be more influenced by this cheaper media than by a nationwide advertisement. This needs to be a sophisticated and carefully planned communication run by specialists who manage spend judiciously and target the intended audience precisely.

Secondly, there is such a deluge of messages aimed at modern buyers that it is hard to make a message stand out. So, advertising campaigns need to be creative and sustained. A well-planned and eye-catching campaign running for, say, two years will eventually catch the attention of the intended audience and influence its thinking.

Finally, the impact of advertising can be hard to measure. It is difficult to understand the return on advertising expenditure, even if it is well remembered or wins industry awards. Firms tend to create their own measures which help them come to a judgement about the return they get for the cost of advertising. One common practice, for instance, is to have one geography reserved as a quiet area, which does not receive the campaign. Differences in the effect on sales can then be estimated.

Direct Marketing

The most common form of modern direct marketing is, of course, via email. This is so routine that it is easy to forget that it is a broad category of work with some common principles. Direct marketing is the practice of communicating to intended buyers through personalised written or digital messages. It can range from a letter with a printed insert to elaborate creative campaigns, such as sending multimedia messages complete with the MP3 or DVD equipment on which to play them. Email is so much the norm that a printed or mailed piece can stand out from the crowd. One professional membership organisation, Europe's 'Managing Partner Forum', was planning a conference for busy leaders. This event needed to be high profile and stimulate debate. Rather than the usual email invitation, the organisers printed and posted a high-quality card, similar to those used on important social occasions like weddings. Attendance at the event was, perhaps surprisingly, up by nearly 100%. One of the attendees, the managing partner of a large professional partnership, said that the invitation stood out from others and stayed on his desk until he dealt with it; effective retro marketing in a modern age.

Direct communication requires an accurate and up-to-date database of contacts (something harder to achieve than might be imagined), a relevant offer, clear written technique and a concise message. It is most effective when part of a relationship with customers or when the customer has given permission to be contacted in this way. This, though, is an act of trust. Most people detest direct mail, whether physical or email. It irritates and frustrates them to receive unwanted mailers, however attractive. By giving a supplier permission to mail to them, the customer is acting on trust that the firm will take the trouble to filter out unwanted communications and make their written offers relevant. Yet, this is not always achieved, even by leading firms, due to the inadequate attention given to systems and planning.

Direct communication is used extensively in the service industries. For instance, firms routinely invite customers to hospitality events and to seminars on technical subjects. They also send research reports or other material. Some use very effective Internet-based communication. More than one merchant bank, for example, sends an Internet-based newsletter on mergers and acquisition activity in their clients' industries. These are widely appreciated by people at all levels in their customers' organisations, even chief executives, because they are timely, relevant and give valuable insight.

Any service firm which engages in these activities should ensure that they are effective and well planned. If not, they will be very costly in terms of data manipulation, design and management time. Worse, they can undermine the prestige of the firm in the minds of customers and cause dissatisfaction. Even leading firms have sent out poorly written communications; or written to dead people still caught on their poorly managed database; or inundated their customers with so much communication that they become annoyed. Most leading firms now invest in a client database with a campaign management facility. The data in this system should be seen as an asset of the firm. It needs to be kept up-to-date by internal disciplines (ensuring that sales staff keep records current) and by a, perhaps six-monthly, check with customers. It also needs to comply with local data protection laws.

Public Relations

PR is a generic term for a range of specialist and sophisticated skills. Media relations, for example, is the process of managing a firm's relationship with public media like newspapers and news channels. This will involve regular contact with journalists and editors, fielding a range of enquiries from new business ventures, through to the chief executive's press programme and on to handling embarrassing aberrations or professional errors. It is rare to find a service firm that does not have at least a 'press officer' dealing with these enquiries. As specialists, they develop their own skills and their own professional networks to handle the media. In fact, their expertise is often better judged by what they keep out of the press than what appears in it.

It is in the field of proactive media relations that many service firms fall down. Some press officers, particularly in smaller firms, find it difficult to get time away from the demands of ad hoc events and enquiries to take any initiative. When they do, it is often to issue press releases that go out to general circulation lists. Unfortunately, journalists receive so many press releases that they rarely give these much attention. It is better to create a proactive media relations plan. This will identify which messages will be emphasised by the firm and which media it will focus on. The aim of the plan is to develop good relationships between principals in the firm and the editors of key media. In fact, some take an approach similar to client account management with important publications.

A typical PR plan aims to influence opinion formers through carefully targeted communications. It should include:

- News releases on new business or contract wins.
- News releases on new services.
- Favourable journalistic coverage in relevant publications.
- Publicity profiles on people of note within the organisation.
- Trying to get employees speaking on industry conference platforms that the intended customers attend. This puts the audience in the position of pupil to teacher and will increase respect for the suppliers' personnel.
- Relationships with influential professional associations. Company representatives can help them to explain the purchase of services to their members.
- Open communications with management institutes or groups representing opinion formers.
- Feature coverage inside the internal magazines of major customers. If features on the company's strategy appear in the internal magazines of major organisations then there will be strong implicit endorsement.

Clearly, though, the messages developed and promulgated through these mechanisms must be a part of the overall communication strategy. The style and nature of the message might be adapted to fit the needs of the journalists or specific titles, but it needs to be an articulation of exactly the same message as the marketing department is emphasising through other media.

More recently, a complementary activity has developed alongside PR to reach other influencers such as industry analysts or third-party procurement advisors. Called analyst or influencer relations, these activities often fall into the remit of the PR group, with specialists focused on these different audiences. The same group of activities is often directed at these

audiences, including direct marketing to announce company news or provide in-depth information on specific issues or topics, individual briefings by senior personnel and annual events
to update the audience on recent performance and future plans.

Online Marketing Communication

Marketing communications has, as with many things, been dramatically affected by the advent
of the Internet. At its most basic, no firm can now ignore the need to create an impressive
and representative website. Many people search for information on services and products on
the Internet (whether at home or at work) before starting their buying process. In business-
to-business markets it is quite common for top management to ask more junior employees
to research a subject, normally on the Web. Firms must, therefore, ensure that they have a
Web presence and produce materials with key words that reflect the culture and aspirations of
their firm. This is, in many ways, the very heart of their business and they must ensure that
it will be found by search engines. As a result, search engine optimisation is a very common
communications strategy used by marketers to ensure that their firm's offer is easily seen by
Internet searchers.

In addition to the generic website, many firms now have websites created for specific
customers, segments or campaigns, where messages can be more targeted to the audience and
objective of the campaign. Often called 'landing sites', these are easier to integrate into a
campaign that takes buyers through a relationship from awareness to purchase.

The wider range of Internet-based media, such as blogs, RMS feeds and social networking
sites (e.g. Linked-In), need to be considered as part of a wider online strategy. Yet, just because
it is relatively new, this form of communications does not need to be given immediate, overdue
importance. It is not necessarily the only means of communication, nor is it necessarily
the most effective. It should certainly not be used solely because it is relatively cheap. Online communication techniques need to be just one part of a well rounded, hard-headed
communications plan. Marketers need to take into account the behaviour of the audience
much more than the aspirations of the online marketing team to do something innovative.

Events and Hospitality

Customer hospitality ranges from large customised events, through packages bought at large
public events (such as the Ryder Cup or Wimbledon) to smaller, more intimate occasions with
like-minded groups of customers, and individual customer entertainment. The philosophy
behind it is that the informal situation will help to create a relationship with the customer,
which will encourage further business.

A well-managed and customised event can certainly contribute to closer relations, especially
if part of a wider communications programme. There can be, however, a lot of waste and
indulgence. Such events need to be well planned, with the objectives, suitability and intended
audience carefully thought through. For example, as many companies select their events
programme before considering the personal interests and preferences of their customers and
prospects, it is easy for attendees to end up being customer contacts with little influence,
or worse, predominantly 'friends' or internal people. In addition, some people, particularly
public sector organisations, cannot accept hospitality.

Over recent years, with the increasing experience and sophistication in hospitality and the
plethora of options open to customers from competing suppliers, hospitality programmes have

changed shape. Modern events are, for example, often either 'issue-led', bringing customers with similar questions together to share knowledge and approaches, or they are events that customers themselves cannot buy. Several companies have taken advantage of the popularity of the Harry Potter films, for example, offering advanced screenings to customers and their families before the latest film goes on general release. When BT used this approach, it even hired actors in costume to make the event more memorable for important customers and their families.

Collateral

Collateral is the expression of marketing messages in brochures, case studies and other physical handouts. It also includes digital collateral such as websites and videos. The website can act, for instance, as the welcome mat of the firm. When customers meet a service provider for the first time, it is very likely they will look up the firm on the Internet, when back at the office, to read about its offer and its expertise. So, some put an individual micro-site on their business cards containing the biographies of employees, articles and examples of expertise.

In retail services such as mobile stores, collateral is generally 'point of sale' (POS) which, properly devised and presented, helps a customer to choose and reassure themselves about the purchase. It is frequently provided by manufacturers, but in large chains is self-generated. It needs to reflect the brand of the firm and the objectives of the campaign. It should also be easy to assemble, durable and have high impact. The chain needs to control the number of POS campaigns and their effect on shop space, to ensure there are no adverse consequences.

Collateral is also important in business-to-business markets, particularly when aimed at those buying for the first time. It may be that a customer is in emotionally distressing circumstances, such as a significant business problem, or it may be that the service seems expensive. Collateral plays an important reassurance role in these circumstances. For example, if the firm is presenting its credentials during a 'beauty parade', those involved should leave behind a bound copy of the presentation in a format acceptable to the customer, so that they can leaf through it and consider it after the meeting. Also, at the point where the customer is about to make a decision but hasn't yet signed a contract, it makes sense to send a case study of a company which has purchased a similar service from the firm, so that there is something physical to refer to. Finally, after signing the contract, it is sensible to leave behind a directory of the people involved in the project, their experience and contact numbers in a format that the customer can use internally to show others. At the very least, a 'thank-you' note should be sent to the customer for their business.

Collateral needs to be carefully designed and crafted to communicate benefits clearly. It is normal to create a collateral strategy for the whole firm, covering: intended use, design structure, renewal and reprinting methods, plus the types favoured for particular customers. Brochures in particular can be an exorbitant waste of money. Many consultancy firms produce expensive annual reports with photos of their own people. There are several problems with this approach. Firstly, there are frequently few differences between competitive brochures. They are often filled with the same pictures of people and descriptions of dubious, irrelevant 'values'. Secondly, these brochures are often written to extol the virtues of the firm and its leaders, rather than focusing on customer needs and the benefits the firm gives them. Thirdly, they are fixed in time, printed and bound using expensive, high-quality print and design techniques. (A more effective strategy, if an electronic format is not an option, is to create a 'gate folder', a designed jacket which positions the firm, its values and benefits. Sales people can then insert pre-prepared descriptions of expertise, case studies and CVs.)

More important, though, is whether the firm's representatives understand how to use collateral and the moments in the customer relationship when it is most effective. For instance, it can be used to overcome post-purchase distress (described earlier). A piece of material, delivered at the right time, can allay concerns and safeguard the value of a project.

All of these communications, from the written material through to the micro-site on the business card, are techniques to make the intangible tangible and to manage the emotional discomfort of potential customers. It is all too common for services firms of all types and sizes to get on with the work immediately after the contract is signed without communicating with the client using these considered techniques. They are therefore confronted with days and even weeks of silence in which their anxiety about the project can grow.

Planning the Integration of Media Use

The skill of communications planning is designing the right mix of these different media and different techniques to influence customers' thinking; called 'Integrated Marketing Communications' (IMC). By planning an integrated approach and managing various agencies to deliver it, a concerted approach can be made to communicate an idea to a customer. The customer might receive an invite to an event one day, see a newspaper article the next, notice an advertisement after that and, finally, talk to an employee of the firm. These communications prepare the way for a direct conversation, optimising time with the customer. Marketers should examine all the channels through which people receive messages about the offer and use models to plan the frequency of communications through different media. They should also plan both the message and the mechanism of response from the intended buyer.

IMC stresses the need to plan the media through which the message goes, the message itself and the mechanisms by which the response from buyers can be measured. This must be done by the firm itself to achieve a good balance of effective communication and spend. Normally, a marketing communications director will manage a group of different agencies throughout one campaign to achieve this.

At the same time, IMC improves the cost-effectiveness of marketing communications. To date, marketing promotion has tended to be a supplier-driven industry. External specialist agencies have been hired by companies to carry out one of a wide range of promotional techniques. Developing an integrated approach has been made difficult by the focus these agencies have on their own area of expertise. Most advertising agencies want to do advertising, most direct marketing and public relations consultancies want to do direct marketing or PR, and so on. They all tend to stay in their own silos. (For a time there was a move in the industry towards a 'full-service agency', or the one-stop shop, but that concept has not really taken hold.) Even the big networks, which own a range of different marketing agencies, have difficulty in combining them effectively to produce the best balance of communications vehicles for customers. However, a firm that adopts IMC can itself manage the balance of different agencies, saving costs.

IMC also looks for a response from the buyer. This goes beyond the relatively unsophisticated practice of broadcast marketing promotion in that it plans, from the start, a process of two-way communications between the firm and the buyers. A response, through a database or a response-driven promotional campaign, can give the company an idea of the effectiveness of the communication. It can also point to improvements the buyer needs from the company in the form of either quality of service or adjustment to the proposition. The company can then improve its profits by responding to these messages.

PRACTICALITIES AND MANAGEMENT BASICS

Marketing communications demonstrate that both good strategy and good execution are needed to succeed. Good, practical execution of dull campaigns does not motivate people to buy. On the other hand, the marketing industry is littered with award-winning campaigns that produce no interest in sales because they have been poorly managed. Communications campaigns that impact markets are a good example of the need to combine both brilliant basics and magic, creative moments.

The Marketing Communications Plan

Larger firms have such a complexity of media, market and messages that they employ communications specialists and dedicate resources to communications planning. Some create a dedicated communications plan whereas others include communications in their marketing plan. Whatever its format, this plan should outline all the communications elements shown in Figure 9.3.

Response Models

During the 20th century, marketing academics and practitioners developed various response models designed to understand how buyers respond to broadcast promotion. For instance, the AIDA model (see the Tools and Techniques appendix) describes a learning process whereby buyers move from 'attention' to 'action' (i.e. buying) through the influence of marketing messages. While measurement of marketing effectiveness is erratic and varies in sophistication, there is evidence that marketers in various markets have achieved this.

Communications Aspects of Relationship Marketing

Relationship marketing (RM) is a relatively recent addition to the more established marketing theory that is very relevant to service businesses. It suggests that companies can have a relationship with their customers, whether businesses or consumers. By understanding their customers, in a systematic way, and by managing deepening levels of interaction with them, revenue and profit will increase. For many service firms (large consultancies, for instance) this has codified intuitive approaches that have been at the heart of their business for many years.

1. Business and marketing objectives
2. Related communications objectives
3. Key messages and prime target markets
4. Media
5. Key campaigns and programmes
6. PR plan
7. Collateral strategy
8. Agencies
9. Budget
10. Outline of campaign timetable
11. Success measures

Figure 9.3   The contents of a marketing communications plan

RM seeks to use the networks and moments of contact with customers to enhance reputation. Business buyers, for instance, use their professional networks to gain information on service providers. In fact, these networks are often the most influential source used prior to purchase. A strong reputation and positive word-of-mouth will therefore yield work for service companies. Marketers can use reputation enhancement and viral marketing techniques to increase word-of-mouth about their service or firm. This will increase the propensity to buy.

Marketing communication should be planned and developed as part of RM strategies. For example, an important fundamental tool of RM is database marketing, often based on 'Customer Relationship Management' (CRM) systems. These systems hold details of customers and have a facility (called campaign management) which controls communication with them. They register customers' interests and preferences. They also allow marketing managers or account managers the chance to edit communications from their firm before they are sent. They can edit out communications pieces which are unwanted or irrelevant to individual customers.

This facility alone can solve a very common service problem: erratic and uncoordinated mailings to customers. Once these systems are in place, marketers planning any communications to customers (from hospitality invitations through to generic campaigns) must iron out any conflicts with other communications at the planning stage. Relationship marketing therefore introduces a discipline into the communications processes of the firm that removes potential dissatisfaction.

One IT services company that has used relationship marketing to its benefit internationally is Indian outsourcer Mahindra Satyam. By creating an integrated relationship marketing programme for its most valued customers, it repositioned itself and grew its business outside India.

SATYAM'S 'DIAMOND CUSTOMER-ENGAGEMENT STRATEGY'

Mahindra Satyam (erstwhile Satyam Computer Services) is a leading international business and information technology services company that uses deep industry and functional expertise, leading technology practices and an advanced, global delivery model to help its clients transform their highest-value business processes and improve their business performance.

Acquired by Tech Mahindra in 2009, Satyam is now part of the $6.3 billion Mahindra Group, a global industrial conglomerate and one of the top-10 industrial firms based in India. The company was hit by a corporate fraud scandal at the start of the century but, at the time of writing, is moving rapidly from crisis to rehabilitation under new ownership. The customer engagement programme is one of the reasons that it has held together and recovered.

Satyam Delivers a Highly Integrated Customer-centric Campaign to Successfully Reposition the Company and Build their Global Business

At the start of the programme Satyam was widely viewed as a code cruncher that saved the IT world from the collective embarrassment of the Y2K crisis. So, it needed to evolve its image and communicate its assets to a market that had pigeonholed it as a basic, commoditised service provider. Satyam also had its eye on a bigger prize: 'mission-critical IT initiatives' that would help elevate the company's reputation, develop new clients in different industries and muscle in on the business that had been the traditional purview of IBM, EDS and Accenture.

So how do you move an IT-oriented company from a developing nation onto the global stage, surmount cultural resistance and change the conversation with customers from commoditised services to value-added consulting and ultimately trusted partners? You exploit every current customer relationship, basing your growth strategy on what is working with your top customers and your account management processes. You restrict the use of traditional media and completely surround the customers, exploiting hard-won loyalty and promoting the reasons behind it to prospective new customers.

That was the essence of the 'Diamond Customer-engagement Strategy', a growth and marketing model which Satyam created to tackle this remarkable transformation. The programme was designed to systematically deal with four important areas: Business, Technology, Industry and Culture. These were to be the 'four points of the diamond' at the heart of the campaign. Its goal was to move clients from transactional to consulting relationships, and, ultimately, to 'complete seamless integrated cooperative relationships', as well as promoting the process to prospective customers.

The objective was to push forward all four points of the diamond with equal energy surrounding the client, changing the client's view of Satyam and achieving a fully integrated leading global brand. As Satyam's marketing group began the journey, it understood that its strength was in the technology component and it had to develop the business, industry and cultural components of the plan.

It was a daunting objective, which no other Indian peer had approached as systematically or holistically. Others had chosen to focus on hiring local talent in markets like America to short-cut cultural challenges and appear more 'American'. Some were pushing forward simply on their technological merit and price advantage. Others were using higher-level business messages, but their ability to deliver was being received sceptically by the US market.

From 2001–2003 Satyam worked hard at getting account planning and management right, and that meant understanding the customers' objectives in detail and putting the right managers in place to execute flawlessly. Marketing reviewed all client plans and categorized them into four growth potential levels, allocating more dollars to those with the most potential. To be considered for execution, every marketing initiative had to address at least one of the four points of the diamond.

Early planning and research of the message strategy concentrated on historical cases of foreign companies in other industries entering the US market. It was decided that, to be successful, it was important not to downplay Satyam's Indian culture but rather to align the most compelling aspects of it with those most admired by American business; doing for Satyam what 'German engineering' did for Mercedes and BMW.

It was a bold move that involved a massive shift in market attitudes. As a result, a new streamlined tag-line was launched which communicated a sense of urgency, service and business savvy: 'What business demands'. These were attributes that the marketing department knew, from client interviews, were admired by different customers. They also aligned well with Indian culture. All marketing messages expanded on this tag-line or landed on it, providing a focus for both clients and Satyam. Most importantly, it addressed two points of the diamond, head on: business and culture.

With budgets tight, the main focus was on using every internal asset and that meant using relationships at every level inside Satyam. Marketing met with horizontal (technical) and vertical (industry-specific) practice directors to enlist their support for the Diamond Customer-engagement Strategy. By then it incorporated the best practices of Satyam's long-term client relationships and comprehensive account management.

A number of initiatives were developed to address each point of the diamond:

1. **Technology.** Satyam had proven its technology excellence and was on the road to increasing its high-tech credentials. They were also aggressively cultivating relationships with key industry partners (SAP being the first, with others following in quick succession). At the risk of appearing overly tactical, Satyam relentlessly communicated every new certification and partnership to its market, creating an overall impression that the company was on the move with unquestioned technology credentials.

2. **Business.** Satyam set out to change the conversation from 'technology as competitive advantage' to 'technology as an industry enabler' which would reduce industry-wide costs, enhance revenues and create value across specific sectors. With a need to focus business on competing on core competencies, this component was tackled through exclusive customer events, thought leadership publications and a modest intra-IT industry advertising programme on the subject of business transformation.

3. **Industry.** Satyam had strong expertise in several sectors such as manufacturing and finance. It moved aggressively into new industries, leveraging cross-industry and technology experience, and systematically acquiring expertise to break into new accounts. To build its credentials, marketing used forums, customer 'boot camps', tradeshow participation and industry analysts.

4. **Culture.** Here, marketing had to weave the Satyam culture and the client into one seamless fabric. They needed to elevate Satyam's image. It had to be changed from an outsourcing vendor, or even a strategic partner, to a fully integrated division of the client company. They did this through creating replicas of their clients' business locations, making the movement between client offices and India development locations easier. This was combined with comprehensive visiting programmes to India for customers as well as hosted social events.

Some of its most compelling components were:

- **A Premier Annual Customer Event: Satyam World**

Diamond Points: Business and Culture

Satyam created their own highly successful premier annual customer event, 'Satyam World'. They copied what they learned from 'other major' industry events that attracted senior executives and decision-makers. Although the cost was higher than a shared event, they had the undivided attention of attendees and could control messages for three days. The event revolved around Satyam's clients, carefully selected prospective clients and the Satyam brand.

- **Geo Sourcing Forums: Cross-industry Gatherings**

Diamond Points: Technology and Industry

Each of these 'Geo Sourcing Forums' was a reflection of its industry, technology or region. Satyam demonstrated its expertise through actions in each Forum. These small workshop gatherings provided a crucial venue for those actively engaged in executing IT initiatives. They were a place where Satyam could listen to client needs and industry analysts. It could also present views on intra-industry problems. Satyam clients chose the topics and as each Forum matured, it became an entity in itself, independent of but underwritten by Satyam. Several Forums have grown into 'multiple chapters', 'some global in scope' and video-conferenced to the attendees. The most mature Forums invited executives from companies that were not yet Satyam clients.

- **Customer Boot Camps: Client-specific Gatherings**

Diamond Points: Technology and Industry

Customer boot camps were serious problem-solving gatherings held specifically to solve customer challenges across their business. Satyam often invited an analyst or an independent expert along to the workshop. Brainstorming, problem-solving, planning and 'solution roadmap development' were the main focuses, while customer service and intimacy were the benefits. A Satyam customer-specific intranet provided deep Satyam capabilities and client-specific resources, as well as a forum to continue the work.

- **Location Replication**

Diamond Points: Culture and Industry

The Replication of clients' headquarters at Satyam's offices in India made customers feel at home and surrounded relevant Satyam employees with their brand. Every detail of the client's headquarters was replicated and branded in their livery (from product display cases, corporate colours and office furniture to small details like coasters). This created a seamless experience between Satyam and its clients; making it seem more like the relationship of a corporate headquarters to an operational division rather than a client/customer experience.

- **Comprehensive Customer Visiting Programmes**

Diamond Points: Culture and Industry

Customer visiting programmes provided a comprehensive door-to-door Satyam experience aimed at demonstrating service through action. Programmes were planned down to the last detail from the pre-travel communications packages to the events scheduled while in India. All materials were co-branded to communicate client partnership and intimacy.

- **Customer Appreciation Events**

Diamond Points: Culture

Satyam hosted clients at major sporting events in their home country. Local marketing people ensured that Indian managers were fully briefed in both the sport and cultural nuances, resulting in successful client experiences and a closer bond between Indian and in-country management.

As a result of the programme, in less than five years, Satyam moved from a 'code cruncher' to a business leader, more than tripled its annual revenues and expanded its workforce by 30,000 employees. The four points of the Diamond Customer-engagement Strategy were not only marketing's mantra but, importantly, shared by internal executives across Satyam. Over the five years to 2008, for example, Satyam's financial practice grew 92% (from 150 customers); a direct result of trust built on the Diamond Client-engagement initiatives. A large proportion of clients were moved up the value chain from transactional to deeply cooperative, with sophisticated joint venture relationships as a direct result of efforts.

Using the Diamond Customer-engagement Strategy as their guide, Satyam's marketing group effectively:

- Embraced its Indian culture when competitors were running from it and used it to its advantage, changing the market perception.
- Changed the conversation from technology to business issues through leadership events.
- Built a global brand that is looked to by global enterprises for advice, counsel, leadership and solutions.

Data-driven Marketing Communication

If there is one major difference between industries which value marketing and those which don't, it is the willingness to invest in data to aid marketing decisions. Whereas some industries use weekly market research to aid their decisions, the services industry has been more cautious in its use of research or good-quality data to guide strategy or marketing.

Yet good marketing communication is also grounded on data. Many broadcast advertisers know how many shoppers they are aiming at and their attitudes to the product through systematic investment in research data. There is no reason why service marketers should not have data on: their customers' attitudes, their competitive reputation, the propensity of customers to refer them to others and the play of their brand in the purchase process. All can be obtained through careful, systematic research and all will increase the effectiveness of communications. Leading service companies in technology sectors are steadily investing more in customer knowledge. They are setting up processes to retain 'corporate memory' of customer behaviour. There is no excuse for the arrogance of 'gut feel' based on the experience of a few customers and internal dialogue.

COMMUNICATIONS DURING A CRISIS

An important aspect of marketing communications with which service firms have to deal is when a serious problem attracts significant press attention. This might be when a client takes legal action, or a regulator is publicly critical or, as in the case of Satyam, a significant individual in the firm receives adverse attention. The first step is to plan to avoid this happening. Communications plans should be put in place alongside regulatory or policy guidelines. These should include:

- **Editorial relationships.** When a negative story appears it is the editors who decide how to treat it. The firm should set up mechanisms to develop relationships between key individuals in the firm and editors of print and broadcast media in the major world centres. Such relationships (apart from helping to raise the profile of the firm) will not stop a negative story running altogether, but will minimise damage and enable the firm to put its point of view more easily.

- **Rehearse the leadership.** Evidence suggests that it is the one word ('shredding' in the case of the now defunct but massive accountancy firm Andersen) spoken by a leading figure which can cause catastrophic damage. Leaders should be well-rehearsed in media management and publicists should have control over how to respond to media interest.
- **Preventative management of sensitive issues.** Meetings should be established between the firm's media team and the firm's leadership to discuss how to handle potential issues.

SUMMARY

All companies must communicate with their markets, and the service businesses of technology industries are no exception. Service marketers should get to grips with the range of tools and techniques available to communicate effectively with both buyers and potential buyers. In addition, they have to make sure that the wider community understands their values and brand. They can now call on an array of concepts tailored to their specific needs, from advertising and PR through to understanding how best to use collateral. Even more importantly, to see a return on their investment, they must make sure that this vital function is overseen by qualified and experienced teams; that organisational expertise is developed.

Marketers in service firms must consider how effectively their campaign communicates with the market as a whole and with individual markets within the broad economy. This might entail messages aimed at potential customers about the firm's capability, focused campaigns to existing customers about the firm's approach or broadcast messages to the wider community about its values and brand. All have their place and, managed properly, can contribute to improved margins through enhanced revenue and successful cost control. This is a different method of communication than the one-to-one relationship of the sales force. It requires processes and methods which communicate effectively across different media in concert with other messages in order to influence the thinking, behaviour and actions of a group of people.

Properly done, it is a highly sophisticated and subtle two-way communication with buyers and potential buyers. It can influence thinking, steering potential customers towards a firm's offer, and help to keep the cost of sales down. The function requires experienced, specialist individuals working in a well-managed team to be effective over time. It is poor leadership to allow inexperienced or unqualified people to run this function, since it results in both wasted expenditure and ineffective communication. Even firms conducting large broadcast programmes, aimed at a wide audience, have to have a clear understanding of the people they are targeting. Whether it is a consumer or business-to-business group, the supplier is aiming to reach and influence people who will make a decision to buy the services on offer.

10

Service Quality

INTRODUCTION

The concepts and policies of service quality have several different generic names: 'customer care', 'after-care', 'service recovery' and 'customer experience management'. All capture the fact that the experience of a service can contribute to how well or badly a business does in its market, and this is as true of technology-based services as it is of any others. There are a number of quality issues that need to be understood and acted upon, including the type of after-care being offered and how it resonates with customers, the determinants of customer expectations and all the different aspects that make up the firm's relationships with them. This is not a vague or imprecise after thought, nor is it an unrefined, passionate desire to give buyers everything they want. As in every other area of business, developing strategies and processes which result in the design of an appropriate service experience and excellent after-care is essentially very practical. This is as much about setting policy and installing processes as any other aspect of sensible business. Unfortunately for marketers in service businesses, though, several characteristics of a service (like simultaneous consumption) demand that they involve themselves in this field. It is a marketing issue because it affects the growth of the firm. A poor service experience will undermine any marketing, no matter how good. So, this chapter examines service quality issues, focusing on both strategy development and the practicalities of the most important aspects to influence.

THE IMPORTANCE OF SERVICE QUALITY

Customer care and service quality are important for several reasons. Firstly, service quality affects the attitudes of buyers towards repurchase. If they have a good experience they are more likely to buy again and, if a poor experience, less likely. This simple fact can be surprisingly neglected and mismanaged by even the largest and most well-intentioned firms. Secondly, customers may judge the service they receive on different criteria from that of the provider. Whereas the marketer might focus on technical expertise or speed of execution, the customer might value 'bedside manner', the way they are treated and the perceived attitude of employees, just as highly. In fact, the latter can cause customers to question the former, if unsatisfactory.

Some recent writers have even suggested that customers' reactions to service quality induce loyalty to the supplier, which can be measured and managed. Frederick Reicheld (Reicheld, 2003), who has made a career out of this, has even suggested that it is the 'only measure necessary'. Whatever the reality of these claims, service firms need to determine which issues influence their customers' views sufficiently to enhance or degrade reputation. They have to understand the components of the service which lead to repeat business and referrals (and thus future revenues), and which processes or techniques should be employed to improve them.

SERVICE QUALITY AND CUSTOMER CARE: A RECENT HISTORY

Proponents of service quality concepts normally suggest that an emphasis on customer care was not necessary before the 20th century and only developed as consumerism grew. That is not the case though. For instance, British Potter Josiah Wedgwood routinely used international direct marketing in the 18th century. Each of his offers included a 'money-back guarantee' and free replacement of breakages, despite there being such poor infrastructure for transporting dinner services. American processed food producer Henry Heinz saw lobbying for quality as a foundation for the creation of his brand in the 19th century and the reputation of Chicago's 'up-market store' entrepreneur Marshall Field was built on outstanding courtesy and quality at the start of the 20th century. It seems that difficulty occurred as distribution chains grew and marketing was functionalised in the mid-20th century. As a result, leaders of businesses became more remote from their customers' experience of their products and marketers grew distant from the responsibility for service quality.

In the latter half of the 20th century, this began to change as the result of a number of developments. They included:

1. An evolution in customers' standards and expectations, prompting a rise in consumerism after the austerity of the mid-century worldwide conflict.
2. The then-steady decline in the performance of much of Western manufacturing in the light of the success of certain Asian, particularly Japanese, companies.
3. Emphasis on after-care and service in some sectors, particularly retail and computing.
4. The publicity gained by several populist writers and speakers on service quality, particularly Tom Peters (Peters and Waterman, 1982) who modelled the dynamics of successful businesses, creating general principles out of case studies. His primary emphasis was on quality of service and customer care. He thought that much Western business had moved away from an emphasis on serving their buyers and had lost world leadership as a result.
5. Certain well-publicised, dramatic improvements in service which affected the share price of the firms involved. One of the early examples was British retailer Marks & Spencer. They gained market leadership and the affections of the entire nation for many decades because of a commitment to service. In the mid-20th century, a time when consumer spending power in Britain was just beginning to increase, they met one of the emotional needs of their buyers with a money-back guarantee. This meant that consumers were happy to buy and felt a warmth and loyalty to the company, which underpinned their profits for many years to come.

Two, further examples were in the European airline industry. In 1983 the then-chairman of British Airways (BA), Lord King, and his chief executive, Colin Marshall, created a radical improvement in the market position of the newly privatised airline by engaging front-line staff in a massive improvement prioritisation programme across the whole firm. Called 'putting people first', the programme engaged many thousands of staff in workshops where the importance of excellent service to passengers was stressed and barriers to such service were removed from the employees. This was one aspect of a broader marketing programme which encompassed a new brand initiative and new quality measures, based on extensive market research. Before the programme, BA had a terrible reputation for punctuality, was beset by industrial disputes and was losing substantial money (£140 million in 1981). A decade later, it was not only popular with customers but was the world's most profitable airline. Although there were clearly other factors which contributed to this performance (see, for example, Gruglis and Wilkinson, 2001), the impact of the programme caught management attention throughout the world.

At the same time the new chief executive of the Scandinavian airline SAS, Jan Carlzon, introduced a similar programme under the banner 'moments of truth'. Carlzon introduced a company-wide programme to enhance all moments of interaction with customers which had a major impact on the company's market position. In one year, the airline became the most punctual in Europe and the first choice of its intended segment (business travellers). Just as important, though, corporate overheads were reduced by 25% and before-tax profit went from a loss of $10 million to a profit of $70 million (Vandermerwe, 1988).

In business-to-business, IBM made similar strides through quality commitments in the early computer industry. It met an emotional need amongst business buyers when the technology was new and unstable by using leasing to finance many of its early sales. This made it easy and low risk for buyers to invest in the new technology. It also put enormous emphasis on the quality of its maintenance and installation work. This approach gave the company worldwide market leadership for nearly 30 years.

6. The publication of a number of influential research reports. One was, for example, conducted by TARP (Technical Assistance and Research Programmes) in the United States, for the White House Office of Consumer Affairs. Interestingly, this research has become part of management rhetoric and is still, often unknowingly, quoted at conferences because, at the time, it was given so much attention. The work, which interviewed 200 companies, showed that:

- The average business did not hear from 96% of its unhappy customers.
- For every complaint received, there were 26 other people with problems and six with serious problems.
- People with problems who failed to complain were far less likely to repeat their orders and were likely to stop completely their business with that supplier.
- People who complained and whose problems were handled well were much more likely to continue doing business.
- People with bad experiences were twice as likely to tell others about it as those with good experiences.

7. This combination of factors attracted the attention of management throughout Western businesses and put a spotlight on quality of service principles; so, for a period of time, service quality and customer care became a popular management fad. Some of the stories told of employees delighting customers were so anarchic that a number of concerned academics set out to research the field and develop credible principles that could be applied within it. For instance, Sri Parasuraman and a group of academics that specialise in services marketing produced their 'GAP model' (Parasuraman et al., 1985) as an analytical tool; Heskett and Sasser applied value chain analysis to service businesses with their 'service profit chain' (Heskett et al., 1997); and Fredrick Reicheld published on his 'loyalty effect' (Reicheld, 2006). Although each has been evaluated, tested and critiqued since, they provide a range of practical management concepts which ensure that service quality issues can be tackled in as pragmatic and realistic a way as any other aspect of business.

8. Public metrics of service quality were then developed. The American Customer Satisfaction Index (ACSI) is an economic indicator that measures the satisfaction of consumers across the US economy. Set up in 1994, it is produced by the National Quality Research Center, University of Michigan. The ACSI interviews about 80,000 Americans annually and asks

about their satisfaction with the goods and services they have consumed. Respondents are screened to cover a wide range of consumer products and services, including goods, services, local government services and federal government agencies. The results are published each quarter. The programme measures customer satisfaction for more than 200 companies in 43 industries and 10 sectors. The ACSI aims to represent the satisfaction of the 'average American consumer'.

The ACSI was based on a model originally set up in Sweden in 1989, called the Swedish Customer Satisfaction Barometer (SCSB), and versions now exist in other countries. The UK version, for example, is based on a representative sample of 25,000 adults surveyed over the Internet. The British Institute of Customer Service calls its version the UK Customer Satisfaction Index (UKCSI).

TERMINOLOGY AND PERCEIVED WISDOM

As a result of this history, a number of concepts, policies and, frankly, ill-founded beliefs have built up around the concept of service quality and customer care. There is lexicography around the subject which rests in the back of the mind of many senior executives, affecting policy formulation. Before constructing a strategy or integrating service quality concepts into marketing programmes, service marketers should take a clear-headed view of these concepts.

Customer Service and the Propensity to Repurchase

Researchers have explored the effect of quality of service initiatives on the propensity of people to repurchase from the same supplier. Their particular initial area of interest was in transactional measures such as satisfaction surveys, which are often sent to buyers asking for an opinion on the service performed or, say, left in hotel rooms for visitors to complete. They found that these are, generally, poorly designed research questionnaires, asking the wrong questions about the wrong issues and giving no manageable data which the supplier could use to improve service still further. Worse, they found that it cannot be assumed that satisfied buyers will return to buy again.

They found that there was very little evidence that good quality of service increased the propensity of people to buy again. It was quite possible for a buyer to be entirely satisfied with service and yet next time buy from an alternative supplier because there is a better value proposition. Many now argue that quality of service feedback mechanisms must therefore look at underlying motivations and tease out the rational and emotional needs that buyers want satisfied.

The Concept of Loyalty

The concept of 'customer loyalty' relates closely to service quality. It suggests that, if buyers think they receive good service, they will be loyal to the supplier, returning again and again to buy. Some have asserted that it is the primary determinant of profit and growth because loyal buyers produce greater cash flow, cost less to service and spread positive word-of-mouth (see Reicheld, 2003). Loyalty, a long-term feeling of attachment to a supplier, is thought to occur when buyers are satisfied and have investment in the relationship which is too great to sacrifice for a cheaper or lower-quality alternative. It has been the thinking behind many discount retail schemes to repeat purchasers and huge investment in customer relationship management (CRM) systems.

There is evidence that people do become loyal at the point of purchase, feeling warmth to a supplier and returning to buy. Yet loyalty is far more than a revamped direct marketing programme or the launch of a consumer magazine. People need something to be loyal to; they need emotional resonance. For instance, as explained in Chapter 5, by gaining emotional allegiance to some features of a brand, product managers have been able to earn incremental revenue over many years. Their brands have remained high in their category's sales leagues due to repeat purchase. The phenomenon also occurs with some corporate brands, such as BA or Virgin. Buyers find that these services offer them an experience that is reliably the same. It saves them time and effort, meeting a need in their life, time and time again.

There are, then, some difficulties with this concept. Some corporate names and some service firms, for example, have no brand value and induce no emotional allegiance. Also, there is evidence to suggest that a number of customers regard the service of some competing service suppliers as technically similar. Their reluctance to change is more apathy than loyalty. It is risky to try to create loyalty schemes in these circumstances or to rely on the assumption that the business is 'all about relationships'. If another supplier were to come along with a better scheme or, more effectively, a better value proposition, it would attract these buyers away. Suppliers need to take a hard-headed view of this issue. They need to understand the emotional allegiance of their customers in terms of the relationship and the value of any goodwill in terms of repeat purchase.

Some academics think that the studies on which loyalty work relies have not, at the time of writing, been thoroughly tested. There is evidence, for instance, that some people actually want anonymity. There are some circumstances where it is delightful to be known by name and given special treatment, but in others buyers just want to get on with the purchase and not to be interfered with at all. They want some transactions to remain insignificant in their lives and the supplier should recognise that. Conversely, there are some buyers that a supplier does not want. Satisfaction is not an end in itself, but ultimately the basis of shareholder value. It is quite possible to have satisfied buyers who are not at all profitable. Logically, suppliers should choose which buyers they want and discourage those that they don't. This makes segmentation, as outlined in Chapter 3, a key strategic issue behind loyalty investment.

The Concept of Holistic Service

The concept of holistic service, introduced in Chapter 6, argues that all the components of service need to work together in a way that the buyers perceive to be a fluid uninterrupted experience. It suggests that the core service of the company should be designed, like a product offer, to meet customer needs. Service quality and customer care often refer to more than the technical components of the work. They refer to the way in which the customer is handled by the firm.

In service firms, the interactions with the customer are continuous and need to match the technical quality. So, service marketers need to ensure that their customer care and their technical performance are integrated into one seamless performance which meets customer expectations and creates one value proposition. This will have an enormous effect on reputation, and then on repeat business and referrals.

The model shown in Figure 10.1 was proposed by an early customer care consultant, John Humble (Humble, 1989). Interestingly, it emphasises the balance between business basics and human interaction that leading companies in service marketing are trying to achieve.

A model for managing holistic service

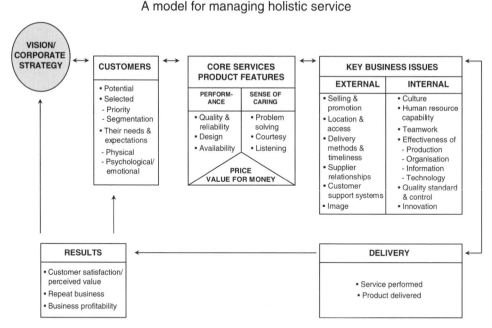

Figure 10.1 John Humble's model for integrated service
Reprinted with permission from *Harvard Business Review*, Humble, July/Aug 1989.

It suggests that value for money from the 'core service' is a balance between technical performance and a 'sense of caring'. The model prompts the supplier to think through the integration of service quality into the total mix of their business to ensure a streamlined and enjoyable service experience. This concept has evolved into a more modern planning approach to service quality, called 'customer experience management' (CEM).

Customer Experience Management

This is a relatively new term that seems to have no clear definition, leaving room for a number of different interpretations. Some think that it is quite simple: every time a company and a customer interact, the customer learns something about the company that will either strengthen or weaken the future relationship. This accumulated experience will affect that customer's desire to return, spend more and recommend it to others.

Others distinguish it from generic customer service. They point out that even if customers are served well, they might still buy from competitors next time. They see customer service as limited to the quality of the last transaction but CEM as a framework for building an organisational capability to deliver a distinctive service experience. Much of their writing emphasises the need for the service experience to deliver the promises raised by customers' perception of the company's brand; the service equivalent of brand integrity.

The concept prompts service marketers to identify touch points, similar to the 'moments of truth' used by Jan Carlson to revolutionise the service of SAS. They should conduct a 'touch point audit', very similar to the contact audit, used to understand all the points of communication between an organisation and its customers.

Total Quality Management

This philosophy and approach to quality came to prominence in the West during the 1980s, led by prominent specialists such as Edwards Deming, Joseph Juran and Philip Crosby. Analytical in emphasis, it suggests that quality issues should be resolved prior to affecting operations, not checked and counted after they occur. It calculates the 'cost of quality' in the organisation and sets out to reduce failure in a systematic way, using multi-disciplined teams of people.

In fact, the prime aim of a 'total quality management' (TQM) system is 'zero defects', achieved by progressively examining smaller and smaller errors in an objective and numerate way, using a range of specialist techniques. Its 'kaizen' principle of continuous improvement based on analysis, 'quality circles' (mixed discussion groups) and open suggestion boxes is fundamental to its success and must be supported by senior management.

Where TQM is particularly powerful is in its philosophy of engaging the whole organisation in guaranteeing and improving quality. The quality activities of the organisation are not seen as a detached and specialist function. As a result, it engages the total operations function and, although, in reality, the target of 'zero defects' might ultimately be unachievable, the approach has radically improved the performance of many companies, saving some from extinction. Although TQM came to prominence in manufacturing companies, it has been enthusiastically adopted by large, process-dominated service companies such as telecommunication operators and utilities.

'Six Sigma'

Six Sigma, an extension of TQM, was pioneered by Motorola and Allied Signal, although it rose to prominence when implemented by GE in the 1990s. It is a measurement-based strategy that focuses on process improvement by applying a model of: 'design, measure, analyse, improve and control'. It appears to have helped many manufacturing companies improve the quality and effectiveness of their operations.

It has also been applied to service businesses. For example, in an interview in the Spring 2005 *McKinsey Quarterly*, Claude Brunet, the management board member of insurance giant Axa who was responsible for operations, discussed the lessons that service companies can learn from manufacturers, having embarked on a Six Sigma programme several years before. He said:

> The biggest challenge is getting service employees to understand that they use processes. After 30 years of Total Quality management and other improvement programs, people in manufacturing already have this perspective. They know what a process is, how to analyse it and how to improve it. This is all new for people in service companies. When we started implementing Six Sigma, early in 2002, we had to get our people to understand existing operations from a process perspective. And when you look at a service company's processes in detail, they may not be as repetitive as the processes and activities at a manufacturing plant. (Reprinted with permission from *The McKinsey Quarterly*, Monnoyer and Spang, May, 2005)

Moments of Truth

A moment of truth is any moment of interface between a buyer and the firm. These range from interfaces with customer service staff, reception and support staff through to administrative processes like invoicing or contracts. All can positively or negatively affect their impression of the service, enhancing or damaging reputation. It is sensible to conduct a periodic review of all moments of truth to ensure that all contribute to the health of the firm.

The Perceived Transaction Period

The perceived transaction period is the time that the buyer thinks they are engaged with the supplier. Service issues arise if the customer thinks that their engagement with the supplier is longer or shorter than the supplier does.

Employee Behaviour and Emotional Empathy

British comedy *Little Britain* caught the mood of the nation when they portrayed a young employee in a travel agency trying to serve an elderly couple. The young girl barely looks up from her screen and offers a range of totally unsuitable vacation ideas, like white-water rafting. She completely fails to listen or empathise with them. The encounter is full of the metallic and robotic phrase: 'the computer says no'.

'Emotional empathy' is the term used by service strategists to ensure that this kind of mechanical human interaction does not happen; that employees who serve customers are responsive, competent and empathetic. This not only means that they listen to the needs of people but also that they demonstrate empathy; that their tone of voice and demeanour show that they sympathise. Interactions need to be warm and flexible, encouraging positive interchanges between people. If a dull, mechanistic approach is taken, buyers will not respond positively to their experience and will, ultimately, move away. As important, but more difficult, this sense of spontaneity needs to continue after the honeymoon period of the initial launch of any new or improved service.

There is evidence that this is a very important factor in customer satisfaction. OgilvyOne published, for instance, a major study into customer experience management in which they found:

> Both emotional bonding and operational attributes must work together to produce customer experience. However, while the study shows that 78% of companies measure customer experience through functional information they captured, only 64% measure emotional (bonding) aspects – even these may be confusing 'customer satisfaction' for bonding, which is more powerful.

Although they reported it as a negative, it is interesting that over 60% of companies had tried to measure emotional empathy and had given some importance to it. British building society Bradford and Bingley, for example, have used the Myers Briggs categories to understand the approach of customers and how to handle their requests. They taught their employees to distinguish between a 'results orientated controller' and a 'social orientated entertainer' (and others). The body language, tone of voice and method of resolution had to be different for each. Others have used 'transactional psychology' (Harris, 1973) to similar effect. People can choose to relate as a 'parent', 'child' or 'adult' and the best transactions are 'adult to adult'. Employees are taught to recognise in what mode a customer is approaching them and to use various techniques to turn an encounter into an 'adult to adult' conversation. Several companies have reported improved satisfaction, particularly with complaint handling, using this approach.

There is evidence that this is very difficult in a call centre where body language and gesture are not available clues to either party. Employees have to judge the tone of a caller very quickly, to pick up on emotional needs and respond correctly. Nevertheless, some companies do report that it is possible to put in place sensitive questioning techniques, which allow their people to detect emotional states and respond accordingly. Successful operational managers (in places like First Direct, for instance) report that they have to ensure their staff concentrate carefully on these first encounters. They take care to use coaching, breaks away from the phones and refresher training.

Some, though, have taken this much further. They think of the numerous encounters within the service reception facility as dramatic moments, with the elements of a theatrical performance and following an explicit or implicit 'script'. As Professor Leonard Berry said, after years of research into the service industry (Berry, 1995):

> Services are performances and people are the performers. From the customers' perspective, the people performing the service are the company. An incompetent insurance agent is an incompetent insurance company.

The components of this 'performance' can, as in the theatre, be planned and designed to achieve any desired outcome and maintained over a long period of time. What the supplier sees as a set of operational procedures can be thought of as a 'customer script', guiding interactions between the customers and the supplier's employees or equipment. In fact, there is evidence that suppliers with a creative or theatrical heritage, like Disney resorts, have achieved success by intuitively applying dramatic techniques to service businesses. (In the Disney resorts, for example, employees are largely resting actors who see signs reminding them that they are going 'on stage' as they start work.) They have demonstrated that, if a 'service script' is followed, it is likely that the quality of service will match the expectations of the buyers and the intentions of the supplier. So this script needs to be worked out carefully, communicated to buyers and, where necessary, form the basis of training for both employees and customers.

It seems, then, that concepts and paradigms do help employees who have to serve customers to demonstrate emotional empathy. A recognised framework puts words, concepts and language around the intuitive mechanisms that all adults develop to interact with people in distress.

Empowerment

Uncertainties about script and boundary can be very damaging to the relationship between customer and supplier. One of the roles of employees is to know when and where they can deviate from the script and break rules to give buyers better service without affecting the overall efficiency of operation. So, operational procedures in a service company should 'empower' front-office workers to meet any unforeseen needs by giving them the right to make concessions, within a controlled framework.

'Empowerment' is the discretion given to employees to respond to individual customer needs. This concept arose as service organisations needed to communicate their desire to respond to service needs and in response to strong evidence that any inflexibility in employees' behaviour damages both reputation and future business. John Bateson (Bateson and Hoffman, 1999) suggests that there are three levels to it:

- **Routine discretion**, where employees are given a list of alternative actions to choose from.
- **Creative discretion**, which requires an employee to create a list of alternatives as well as choosing between them.
- **Deviant discretion**, which expects people to do things which are not part of their job description or management's expectations.

The latter are normally most noticed and appreciated by buyers. Yet, if good service depends on experienced employees consistently breaking the rules, the service marketer has not done their job properly. They should ensure that concessions are recorded, so that the fundamental processes of the organisation can be re-examined in the light of the type and frequency of concession.

Delight Factor

A term closely related to Tom Peters (Peters and Waterman, 1982) which sets out to measure the emotional pleasure or pleasant surprise given to customers by a service experience; some even claim that they have 'customer delight measures'. The long-term problem with this is twofold. Firstly, it builds cost into the business. Secondly, it introduces an upward spiral of exceeding expectations, with expectations being increased and delivery having to be increased further still. This eventually becomes unsustainable, and the momentum of the company to provide excellent service eventually runs out. Nearly all the examples held up in Peter's original book have, for example, had massive difficulties since publication, damaging the crux of his thesis.

A practical alternative would be to use communication and marketing techniques to lower expectations. The chief executive of cut-price airline Ryan Air is continually making controversial announcements about planned changes to their policy. They considered charging overweight people more, charging to use the toilet on flights and for very precise excess baggage. Many of these ruminations do not actually get implemented, but it sets the expectations of potential passengers very low. Some are pleasantly surprised when the experience of such cheap travel is not as bad as they feared.

Personalisation

Amazon has driven personalisation of service on a technical infrastructure to new heights through its Amazon.com experience, as discussed in the case study below. With the combination of technological sophistication and proliferation of messages and products competing for consumers' attention, the idea that 'you no longer find products and services, they find you' is becoming reality through Amazon's approach. Even consumer product companies are jumping on the bandwagon, such as Nike with their NikeID facility, through which customers can design their own trainers, adapting the colours and features to suit their personal preferences.

A PERSONALISED SERVICE FROM AMAZON.COM

The online shopping giant Amazon.com opened its virtual doors on the World Wide Web in July 1995 and today offers 'Earth's Selection'. A Fortune 500 company based in Seattle, Amazon strives to be Earth's most customer-centric company where people can find and discover virtually anything they want to buy online. By giving customers more of what they want (low prices, vast selection and convenience), Amazon continues to grow and evolve as a world-class e-commerce platform.

Founded by Jeff Bezos, the Amazon.com website started as a place to buy books because of the unique customer experience the Web could offer book lovers. During the first 30 days of business, Amazon.com fulfilled orders for customers in 50 states and 45 countries;

all shipped from his Seattle-area garage. Over time, the company's evolution from website to e-commerce partner to development platform is clearly driven by the same spirit of innovation.

Achieving customer loyalty and repeat purchases has been key to Amazon's success. Many dot.coms failed because they succeeded in achieving awareness, but not loyalty. Amazon achieved both:

> We work to earn repeat purchases by providing easy-to-use functionality, fast and reliable fulfilment, timely customer service, feature rich content, and a trusted transaction environment.

Amazon's focus on its technology has enabled a smooth shopping experience, which is all about the customer and ease of use. Among its many technological innovations, Amazon offers a personalised shopping experience for each customer, book discovery through 'Search Inside This Book', convenient checkout using '1-Click® Shopping', and several community features like Listmania and Wish Lists that help customers discover new products and make informed buying decisions.

Personalisation is considered fundamental to Amazon's customer-centric experience. Although personalisation broadly defines a variety of eTail tools, it is most commonly associated with the use of recommendations. 'Customers who bought X . . . also bought Y' is Amazon's signature feature and remains the industry benchmark in web-based business. It revolutionised the online customer experience and continues to drive service quality today.

Getting Personal

In an age of information explosion, personalisation offers a way to focus customers. It helps to better serve the customer by anticipating needs, making the interaction efficient and satisfying for both parties, and building a relationship that encourages the customer to come back and buy again.

In a 1998 interview, Jeff Bezos said:

> In the online world, businesses have the opportunity to develop very deep relationships with customers, both through accepting preferences of customers and then observing their purchase behaviour over time, so that you can get that individualised knowledge of the customer and use that individualised knowledge of the customer to accelerate their discovery process.
>
> If we can do that, then the customers are going to feel a deep loyalty to us, because we know them so well. And if they switch to a competitive website – as long as we never give them a reason to switch, as long as we're not trying to charge higher prices or providing lousy service, or don't have the selection that they require; as long as none of those things happen – they're going to stick with us because they are going to be able to get a personalised service, a customised website that takes into account the years of relationship we've built with them.

Right from the beginning, it's clear how Amazon's business strategy has been manifested in their design decisions. They build tools that accept and observe preferences (browsing behaviour, search queries, wish lists, purchasing history, etc.) and design features to take advantage of that knowledge by finding related and closely relevant products.

Indeed, it was web-based business that gave Amazon this rare window into human behaviour, where they could record every move a visitor made at every click of the mouse.

From this data, all sorts of conclusions could be drawn about the consumer. In this sense, Amazon was not merely a store, but an immense repository of facts. All they needed were the right equations to plug into them.

Fuelled by advances in technology, personalisation essentially entails the ability to adapt a product or service continually and independently, whether by altering services or by changing product configurations or applications. It enables Amazon to:

- Identify customer characteristics by initiating interaction and two-way communication.
- Predict customer needs and purchase patterns.
- Provide a customised online shopping experience that reflects in-person shopping.
- Target each individual customer differently with one-to-one marketing.
- Build a relationship that encourages the customer to return.
- Make the interaction efficient and satisfying for both parties.
- Promote sales of a variety of products, up-sell and cross-sell.

Making Recommendations

Amazon's personalisation system is based on recommendation algorithms that rely on the huge amount of data available through the Web. There are three common approaches to solving the technical aspects of the recommendation problem. But since none of these scaled to Amazon's tens of millions of customers and products, they developed their own: item-to-item collaborative filtering. This produces recommendations in real time, scales to massive data sets and generates high-quality recommendations.

Rather than matching the user to similar customers, Amazon's filtering matches each of the user's purchased and rated items to similar items, then combines those similar items into a recommendation list. It relies on acquiring and then crunching massive data volumes; every page view, every search, every purchase.

Recommendation algorithms are best known for their use on e-commerce websites, where they use input about a customer's interests to generate a list of recommended items. Many applications use only the items that customers purchase and explicitly rate to represent their interests, but they can also use other attributes, including items viewed, demographic data, subject interests and favourite artists.

For large retailers like Amazon, a good recommendation algorithm is scalable over very large customer numbers and product catalogues, requires only sub-second processing time to generate online recommendations, is able to react immediately to changes in a user's data, and makes compelling recommendations for all users regardless of the number of purchases and ratings.

Targeting Marketing

Amazon uses its personalisation system as a marketing tool in many email campaigns and on most of its websites' pages, including the high-traffic Amazon.com homepage. The online store changes radically based on customer interests; for example, showing programming titles to a software engineer and baby toys to a new mother. The click-through and conversion rates, two important measures of web-based and email advertising effectiveness, vastly exceed those of untargeted content such as banner advertisements and top-seller lists.

Clicking on the 'Your Recommendations' link leads customers to an area where they can filter their recommendations by product line and subject area, rate the recommended products, rate their previous purchases and see why items are recommended. Using this feature, customers can sort recommendations and add their own product ratings. Further, the shopping cart recommendations offer customers product suggestions based on the items in their shopping cart: 'Customers who bought items in your shopping cart also bought . . .' The feature is similar to the impulse items in a supermarket checkout line, but Amazon's impulse items are customised to each customer.

Amazon also used the personalisation enabled through technology to reach out to a group that was difficult to reach, which Bezos originally called 'the hard middle'. Bezos's view was that it was easy to reach 10 people (you called them on the phone), or the 10 million people who bought the most popular products (you placed a superbowl advert), but more difficult to reach those in between. The search facilities in the search engine and on the Amazon site, together with its product recommendation features, meant that Amazon could connect its products with the interests of these people.

Amazon marketing works to direct customers to their websites primarily through a number of online marketing channels, such as their Associates programme, sponsored search, portal advertising, email campaigns and other initiatives. However, Amazon believe that their most effective marketing is a direct consequence of their focus on continuously improving the customer experience; creating word-of-mouth promotion that helps to acquire new customers and encourage repeat customer visits.

In Summary

Customer service is at the very core of Amazon's mission, to become Earth's most customer-centric company. Their focus on personalising the customer's online experience has translated to an excellence in service for which Amazon is consistently ranked top in customer satisfaction.

Personalisation is becoming more deeply entrenched on the Web and beginning to move beyond the consumer, with business-to-business techniques evolving. Amazon believes that while e-commerce businesses have the easiest vehicles for personalisation, the technology's increased conversion rates, as compared with traditional broad-scale approaches, will also make it compelling for offline retailers' customer communications.

ANALYSIS

In view of the fog of confusion around such a simple subject, it is sensible to undertake pragmatic clear research and analysis before embarking on a service quality strategy. There are a number of tools used by service marketing specialists in leading technology companies. They include the following.

Understanding Customers' Expectations

Many different researchers have demonstrated that, across a range of industries, buyers' views of quality of service depend on their expectations. If the performance of the service meets their expectations then they regard it as good service. If it does not, it is bad service; and the degree of

The Gap model

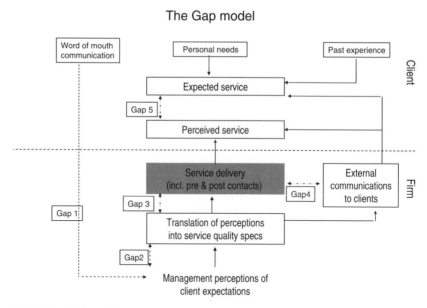

Figure 10.2   The GAP model
Reprinted with permission from *The Journal of Marketing*, published by the American Marketing Association, Parasuraman, Zeithaml and Berry, Volume 49, Fall 1985.

negative reaction depends on the degree of difference between the two. Their research has also shown that different degrees of satisfaction occur in different conditions and in different product or service groups. For example, there appear to be different expectation/satisfaction dynamics between products and services and also between 'high-involvement' and 'low-involvement' products.

Sri Parasuraman and his colleagues (Parasuraman *et al*, 1985) have been responsible for creating one of the most significant diagnostic tools: the well-regarded 'gap model' (see Figure 10.2 and the Tools and Techniques appendix). This takes a perspective on a firm in its environment, the management process, the expectations of its customers and the effect upon their thinking of messages from the firm. It identifies five gaps to which management should address attention:

1. **The 'management perception' gap.** This focuses upon any difference between management's views and those of customers.
2. **The 'quality specification' gap.** This exists when quality standards, strategy or plans do not reflect management objectives or views.
3. **The 'service delivery' gap.** This exists if there is a difference between quality strategy or plans and the firm's delivery to customers.
4. **The 'market communications' gap.** This tracks any difference between marketing communications and service delivery.
5. **The 'perceived service' gap.** This exists if there is a difference between the service delivery perceived or experienced by customers and their expectations.

The model has been tested in a number of research and real business projects. It has held up well and yields useful, practical insights for leadership teams about the customer service

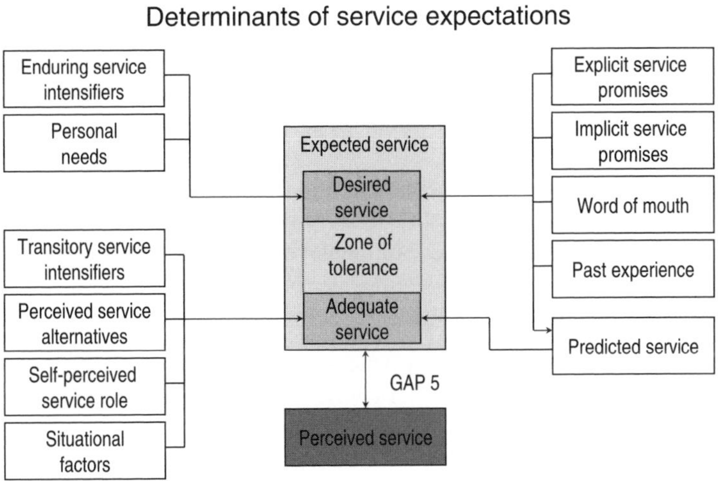

Determinants of service expectations

Figure 10.3 Determinants of expectation
Reproduced with permission from Sage Publications, Zeithaml, Berry and Parasuraman, Winter 1993.

performance of their firm. It helps managers understand both how and where they need to improve.

Professor Zeithaml (Zeithaml and Bitner, 2003) undertook research to understand what influences expectations in much greater detail. Her model, in Figure 10.3, is a good framework for research projects trying to understand this issue in detail.

Dimensions of Service Quality

When trying to plan service improvements, particularly in a large firm, it is possible to break the service offer into recognisable components, or 'features', which can be individually improved. By doing so, suppliers can understand which aspect of their service is deficient when compared with customers' expectations or competitor performance. They can then rank these in importance and calculate the real cost of any improvement. Service improvement then becomes a practical and manageable programme, rather than a vague wish list.

Service marketing academics have made a very valuable contribution in this field. For example, Professor Christian Grönroos suggests (Grönroos, 2003) that services have both a 'technical or outcome' dimension and a 'functional or process outcome'. Both affect customer perception of value and need to be studied. Harte and Dale (1995) produced guidelines based on six pieces of research in service firms. The attributes required by most customers were:

- Timeliness: the service is provided promptly.
- Empathy: the organisation understands the customers' needs.
- Assurance: technical correctness of the work.
- Fees: providing value for money.
- Tangibles: providing evidence that the work is performed correctly.
- Reliability: the firm does what it says it will.

On the other hand, Professor Leonard Berry and his colleagues (Berry, 1995) have demonstrated in their research that there are five general criteria by which buyers judge service quality. Not dissimilar to Harte and Dale's guidelines, these appear to be largely constant across any sector and can form the basis of discussions with customers on service quality. They are:

- Reliability: the ability to perform the promised service dependably and accurately. This is thought to be the most important of all.
- Tangibles: the appearance of physical manifestations of the service.
- Responsiveness: the willingness to help with prompt attention.
- Assurance: the knowledge and courtesy of employees plus their ability to convey trust and confidence.
- Empathy: caring and individual attention.

Within this group of criteria, reliability (in the sense of doing what is promised) is by far the most important. They found that these criteria were consistent across many service industries with only a few minor variations.

Professor Parasuraman (Parasuraman *et al.*, 1991) also developed a detailed technique called 'zones of tolerance'. This suggests that people's expectations have two levels: a desired level and an adequate level. If the service delivered falls between these two standards then it is tolerated by the customer; properly used, this very practical research technique allows service marketers to gain very detailed information for planning and for flexibility of investment in service quality.

Tolerance levels vary by customer and by service attribute. In fact it seems that customers are less flexible on technical outcomes of service than customer service processes. The concept can be used as the basis of research projects that can be designed to understand first, the customer's ideal preferences, second, acceptable levels of service and third, unacceptable levels of service. This will help management to understand where improvements need to be made.

Figure 10.4, for example, represents detailed research of a fictitious maintenance service. The analysis has revealed the customers' most important features, of which 'speed of response' is the highest. Their ideal requirement for each feature has been identified and contrasted against both the supplier's performance and their nearest competitor. Some might want to

A fictional example of graded service dimensions

Hierarchy of Feature (Prioritised)	Requirement in Hours	Our delivery in Hours	Competitors' Delivery in Hours
1. Speed of response	4.0	8.0	24.0
2. Speed of diagnosis	0.5	2.0	0.5
3. Time to clear problem	2.0	6.0	2.0

Figure 10.4   Detailed analysis of competitive service features

invest to improve feature 'response'; perhaps moving it nearer to the customers' ideal or even investing to exceed their expectations. Others might settle for remaining ahead of competitors in this important area and seek to improve 'time to clear' problems; reflecting their performance in front of the customers once having arrived. Different service marketers will make different choices, but the analysis allows them to understand these choices in depth and use them as a basis of calculating the investment needed to take any chosen course of action.

Understanding Lifetime Value and Customer Profitability

This important concept changes the perspective on a customer and prompts investment in customer care. Although he didn't give it this name, one of the most practical applications of lifetime value was demonstrated by Lord Marcus Sieff (Sieff, 1986), chairman of Marks & Spencer in the 1970s. He said that this was behind the commitment of the retailer to service quality; pointing out that, if a young couple married and moved into an area near one of his stores, they might spend a lot of money over, say, 10 years. He didn't want to cut off this revenue stream in an argument over one jumper.

The concept of lifetime value of customers suggests that firms know four things about a customer:

- The total revenue from all work done in any given year.
- The costs of service to those customers, including proposal and prospecting costs.
- The anticipated duration of the relationship of the customer to the firm.
- The profit in any given year and the total profit.

This analysis provides the total value or equity of customers and can change perspectives. It may be, for example, that the cost service to major accounts is high, making them relatively unprofitable, and needs to be adjusted. On the other hand, a simple analysis of acquisition costs in comparison to service costs has caused many firms to put more emphasis on quality of service to existing buyers rather than gaining new ones.

The Contact Audit

The contact audit (see the Tools and Techniques appendix) is borrowed from the direct marketing industry. It identifies all the points of contact a customer has with a firm and reaches a judgement on the level of influence they have on customers' perception of service quality. The technique allows the leadership to identify all the 'moments of truth'; those times when customers interface with the firm and form judgements of capability which influence reputation and, as a consequence, future revenue.

SERVICE STRATEGY

Like all other strategic issues, quality of service is so important to service companies that an explicit service strategy should be developed. Leaders should use the relevant concepts, analysis and techniques to construct a specific service strategy. This, once again, ought to be produced in a manner and style suitable for the firm. In some cases a detailed, written document; in others, a general statement of strategic intent, emphasising one or two important priorities. The strategy itself should be based on market and customer insights and address several key issues.

Format for a service strategy

- Strategic context
- Summary of customer and market insights (including customers' views of service attributes contrasted with competitor services)
- Service style
- Improvement or change programmes
- Marketing communications on service issues
- Resources
- Measurement
- Service recovery policy and practice

Figure 10.5   Elements of service strategy

The first is the style and ambience of the service, which should reflect the positioning of the firm. The service of a market leader should be substantially different from that of a niche supplier trying to command attention on the basis of differentiation. Positioning maps (see the Tools and Techniques appendix) can be used to clarify this.

The second issue is strategic impact; understanding whether the market is at one of those moments when quality of service is strategically significant and can enable a supplier to gain real competitive advantage. If it is anticipated that a specific service strategy will revolutionise its industry (as per IBM and BA, each in their time) then it ought to be a major emphasis of the marketing strategy, demanding priority and serious investment, shouldering aside all other competing plans. In these circumstances, service quality becomes the marketing strategy.

Continuous improvement or change programmes should be specifically identified and specified in the strategy. If not, the reputation for quality will falter. A number of companies which have had a reputation for service quality, or have been used as case studies, have since suffered real business difficulty. IBM and BA have had, for instance, times in their recent history when their survival has been threatened and service quality seriously disrupted. This gives lie to the suggestions by many customer care gurus that service quality, 'customer delight' and loyalty are the only issues to focus on. It also suggests that the immense gains from their exploitation of the strategic significance of their service strategy were dissipated by later management teams who failed to institutionalise service improvements. Detailed research into competitive service performance using, say, conjoint research (see the Tools and Techniques appendix) or zones of tolerance should be regularly used by service marketers to prioritise and specify systematic improvements in the operations of the firm.

Another aspect of the service strategy should be the adjustment of the service features in order to meet changed buyer values. The firm should know those values which are important to key customers and how they match its competencies. It should also know customer views of competitor services and their service strengths or weaknesses. These should be integrated into a view of how technical performance and customer service features need to be changed, if at all.

The strategy also ought to address resources, measurement and service recovery policy. A typical format is shown in Figure 10.5.

PLANNING SERVICE QUALITY

In the midst of the market planning process, service issues need to be planned and resourced. Marketers need, in the light of market analysis and strategic imperatives, to specify the service quality plan. It might be implemented or even written by others, but it needs to be given direction by service marketing skills and market insight. This is a management process, like

Figure 10.6   Leonard Berry's model of service planning components
Reprinted with the permission of The Free Press, a Division of Simon & Schuster, Inc from *Marketing Service: Competing Through Quality* by Berry. Copyright © 1991 by The Free Press. All rights reserved.

any other, which ensures that service improvements are integrated into changes in the whole business. It ensures that the firm's service evolves with changing market needs. It is also a cyclical activity, progressively responding to data on customer views and market needs. As in other functions, planning might be fast, unstructured and intuitive or it might be detailed, complex and written into an elaborate document.

There are different constructs that leaders can use for this. For example, Figure 10.6 shows Professor Leonard Berry's 'framework for great service' (Berry, 1995). It is based on analysis of different service companies and allows leaders to think through the logical components of a specific service strategy and how it might be implemented in the fabric of the firm.

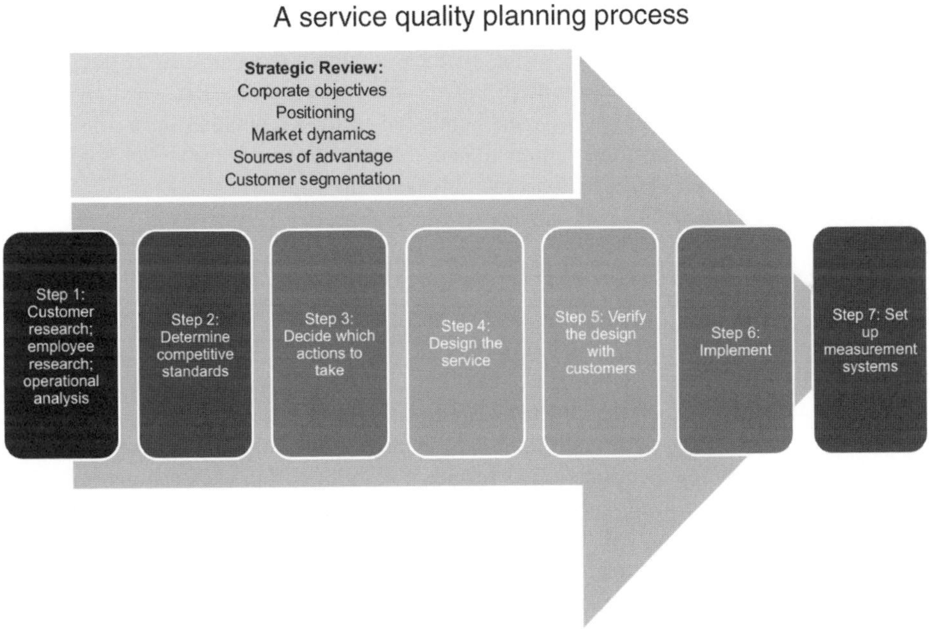

A service quality planning process

Strategic Review:
Corporate objectives
Positioning
Market dynamics
Sources of advantage
Customer segmentation

Step 1: Customer research; employee research; operational analysis

Step 2: Determine competitive standards

Step 3: Decide which actions to take

Step 4: Design the service

Step 5: Verify the design with customers

Step 6: Implement

Step 7: Set up measurement systems

Figure 10.7   A service quality planning process

Figure 10.7 is an alternative structure, covering similar steps but trying to link strategic perspective to the practicalities of service improvement (Stone and Young, 1994).

SERVICE RECOVERY

In even the best run companies, with the most thoughtful management and well-crafted service strategy, things go wrong. The manner and style with which these service aberrations occur can, though, be part of the charm and attraction of an elegantly positioned service. Research even suggests that, if service recovery is properly handled, buyers feel more positive towards a supplier after a difficulty than before. To quote specialists who have studied this area (Hart *et al.*, 1990):

> Service companies must become gymnasts, able to regain their balance instantly after a slip-up and continue their routines. Such grace is earned by focusing on the goal of customer satisfaction, adopting a customer-focused attitude, and cultivating the special skills necessary to recover ... recovery is fundamental to service excellence and should therefore be regarded as an integral part of a service company's strategy. (Reprinted with permission from *Harvard Business Review*, Hart, Heskett and Sasser, July/Aug 1990)

So, service recovery should be a clear and well-managed aspect of service management. A number of general principles about managing service aberrations have evolved which are sensible and practical. They need to be put in place, institutionalised across the firm, to avoid damage to reputation and to future growth.

People are often as anxious about making complaints as much as they are distressed by the service aberration itself. So, the supplier should, first, ensure that it is easy and unembarrassing to complain. There should be mechanisms to ensure that someone immediately apologises and sets about putting the problem straight. They should then find some adequate compensation for the distress and inconvenience caused to the customer. Finally, measures should be put in place to ensure that the processes of the business stop any reoccurrence of the problem. As shown in the case study, these principles were put into action by BA when Heathrow's Terminal 5 opened to chaos in May 2008, damaging the airline's reputation. As a consequence, the airline recovered much customer goodwill and, through an effective advertising campaign, changed public perceptions back to 'Terminal 5 is working'.

Professor Hart and his colleagues suggest service companies should:

> ... measure the costs of effective service recovery, break customer silence and listen closely for complaints, anticipate needs for recovery, act fast, train employees, empower the front line and close the customer feedback loop.

The main priority, though, is to make sure that leaders are aware of service aberrations and maintain a clear perspective on them. Both individuals and the culture of a firm can remain in denial about service difficulties. It is possible for the leadership, by their attitude, to inadvertently suppress the identification of service aberrations. If they create an antagonistic climate, customer service staff may not be prepared to take the risk of exposing problems and even written complaints will be suppressed. The analysis and measurement processes described in this chapter should help to ensure that general trends in quality aberrations are identified and rectified. However, a blameless and open culture is also important. Although leaders should encourage the very highest standards of technical work and customer care, they need to acknowledge that no organisation and no individual are perfect. They need to encourage an honest approach to service aberrations.

BRITISH AIRWAYS ENSURES T5 DISASTER IS NOT TERMINAL

British Airways (BA) is one of the world's largest international airlines, carrying more than 33 million passengers worldwide in the 12 months to 31 March 2009. Together with its partners, BA flies to more than 300 destinations worldwide.

The airline's two main operating bases are London's two principal airports, Heathrow (one of the world's biggest international airports) and Gatwick. In 2008/9, it earned nearly £9 billion in revenue, up 2.7% on the previous year. Passenger traffic accounted for 87.1% of this revenue, while 7.5% came from cargo and 5.4% from other activities. At the end of March 2009, it had 245 aircraft in service.

As one of the world's longest-established airlines, it has always been regarded as an industry leader. The company's commitment to customer service and its strong brand were perhaps two reasons why it was able to ride out the storm that occurred around the chaotic opening of Heathrow's Terminal 5 in 2008, and why it has since been able to recover its position and the goodwill of its customers.

Disaster Strikes as T5 Opens for Business

2008 was a difficult year for everyone, not least for airlines. In addition to the credit crisis that spread quickly from the USA, the value of sterling plunged, consumer confidence collapsed and companies like BA faced record fuel prices. Many went bust, including the UK's third-largest travel operator, XL Airways, leaving 90,000 holidaymakers stranded abroad. The disastrous opening of Heathrow's Terminal 5 came in sharp contrast to an industry-defying, successful year for BA, as the airline reported great financial results with profits up 45% and operating margins up to an industry-leading 10%.

The opening of Terminal 5 was the result of 20 years of planning, the largest construction project in Europe and over £4 billion of investment. BA saw it as a chance to redefine air travel, 'replacing the queues, crowds and stress with space, light and calm'. It was the largest freestanding building in the UK, designed with both customers and sustainability in mind, and the promise that check-in would take less than five minutes.

A press release, embargoed to coincide with the touchdown of the first flight at the terminal, BA026 from Hong Kong, announced the opening of the terminal with its headline, 'Celebrations as Terminal 5 opens for business'. And yet, rather than the celebration envisaged, the opening gave way to one service failure after another, until images from Terminal 5 were transmitted around the world showing the chaos ensuing and damaging the reputation of BA.

Passengers became frustrated by a lack of communication around the service failures. Baggage systems crashed and flights were cancelled and yet only two of the 26 information desks were operational. Some passengers arrived at the airport to be told their flights were delayed, while others were told their flight was cancelled when it was actually scheduled to take off.

By mid-afternoon on 27th May BA had cancelled 33 of the 534 services it planned to operate from the terminal. In the early evening, passengers were told that no one would be allowed to fly with their hold baggage for the rest of the day. At one stage during the day, 28,000 bags were separated from their owners and 19,000 were sent to Milan to be sorted.

As a result, five days after the crisis, 250 flights had been cancelled and there was a backlog of 15,000 bags. Thousands of passengers had their travel plans disrupted, with some stranded for days, and BA were unable to find them hotel rooms.

A number of factors contributed to the service failures on the day, including:

- Baggage handlers and staff not being able to get into BAA car parks, consequently arriving late for work.
- Poor signage making it hard for BA staff to navigate the building.
- Terminal 5's computer system not recognising staff identification cards, meaning that doors that should have opened remained locked.
- 17 out of the 18 terminal lifts jamming.
- The transit system moving passengers from the main terminal to Terminal 5 breaking down.
- A programming error preventing staff from logging onto the baggage system.
- The baggage handling system crashing.
- Staff with already low morale becoming stressed and unhelpful as the problems unfolded.

Alistair Carmichael, a UK member of parliament, described the failure as 'a national disgrace, a national humiliation'. More seriously for BA, with passenger choice only set to increase, particularly on transatlantic routes in the future, the risk of losing customers forever as a result of the fiasco was high.

Work Begins to Recover Both Service and Reputation

In sharp contrast to the celebratory press release issued the day before, BA CEO Willie Walsh apologised in a press release the following day, saying, 'We disappointed many people and I apologise sincerely. I take responsibility for what happened. The buck stops with me.'

Despite many of the service failures having their basis in issues to do with the terminal operator, BAA, throughout the crisis Walsh did not attempt to blame BAA, saying instead '...there were a lot of angry customers out there. They buy their tickets from BA, not BAA, so we had to deal with it.'

BA's Global CIO, Paul Coby, added, 'Everyone in BA recognized that it does not matter to our customers where the problems were – they paid good money, bought tickets from BA and deserved to have a great travel experience, which we did not deliver. Lessons had to be learned . . . We accepted responsibility for what happened to our customers because that was, and remains, our primary accountability.'

Later, when facing a Parliamentary Committee, Walsh acknowledged that risks had been sanctioned by him and that, with hindsight, the opening should have been delayed to allow for more training and familiarisation for BA staff. He also announced that he would forgo his £700,000 bonus.

Walsh apologised yet again in the company's 2007/8 annual report, saying, 'I want to put on record yet again my deep regret for the inconvenience and frustration we caused to customers in the days after Terminal 5 opened. Our people worked tirelessly to put things right and, thanks to this tremendous effort, customers will begin to see this tremendous new facility in its true light.'

Work began to put right the operational issues that had caused the failures on the day.

The airline gave away thousands of free flights in a desperate attempt to retain passengers in the wake of the disaster. Walsh and chairman Martin Broughton both set aside time to personally review a significant number of customer letters and emails so that they could drive the improvements needed to service.

Once the service failures were corrected, there was naturally a lag in public perception, with many people still assuming that Terminal 5 was a failure. To that end, the airline launched an advertising campaign to announce 'Terminal 5 is working', backed by a series of metrics, released daily to prove the claim. The campaign initially ran across a seven-week period in July and August 2008, using a wide range of digital formats. Key metrics from the terminal, such as the percentage of flights departing on time, or the time taken to get through check-in, were sent to media owners at the end of the day before the ads were due to run. The emotions of customers were also drawn on, with roving reporters asking customers to say how they felt about Terminal 5, from a choice of 'happy, excited, calm, impressed, frustrated, disappointed or angry'. Their emotions featured on a website, promoted through web and press advertising. Then-head of BA global marketing communications Katherine Whitton explained, 'As a premium service-driven organization, we understand that we need to deliver not only the rational benefits of flying, but also engage our customers emotionally.'

The Hard Work Pays Off

The 'Terminal 5 is working' advertising campaign resulted in 62% of respondents in a Millward Brown survey agreeing with key statements such as 'most flights arrive on time' and 58% agreeing that the ads made the brand seem more appealing. Readers of the UK's *Daily Telegraph* went so far as to vote the opening of Heathrow's Terminal 5 one of the travel highlights of the year, having 'found its feet'. At the end of 2008, 80% of readers polled felt positive about T5.

Some 21 million passengers passed through Terminal 5 in its first year. Satisfaction levels among them have risen steadily through the year to 76%, as over 82% of flights departed the terminal within 15 minutes of their scheduled time, and BA achieved more than 99% regularity.

BA got through its difficult times, in one of the worst trading environments for the airline industry, and it is now focused on the long-term vision for its business: to be the world's leading global premium airline. The rolling three-year business plan developed by the airline set out its agenda for 2008/9, including its aim to build on Terminal 5's strengths. The terminal has transformed the company's operational performance and customer service, with record-breaking punctuality and customer recommendation scores.

Meanwhile, fascination for Terminal 5 itself continues, and in the summer of 2009 the writer and television presenter Alain de Botton was contacted by Grupo Ferrovial, the Spanish multinational that owns the airport. It commissioned him to become the airport's first 'writer in residence' by spending a week working from a desk in the departures hall. The result, titled 'A Week at the Airport: A Heathrow Diary', was published at the end of September 2009, bringing the terminal to life with more observations and stories of those who pass through it.

SATISFACTION MEASUREMENT

There are several types of measures that are important to understand and have in place in order to manage ongoing customer satisfaction. These are:

- **Event-driven feedback.** Sometimes called 'transaction surveys', because they follow a specific transaction between the supplier and its customer. This is feedback from the buyer in response to a specific event or project (a completed consultancy project, for example, or the installation of a new home broadband service). These feedback mechanisms can either be through some form of questionnaire or response to a telephone survey. Their purpose is to capture the customers' views of their specific experience while it is fresh in their mind. It is therefore important to gather customers' responses as soon after the transaction as possible.

  The surveys need to be carefully designed using experienced researchers. If not, they are likely to reflect the wrong views and mislead the company. They need to be carefully administered and sent to the correct people. (More suppliers than would admit send these to anonymous departments or invoicing addresses.) The firm should keep careful records and undertake trend analysis as an input to its strategic direction.

- **Generic perception studies.** These are research projects undertaken on a regular basis in order to understand the general view that different groups of buyers have about a supplier. Frequency of surveys varies according to volume (some might be monthly, some might be annual). Quite often suppliers will carry out a number of surveys each month and then have either a quarterly or an annual summary of trends. Techniques for conducting this research vary, although many suppliers have used conjoint research (see the Tools and Techniques appendix) because it yields powerful insights into changing customer views of different aspects of service.

  These are very different in style, purpose and design to event-driven research. They are frequently conducted by independent research companies using independent sampling techniques. The aim is to get a generic view of the firm's service performance which can then be compared to the trends in event-driven measures.

- **Internal measures.** Firms should establish internal measures of service quality (percentage of projects completed to predicted time scales, for instance). These should be prioritised around customer need and the leadership's objectives. They should be communicated to all, set as individual objectives for each executive and used as a basis for reward. However, they should also be compared with external measures to identify any gaps. If, for instance, a firm's internal measures show excellent performance in an area that customers criticise, there are two possible solutions. Either internal results are distorted and need adjustment or customer perception needs to be changed through marketing communication.

- **Explorative research.** Many service suppliers have used explorative techniques as a basis for fundamental design of the service. Some, for example, use observational techniques to watch the behaviour of buyers as they interface with the service.

Stone and Young (1994) propose the method shown in Figure 10.8 of drawing these elements together as an ongoing management system of satisfaction measurement.

Effective measurement types

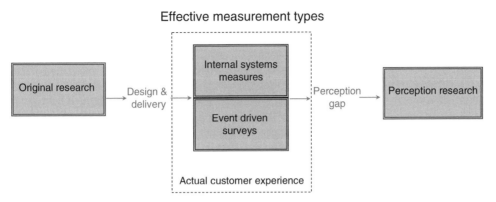

Figure 10.8   A structure for satisfaction measurement

SUMMARY

Customer care is an aspect of service businesses that contributes to growth in revenue and reputation. It has strategic implications for business growth which can, in the right market circumstances, be dramatic. It can be analysed and managed as effectively as any other part of the marketing mix. It is too important for the services firm to approach in an ad hoc way. It needs a specific strategy based on good analysis but, as much as any other field, it needs to have brilliant basics installed in the firm's operations. Every element involved in the customer service process, from the quality of service offered to understanding customer needs and expectations, ought to be carefully crafted and built into the fabric of the organisation. Over the years a number of tools and techniques have emerged which can help service marketers design the customer service processes their companies need. Once set up, they are the foundation of fresh, lively and, even, magical services that give pleasure and pleasant surprises to customers, encouraging repeat purchase.

11

Service on the World Stage

INTRODUCTION

There are a number of different models of international business that employ different growth and marketing strategies. This is important because marketing strategy, which gives shape to international campaigns, is directly related to the approach taken by the owners of the company. Yet, whatever approaches the firm decides to take, there are different organisational and cultural challenges that need to be addressed, not least technical and linguistic differences. For example, will the firm service customers in other countries from the home territory or set up an office in chosen countries? What degree of local expertise will be needed? Does it have a clear international perspective on each market? Moreover, to position a truly enticing service on the world stage, there are a host of human issues that a supplier must address. In fact, some of the most profound and valuable strategic insights into international markets arise from one word: 'culture' (how human beings behave and react within their society). This chapter argues that insight into this aspect of internationalisation is critical to the successful creation and marketing of services; and highlights a number of tools and approaches that help to address it.

THE CHALLENGE OF WORKING ACROSS BORDERS

In the modern world it's hard for any firm to avoid international work, even if it wanted to. Almost any customer, large or small, has international needs, now that virtually all products and services are available on the Internet. So, service firms need to think about how they approach requests for work across geographic borders. Some services (like packaged travel or technical help desks) can be offered from remote locations; others (like consultancy or on-site maintenance) require suppliers to visit in order to meet demand in foreign countries. Some suppliers might have relationships with trusted local representatives to whom they refer work. While others get to the point where they extend their firm into another country, setting up local operations there to meet frequent demand.

Unfortunately, there are a number of complicating factors for service firms moving on to the international scene. The first and most obvious are the differences in regulatory, technical and legal requirements in cross-border work. Although much has been done by many professional bodies to standardise practices across the world (attempts to harmonise international technical standards, for example), there can be large and significant differences in national law, regulation or technical specification. Some firms, for instance, are surprised about the restrictions placed upon the use of customer data by some European countries. So executives need to take extreme care to ensure they have local knowledge if their work is going to be technically correct.

International operations complicate much more than technical content though. They affect almost every aspect of approach to market – whether strategic, tactical or managerial. While another country may look misleadingly similar to home territory, there are underlying attitudes to life and business that mean different criteria are used to make judgements. These differences in lifestyle, values, behaviours and business practices between different geographies and cultures are difficult to identify, digest and manage. While stereotypes and clichés are often exaggerated, there are genuine differences in taste and approach, which can cause real problems.

Language is an obvious example. For example, when Coca-Cola first stepped seriously onto the world stage (in the 1920s) it was astounded to find that the Chinese characters which most closely reproduced the sound 'Coca-Cola' translated roughly to 'bite the wax tadpole' (Pendergrast, 1994), whereas Pepsi's "come alive with the Pepsi generation", translated in Taiwan as "Pepsi will bring your ancestors back from the dead". Although English is generally accepted to be the international language of business, different words have different meanings and the subtlety of intent can be lost in translation. As services are intensely personal and affected by atmosphere, intonation or body language, mistakes can be made in the use of English alone. However, they are even more likely when a supplier is working, through translation, in a customer's local language.

Even worse, culture and upbringing give different meanings to words which can dramatically affect understanding. Disastrous interruptions to service performance can be caused through the ignorance of cultural differences alone. For example, while the English word 'lunch' refers to a meal eaten in the middle of the day, what this signifies varies enormously in European cultures. In Sweden, it is generally a meal at 11:30 am which comprises fish, potatoes and salad accompanied by a 'lättöl' – a light beer. To the English, though, it means a snack, like sandwiches, eaten soon after midday, often at the desk, or possibly in the 'pub '. To the French, particularly in Paris, it is still often a three-course meal at 1:00 pm with red wine and, to the Spanish, a big meal at 3:00 pm. The cultural context gives different meanings to the word, complicating life. If an inexperienced Swedish business person says to a Spanish customer: 'we'll discuss this over lunch', obvious complications arise.

This cultural heritage has an enormous effect on business language and understanding because words are given very different meanings by their cultural context. If technical and linguistic differences can complicate international work, then other factors, explored in this chapter, can cause enormous difficulty if not properly managed. More than one service firm, large and small, has been damaged by ill-conceived international strategies. Leaders should think through the strategic and managerial aspects of international market approaches with real care.

INTERNATIONAL STRATEGIC CONSIDERATIONS

Cultural Differences and their Effect on Customer Expectations

Different cultures create different expectations and attitudes in customers. People in different parts of the world have different languages, mindsets, preferences, attitudes and prejudices. This 'collective programming of the mind' occurs during childhood and adolescence. It is often so subtle that people themselves can be unaware of their own cultural perspectives and biases. Yet it has a profound effect on their attitude to life and business. It also affects their demand for different services. Some cultures, for instance, value strategy and consultancy more than others. The Nordic countries of Europe tend to place emphasis on discussion and debate that leads to consensus. This, in turn, tends to make them value ideas and pay for advice more readily than, say, the British. So, the demand for a consultancy service, and the price people are prepared to pay for it, is different in different cultures of the world.

What is not often understood is the vast difference that cultural heritage makes to quality expectations. 'Quality' to a New Yorker, for example, can mean choice and speed of delivery while to the English it means sincerity and to Germans, thoroughness or accuracy. People approach every service with a set of expectations, which are both emotional and practical. They will only think that a service is good quality if both the technical content and the service performance meet those expectations. Yet they will not make any effort to explain these expectations to the service supplier and, in fact, are often unaware of them themselves. However, they are ruthless in the way they criticise and judge a firm if it does not meet these articulated and unarticulated expectations.

It seems, then, that a desire to give excellent international service forces a company to consider the customisation of their service proposition to meet cultural expectations. This has serious implications. It may be that, in one culture, the delivery of service is a performance designed to serve the customer and take care of all their needs. In another, however, excellent service may be self-service. In yet another, it may be delivered through advanced technology. So, if a service supplier is designing a service that will be received in Scandinavia, Germany, Spain and France, it needs to research the expectations of that service in each of those cultures. The outlook, attitude and assumptions will be different in each culture and the service must be adjusted to suit those different tastes.

Even the ubiquitous American fast-food service, McDonald's, has to be customised in different cultures of the world to deliver satisfaction. Although it is thought to be the same the world over, the franchise dynamic of the firm's international operations has not only ensured that the service is varied around the world but also that the fast-food menu itself is customised to some degree. As John Travolta's character pointed out in the film *Pulp Fiction*, their products in Paris (and other countries) are different from those in the USA. But there are other aspects of the service that have been adjusted over the years. For instance, in the early years of its expansion into the UK, their employees were trained to say 'Have a nice day'. In America this is a polite, unremarkable phrase, which is part of day-to-day life. However, this was not natural to employees in the UK and their uncomfortable tone of voice implied insincerity. Unfortunately, as one of the key service assumptions in the UK is sincerity, this caused offence and affected trade. British people will tolerate many other aspects of poor service if the service supplier is sincere or polite. Eventually, McDonald's publicly apologised to the country through PR and mass advertising media; and dropped the phrase.

It is often wrongly assumed that, in business-to-business markets, attitudes and expectations are very similar across the world and, as a result, cultural differences are less important. However, there is much evidence that differences occur and affect all aspects of business. One of the most useful and extensive studies to demonstrate this was undertaken by Gert Hofstede (see Hofstede, 1984 and the Tools and Techniques appendix). He identified five significant cultural variations and devised matrices by which international business affairs can be planned and managed. Interestingly, his original research was conducted in one company (IBM) that perceived itself as a successful market leader with one consistent global culture. So, even in the perceived monoculture of one large international business, cultural variations have a large influence. This has implications for many aspects of business-to-business practice.

In fact, there is very little business practice that is not influenced by cultural diversity and, as a result, impacts the markets of services suppliers. For instance, different cultures respond differently to colour and this affects any work on international design, advertising and brand. Similarly, there are indications that business innovation occurs differently in different cultures so that the styling of propositions needs to be adapted. Finally, there appears to be a difference in the adoption of new ideas, affecting the pace at which new services are launched and accepted by the market in different parts of the world.

Globalisation

In the latter half of the 20th century, it became common to talk about 'globalisation'. The crux of the globalisation argument is that technology is a powerful force, driving the world towards convergence and commonality. It prompts the free flow of capital and enables big worldwide markets to evolve. It has allowed, for example, suppliers in markets as different as: professional services, vacations, information and commodities (to name but a few) to access much larger groups of customers. A revolution in communications technologies has allowed media like film, music and the Internet to create aspiration amongst massive worldwide audiences; prompting, in turn, the evolution of gigantic markets. In parallel, political initiatives like the GAAT rounds and the emergence of the European Union have added to the creation of common standards in international markets. This has had implications for the capital flows which enable businesses to grow and for the political or legislative frameworks in which they operate. More than one politician and many business leaders have been heard to talk about the inevitability of globalisation; that protectionism cannot be resisted, Canute-like, in this fast-harmonising international community.

So, the globalisation of markets has become a familiar basis for strategy in the boardroom of businesses throughout the world. It is particularly familiar to technology companies as firms like Microsoft, IBM, Oracle, Nokia and Ericsson have become famous brands across the world with offices and factories on almost every continent. So, marketers considering the launch and promotion of a technology-based service on the world stage must consider how global their offer will be. Is it to be a ubiquitous, worldwide entity which is the same wherever it is used and communicated in the same way with the same methods and message? Many voices would argue that it should.

In the early 1980s, for example, Theodore Levitt published an influential article in the *Harvard Business Review* (Levitt, 1983) on the globalisation of markets. He pointed out that people around the world have similar aspirations for their life and family. He argued that this was an opportunity for international firms to save costs. They should be able to create common processes in brand management, advertising and distribution; improving profit through a new international strategy: global marketing. The global approach could be owned by no culture but would serve all equally well with a common offer. While demonstrating his sophisticated understanding of cultural differences, Levitt argued that these were declining in influence in the face of standardised products, which derive from commonality of preference. In fact, he suggested that:

> ...different cultural preferences, national tastes and standards ... are vestiges of the past. Some die and some become global propositions (Italian food, American rock music, French wine etc).

In this view, multinationals which accommodate national preferences in their approach are thoughtlessly accommodating poor marketing which 'means giving the customer what he says he wants'. A global strategy would be to create a cheap, reliable ubiquitous product, which would appeal to a universal need and would create markets for that proposition by investment in marketing communication or brand advertising. Levitt suggested that marketers should set out to convince different audiences to subsume their cultural tastes; that a ubiquitous global offer is what they want. Some well-established global offers, such as Coca-Cola, were cited as leaders in this approach. (The Economist said, for example, "the number of truly global brands are few indeed: the only indisputable mass-market one is Coca-Cola").

Unfortunately, we have been here before. Proponents of globalisation seem unaware that the world was once very global (enabling the direct marketing of numerous offers and free flow of capital) but drew back from it. In 1920, for instance, Maynard Keynes (Keynes, 1920) said of his life in London before the First World War:

> and then in London, a man could order by his telephone, sipping his morning tea in bed, the various products of the earth, in various quantities that he might see fit, and reasonably expect their early delivery upon his doorstep.

In the decades prior to 1910, the world saw an expansion of enabling new technology (telegraphy, steam ships and the railway), which created worldwide networks. Historian Eric Hobsbawm (Hobsbawm, 1999) said of it:

> An all embracing world system of virtually unrestricted flows of capital, labour and goods never actually existed, but between 1860 and 1875 something not too far removed from it came into being.

This led many to think that there would be an ever-closer integration of political structures and greater freedom of trade. H.G. Wells (the scientist and novelist who predicted the Internet nearly a century before its advent) even muted the idea that a 'world government' was inevitable. Yet the world community fractured. Around 1880, countries began to introduce tariff barriers to restrict free flow of goods and (in 1914, with the advent of passports) restrictions on free flow of people. War and the economic depression prompted restrictions on free flow of capital and higher, protectionist, trade barriers.

It is yet to be seen whether the global recession and credit crunch at the start of the 21st century damages global integration. Nevertheless, there has been a resurgence of national identity and a dislike of some Western offers in much of the world which flies in the face of the ubiquitous, global marketing approach. For the service marketer in a technology company this prompts an horrendous strategic dilemma which could, quite literally, destroy the company. Is the service going to be offered as a local service available in some other countries or a truly global offer? The answer to that question needs to be carefully considered and thoroughly analysed. Investment in the chosen direction needs to be assessed, planned and carefully controlled. This, combined with high-profile failures to achieve global positioning, has caused all but a few established firms to attempt risky investment in truly global strategies. Instead, many are adopting a 'think global/act local' strategy, which acknowledges the impact of local cultural differences on marketing (originally attributed to D. Keough and R. Goizeuta leader of Coca-Cola in the 1980s; see Pendergrast, 1994). One is Fujitsu, currently adopting a 'transnational' approach to IT services.

FUJITSU EVOLVES INTO A 'TRANSNATIONAL' ENTERPRISE TO GROW GLOBAL BUSINESS

Japanese IT giant Fujitsu is a leading provider of IT-based business solutions for the global marketplace. With approximately 175,000 employees supporting customers in 70 countries, Fujitsu combines a worldwide team of systems and services experts with highly reliable computing and communications products and advanced microelectronics to deliver value to customers. Headquartered in Japan, with revenues of US$47 billion, Fujitsu is the world's fourth largest IT services provider and Japan's market leader.

Time for a New Global Business Structure

Global expansion is on the agenda of most large enterprises, together with how to best organise themselves for international operations. Developing a new strategy and structure in the increasingly complex environment of international business is an enormous challenge. Yet Fujitsu has taken this head-on in order to achieve its aim of becoming a truly global IT company, within a three-year (FY2009–2011) timeframe.

Fujitsu recognised that it needed to adopt a broader, global perspective, rather than focusing so heavily on Japan, and has undertaken a radical shift in the focus of its operations. This process of structural reform has meant an end to the 'silo approach' to business inside and outside Japan, with integration enabling the group's companies to better work together to accelerate global growth. The strategy to clarify responsibilities and strengthen overall group management is reflected in its adoption of the transnational organisational model: a move that reinforces Fujitsu's ability to view operations from a global standpoint, while still pursuing business that is responsive to the prevailing conditions in each region.

Kuniaki Nozoe, Fujitsu President (2008–2009) explains, 'We will expand business globally by leveraging the base Fujitsu has built in developing its worldwide operations over many years, and by instilling the maxim "Think Globally, Act Locally" in our operations.'

Two business imperatives prompted this strategic change in organisational structure. First and foremost was the need to organise the group's business to better meet the needs of its global markets and clients. This involved improving its ability to: present a 'One Fujitsu' to all customers; take products and services to market together; simplify and transform its regional organisations; derive real, lasting benefits from best practice and standardised offerings and processes across regions. Second, naturally, was future growth and the opportunity to significantly increase the percentage of Fujitsu's business outside the Japanese home market.

The expanding international IT marketplace, where markets are developing at different speeds and in different ways, has been driving this increasing demand for both global and local excellence. In IT markets, analysis of consumer wealth and IT spend alone shows the considerable variations in size and opportunity, particularly between, say, markets in the West and newly emerging economies such as those in Asia. As these markets evolve quite differently, local requirements vary from market to market. Yet clients also demand the benefits of global scale and coherence. It was the need to address this paradox of 'trying to do it all' that was behind Fujitsu's move to become a transnational enterprise; 'thinking globally and acting locally'. With few truly global IT companies, and none that are wholly successful, Fujitsu saw the potential for differentiation through this approach.

Fujitsu has long worked to develop operations globally, and was one of the first IT vendors in Japan to advance operations abroad. Fujitsu Services in the UK, for example, had an impressive track record in securing large-scale outsourcing business deals in Europe. But despite this record, and initiatives such as the Global Service Innovation Programme, Fujitsu's operations outside Japan were limited to an 'Act Locally' mindset, which is why 'Think Globally' was crucial for the group to resolve.

Emergence of the Transnational Model

So, external pressures for local responsiveness and global integration lie at the heart of the international strategy/structure choices considered by Fujitsu. They reviewed four models: *Multinational* (ultimately localised), *Global* (highly centralised), *International* (focused on sharing) and *Transnational* (highly networked). Each differ in their management structures, external approach to the market, and internal lines of communication and reporting. Fujitsu thought that the *Transnational* organisation combined the essential features of the other three models: flexible and sensitive to local conditions like a *Multinational* company, competitive and efficient like a *Global* company and, at the same time, attentive to leveraging learning and sharing knowledge across the local business units as in an *International* company.

Seen as a solution for increasing complexity, the transnational model represented a balance between the pressures for global integration and local responsiveness; centralised control and decentralised autonomy. Described as an 'integrated network', it could deliver simultaneous efficiency and scale benefits for global competitiveness, national responsiveness for flexibility and the scope benefits of corporation-wide learning for competitive innovation. As a concept it was still evolving, but it seemed the obvious route for Fujitsu given the group's multinational heritage, global aspirations and unique culture in the world of IT services and solutions.

Going Down the Transnational Route

Fujitsu is now fully embracing the management principles behind the transnational business model to better serve its clients. A structural reform is underway, that visibly underpins the global delivery of a 'One Fujitsu' to all customers; going to market under one consistent brand, with one single business responsibility. Whereas the 'local' structural component that is the company's heritage helps to simplify and transform their regional operations, providing regional empowerment but not independence. Global coherence has evolved from a small but powerful set of functions, all supported by a shared incentives programme forming the 'collective bonds of performance' for the company.

Clients benefit from the 'think global, act local' approach. The global scale and coherence of the transnational business elevates Fujitsu's global delivery and global client management capabilities. The shared values of the 'Fujitsu Way' embedded across the organisation reinforce the 'One Fujitsu' experience. Further, the global knowledge and experience sharing add considerably to the security and confidence that is a function of scale and reach.

At the same time, the local proximity of focused regional operations helps ensure Fujitsu is fully in-tune with local market conditions and client requirements. An example is the successful roll-out of a client relationship initiative, 'Field Innovation', which converts IT innovation into real business value. Long-term partnerships are also more manageable, while the empowerment of local account teams and management continues to generate innovation and process improvement.

Fujitsu believes the transnational model works well to balance global scale and local touch. From their experience, it's now feasible to reach a higher number of services, sectors and geographies on a 'global' level, yet with the 'local' benefit of greater segment focus and more localised management and delivery systems. Another advantage includes a superior ability to meet key IT purchasing criteria. These include: building a deep understanding of the customer's requirements; getting senior executives involved in the sales and delivery process to build relationships and demonstrate commitment locally; pricing competitively at both a global and local level; bringing all the benefits of a tried and tested solution to bear for the customer; and having the scale and flexibility to offer innovative, outcome-based commercials.

Making it Real at Fujitsu

Fujitsu continues to push ahead on structural reforms and growth strategies geared to each market while strengthening group management, clarifying responsibilities and increasing sales. Where previously each region implemented their own strategy, with significant overlapping of functions, there is now unified planning. R&D, sourcing and supply chain have become 'global' integrated functions, with sales and service positioned squarely at the 'local' level – Japan, Americas, APAC, China, EMEA. Taking 'think global' a step further, the 'Global Business Group' that develops strategy has been organised into four global support functions – namely, Marketing, Delivery, Global Client Management and Business Management.

The specialisation of group companies has been a major component in the restructure. Concentration of resources has put the company in a stronger position to meet the needs of different business markets; large, medium and small. Converting Fujitsu Business Services into a wholly owned subsidiary, for example, effectively created a one-stop shop for medium-sized businesses in Japan.

The acquisition of Fujitsu Siemens Computers, integrated as Fujitsu Technology Solutions in 2009, accelerated the group's globalisation strategy and helped address clients' requirements for IT solutions across the globe. Through reorganisation of this subsidiary, and further realignment of operations outside Japan, Fujitsu intended to double annual IA (Intel Architecture) server sales within two years, and to improve profitability in the longer term by adding services to these deals. Fujitsu says that it is now successfully expanding business globally through differentiated IT infrastructure products and services, with the benefit of enhanced speed, a partnership approach to customers and continuous technological innovation.

The company's global value proposition has evolved noticeably alongside the structural changes. An ambitious transformation that is paying dividends, with the cost efficiencies, strengthened management structure and integrated 'think global, act local' capabilities of a transnational enterprise fast being realised.

As Kuniaki Nozoe explained, 'Fujitsu's goal is to get more than 40 per cent of its sales outside of Japan in fiscal 2011, and it will do this through a "think globally, act locally" strategy.

The Implications of the Characteristics of Services

A number of the characteristics of services outlined in Chapter 2 mean that the international marketing of services has particular issues and problems that need to be addressed. For instance, as services are intangible, much of the investigation, analysis and purchase of them is done in the customers' imaginations; and those imaginations are affected by cultural heritage. Colour, language, intonation and dreamscapes are all affected by childhood and heritage. So cross-border marketing needs to carefully take these issues on board. Also, since a service experience frequently involves an exchange between two human beings, body language and manners affect perceptions of quality. Behaviours which are considered polite in one society can cause offence in another.

A caution: Aspiration and Fame as a Counter-weight to Cultural Differences

The majority of the fans of British soccer team Manchester United are not in Britain or even Europe. Remarkably, a wide range of people in Asia are devoted followers; keeping up with their team's stars and tournament performance through satellite TV. Despite living in the heat of Singapore or Malaysia, in perhaps poor circumstances with, perhaps, Islamic faith, many people are part of an excitable group which charts the course of rich, pampered Western soccer stars. They suspend their cultural heritage and issues for a greater benefit: to be part of an exhilerating tribe. A similar phenomenon occurred just after the Second World War. At the time, America was seen as energetic, modern and rich. People across the devastated globe wanted to feel part of its exciting modernity. They wanted to belong and, as a result, willingly bought into American exports like Coca-Cola with its promise to a young generation.

It is possible, then, to create a service that has appeal and fame across the world. It is so associated with success that it prompts people to desire it and suppress or suspend their cultural predilections. They want to get in on the experience. This desire explains, to some extent, the relentless rise of Richard Branson's Virgin companies.

INTERNATIONAL MARKET ANALYSIS AND PERSPECTIVE

Any debate about international expansion is better conducted in the light of both analysis and insight into that market. The processes and techniques described in Chapter 2 (market research, market audits, etc.) can help with this. All apply as well in international market analysis as they do in domestic markets. However, there is increased risk of error with international markets. Whereas published research reports and industry data can be freely available, a lack of real local knowledge can be very costly if the firm is considering a serious investment. At the very least, there should be a fact-finding visit to any intended countries, discussion with local specialists and even a visit to the commercial representatives of embassies in that country. Each will round out knowledge and yield insight prior to any strategy debate.

For instance, it would be dangerous and wasteful for any firm to attempt to 'target the world' with a broad and amorphous approach to expansion. All have to start by choosing a portion of the international market, a group of customers, to concentrate upon and the most obvious international segmentation method is geographical. This has allowed many firms to grow businesses inside national borders until they are flourishing mature businesses. Yet choosing a country or a geographic market can also be a limitation. The firm is, in fact, restricting itself to agreed borders and legal frameworks (many of which resulted from long-forgotten peace conferences which followed centuries-old wars) that are irrelevant to some modern business needs. Why should a 21st-century service business define itself by the failures of 19th or 20th-century dictators? Moreover, within the boundaries of many nations, like Spain, Germany or the USA, are strong regional structures and differences that also have to be acknowledged.

Some firms have therefore created a segmentation based on cultural groups. For example, a Hispanic emphasis is likely to appeal in Spain, South America and certain Spanish-influenced areas such as Miami in the US state of Florida. This segmentation approach might also lead to an emphasis on 'Anglo-Saxon', Germanic and Chinese groups, which involve similar cross-border opportunities. Some firms have even found that the successful penetration of cultural groups in this way has led them, through trade relationships, into valuable groupings of diasporas which have developed from previous migrations of trading peoples. Others have followed ancient trade routes into new geographies. For example, some service firms relate to customers in Southern America through Spanish offices.

In fact, many of the segmentation approaches outlined in Chapter 3 can apply in the international context. If, for example, a consultancy firm has decided to grow by penetrating a few international clients, it might concentrate on managers or firms with cultures that make them 'early adopters' of new management concepts. There is evidence that certain countries (such as Sweden and Finland) and certain regions (California for example) are 'bellwether' markets, attracted to whatever is new. Large firms in these cultures are likely to be attracted to leading conferences and thought leadership on emerging ideas. So, leaders should define the most relevant segmentation for their own firm using a method similar to that described in the Tools and Techniques appendix.

Yet this is just one example of the greater complexity of international market analysis. Almost every aspect of analysis and research, from environmental analysis to market definition, is more extensive and more risky in the international context. It must be crafted carefully and yield real insight rather than just tables of data and pretty PowerPoint slides.

INTERNATIONAL BRAND STRATEGY

There are several strategic decisions to take with regard to the positioning of a service brand internationally. Each has serious implications to the health of the firm, the development of brand equity and the ease with which revenues will grow.

Is, for instance, the brand going to be perceived to be indigenous to each country in which it operates? It is possible to create a service brand which makes the firm appear native to each country or culture in which it operates. It can be American in America, British in Britain and Australian in Australia; familiar to locals in every country in which it becomes established. This multinational brand strategy is possible because reputation evolves naturally from service experiences. Companies which have strong local business units, established in countries that are staffed with good local professionals, cause a local reputation to evolve through word-of-mouth and this can be developed into a warm, familiar local brand. This strategy can be maintained by following the same approach as new countries are penetrated. There is difficulty, though, in maintaining international consistency because each adaptation to local culture risks creating another variation of the brand. Some companies get around this by following a 'family' branding strategy. The Japanese technology company Fujitsu has often used this approach, particularly as a transition device following acquisition; as in: 'ICL, a Fujitsu Company'.

An alternative is for the brand to reflect the culture of its corporate or domestic headquarters. The brand can simply reflect the values and image of the country in which the firm originated or is headquartered. IBM Global Services is unquestionably American, while the service teams of Ericsson are unquestionably Swedish. This has both advantages and disadvantages. Firstly, it is relatively easy to maintain the integrity of a brand that reflects national identity. It also resonates with some cultures, creating opportunity. In recent years, for example, the countries of Eastern Europe have been very open to American service firms because they have thought that they needed to catch up after years of communism. On the other hand, some Asian countries have been cautious of Western advisors after the last collapse of their economies but warm towards Swedish brands because they are seen as fair-minded.

The brand can, by contrast, be presented to people as that which they perceive a national culture to be. People in different countries have perceptions of foreign cultures and their business practice, some of which are unrealistic images based on media received in their home country. If the brand reflects a dominant culture, the firm needs to decide whether the positioning should embody the reality of that culture or what the local market thinks it is. For example, in some markets American business is perceived to be efficient, process-driven and modern. This may not actually be the case, but the brand positioning which exploits the indigenous market's perception of American business is a strategy open to the firm.

The British national carrier, BA, made a remarkable mistake by misreading this phenomenon. In the latter half of the 20th century it had transformed from a public service airline by radically improving its service quality (when no others did) and then positioning itself as 'the world's favourite airline'. It was the unrivalled and unquestioned market leader, admired by customers and competitors in study after study. It then sought to become a global service supplier changing, en route, its livery. Premier Margaret Thatcher expressed her intuitive dislike of this approach by covering a model of the new tail-plane designs with a napkin. The world took an equally negative view. Research showed that foreign travellers were put off because they had sought out BA for a perceived genteel, English style of service (which may not necessarily be true of the experience or true of Englishness). Yet the airline suffered real revenue loss because of a perceived leap to global blandness.

Finally, is the firm going to attempt to create a real 'global' brand; an entity that commands allegiance from buyers in many parts of the world? An effective global brand has equity and the same identity in many markets. It simply is. Proponents of the approach argue that the commonality of design, distribution and advertising allows its owners to reap efficiencies of spend which competitors are unable to achieve. This is based on a number of global brands (like Coca-Cola and Microsoft) that have achieved enormous benefits from scale. There appear to be, however, very few clear definitions of a global brand. If simple criteria are applied (such as being present in at least three continents of the world and having more then 40% of revenue outside its originating country) then there appear to be very few in existence. The risk here is that the brand will be perceived to be Western rather than truly global or the lowest common denominator with little emotional appeal. This strategy also requires realistic initial investment and can be hugely expensive because the firm has to create a consistent impression in all markets in which it operates.

INTERNATIONAL MARKETING COMMUNICATION ISSUES

As discussed in Chapter 9, a firm working on the international stage has to communicate with both individual customers and with markets. Just as with domestic marketing communication, it needs to use communication techniques to plan and execute its messages so that they pave the way for personal dialogue between customers and sales staff. It needs a strategic communications plan, good message management, clear media selection, programme management processes and response mechanisms. In the international context, however, differences in language, taste, colour preference and humour need to be accommodated. The design and message of all communication pieces need to be checked with internal representatives and external creative agencies in different cultures at the planning stage. There need to be clear processes which engage communications representatives of all the firm's organisational units in the development of both strategy and programmes. These need to influence the thinking of local customers.

INTERNATIONAL SALES AND ACCOUNT MANAGEMENT STRATEGY

Many of the approaches to service sales described in Chapters 7 and 8 also apply in the international context. For instance, it is possible to use the pipeline management tool (see the Tools and Techniques appendix) to focus thinking on the internal management of sales processes. It is also possible to build a personal network of international clients and handle them effectively while, at the same time, managing the implementation of a personal marketing plan. However, 'relationship management' is one of those phrases that, although it uses the same words, is heard differently by different cultures.

The way an American or British executive manages business relationships is different from the way, say, a French specialist approaches them. In the latter, the intimacy of *Grande École* and the expectation of an elite's ability to transfer between public and private sectors creates a different business climate to the freewheeling and open dialogue of, say, New York. Yet this is different again from the way some Asian firms work. 'Anglo-Saxon' consultants, for example, feel at ease approaching all levels of a client organisation and would expect to receive honest, clear feedback about the firm. In fact, consultants will often recommend to the senior executives of their clients that they visit and collect data from front-line employees in order to get a true picture of how a company is performing in its market. However, in some Asian cultures there is a sense of people having their place at definitive levels of society and a respect for order or hierarchy. So, it is not acceptable to criticise senior people. In those countries junior employees will generally not express any view that appears to be negative, even if those views are strongly held. Very often they will be uncomfortable spending any time or developing any relationship with professionals serving top executives. Any project predicated on the need to gather data across the organisation (like, for example, a GAP analysis to diagnose service quality issues) will be very difficult to execute in, say, Thailand. Relationship management must therefore be different when dealing with large customers in different cultures.

INTERNATIONAL SERVICE DESIGN

Innovation and new service design processes also have to be adjusted when working on the international stage. For instance, innovation is most likely to occur in local firms or countries because they are close to the customer. Local leaders are likely to spot developing needs and to deploy resources to meet individual client opportunities. They are unlikely, however, to be successful at spreading that innovation across the firm because they will be limited by their own time constraints, by a lack of local perspective in foreign cultures, by internal politics and, sometimes, petty jealousy.

It is therefore sensible to have a small innovation team who are responsible for new service introduction across the international service firm. They must have a budget to create or stimulate new ideas and to undertake feasibility studies. Their prime task, however, is to legitimise local innovation and spread it across the organisation. They need a clear mandate and international processes to do so. Similarly, the firm's portfolio management and design process must be managed internationally. Normally, these processes are managed by a representative team of senior leaders who ensure that proposals reach the various criteria to pass through the 'gates' in the design process (see Chapter 6). They are responsible for both the encouragement of spend on new idea development and the control of costs. Both are important to success.

There is, however, a local dimension to the design of service components. As different cultures respond differently to process, people, colour, image and technology, these differences need to be built into the innovation process. The international team needs to decide the degree of standardisation or customisation that will occur for each service and in which parts of the world. This will be based on a balance between research into client needs and affordability. Once the policy is decided for each service, tools such as features design or blueprinting (see the Tools and Techniques appendix) can be used to modify the service for local cultures.

DIFFERENT INTERNATIONAL STRATEGIC CHOICES

There is a range of strategic choices open to the leadership teams of service businesses that move beyond their own geographic borders. Each affects the nature of their service business; and particularly the approach to marketing. They are as follows.

The Exporter

The exporter is primarily a domestic firm. Its offer, approach and underlying business assumptions are a reflection of its domestic culture and perspective. In moving abroad it is normally responding to unsolicited demand or obvious opportunities. Its strategy is to reap new revenue streams with little customisation or cost. It often chooses local representatives (whether individuals or firms) who can service demand. Many service firms export through links with trusted partners into countries where there is healthy demand.

The marketing of the exporter normally involves the use, in a foreign country, of campaigns created and executed in the home country. Printed adverts will, for example, be reproduced with the local language or television campaigns with a local 'voice over'. Yet, they will normally look decidedly odd to local audiences (even if they cannot clearly articulate why) because they conflict with subtle conventions of tone and culture. They are therefore likely to have limited impact. So, this form of marketing is appropriate for tactical opportunities only. It is unlikely to contribute to the creation of an effective, beguiling international service.

The 'Follow the Client' Strategy

The international growth of many small firms begins with large international customers. They might, for example, provide service to one part of a large international business. As the work and relationships develop, the supplier might find that it stimulates demand for work in many parts of the world from the same customer. This, in turn, might become the foundation for international offices and employees.

There will come a time when the leaders will judge that there is so much continuing demand from a part of this large client in one location that they can risk establishing a more extensive base there. In fact, in some cases, it is possible to first establish a presence in the client's offices at that location. The supplier then needs to plan how it intends to stimulate demand from other customers in that location and how it will provide local resource to service that demand. It may set up an office using expatriate employees from the home firm. The intention in these circumstances is to eventually hand over to talent developed locally and reap earnings from the profit of the new entity. Alternatively, it may combine recruited talent with an acquisition to create a new firm in that country.

With this strategy, marketing must be aimed specifically at the customer and its vast organisation, while trying to avoid it becoming too dominant a portion of the firm's revenues. Best practice is to craft marketing strategy to develop the relationship with the large organisation as outlined in Chapter 8. However, at the point where leaders judge that there is sufficient momentum for a local country business to be established, a 'market penetration plan' will need to be developed. Sales activities, service design, campaigns and brand work must be specifically crafted to build demand and revenue from other customers in the new location.

The Network Firm

An international network is a group of companies or individuals working under a similar brand or proposition. This might be a holding company owning many different decentralised firms or a federation of locally owned companies which contribute part of their earnings to a central organisation for some greater benefit (usually a shared brand or the ability to service large international customers). The owners of this type of organisation are after synergies from the presentation of one united front on the world stage. They expect this approach to give them more impact than their individual domestic companies could possibly achieve alone; and marketing is at the heart of that activity even if, as in some circumstances, they do not value the function or call it by another name.

There should be strong brand development programmes and clear marketing communication campaigns designed by a central specialist team working on behalf of the whole network. Yet, ironically, marketing in a federal network can be very weak with few effective central guidelines and marketing policies which need not be followed by local countries. The approach is frequently diverse and defuse, driven largely by the local businesses.

The International Franchise

International service franchises are a fast-growing and successful aspect of the service economy, used too infrequently by technology companies. A franchise normally has a clearly defined offer and image. Franchisees pay a large sum to own a version of the business in a geographic region. The business model has already been proven in one market. Its service, its appeal and its intended customer type are clear. Franchisees see themselves as entrepreneurs and small business leaders, working within a proven framework.

Marketing is funded primarily from franchisees' contributions but is centrally prepared, packaged and launched. Franchisees expect detailed, prescriptive and highly effective marketing campaigns to be developed. They understand that their revenue generation and sales success depends upon it. So, marketing is highly controlled. The franchisees' role is normally limited to local implementation only. They are allowed very little leeway and so there is, usually, very little customisation. There must be, however, carefully crafted, effective and well-formed international campaigns.

The Multinational or Transnational

A multinational is a large firm operating on several continents. It has one consolidated set of accounts, one profit pool and a corporate function that directs or leads policy. While it seeks to optimise costs (by streamlining processes, locating facilities in areas of cheap labour and standardising offers) it also tends to configure its approach to local nationalities. America's venerable and massive engineering firm, GE, has, for instance, announced a new strategy for the products it offers in developing countries. As part of its innovation programme, certain products are stripped down and simplified for those markets.

The proponents of a global approach cite the degree of product standardisation and similarity of marketing as a fundamental difference between the multinational and global firm. They are different in both policy and approach. In fact, muddled thinking, which has confused multinationals with global policy, has damaged the earnings of some large firms over the past few decades.

For the multinational, consistent worldwide marketing appears efficient. Yet, most have learnt to set up mechanisms to adapt centrally designed campaigns to local taste. Even in the most centrally run companies, local subsidiaries can put enormous effort into customising international policies to local needs; even if not officially encouraged to do so. In France, for example, it is quite normal for people to work around rigid and clearly defined laws. There is a battery of regulation for many aspects of life captured in *Code Napoléon.* So, people will routinely set out to flex or avoid rules. It is natural, then, for employees of a foreign company there to put enormous effort into identifying or even helping to codify corporate rules; whilst at the same time ensuring that these rules are blithely ignored where they conflict with local business sense.

The Truly Global Firm

The concept of the global firm grew in the 1980s. The proposition is that this entity is based in no dominant location and has no dominant culture. It might have a very small headquarters, created for legal and administrative reasons, but corporate officers can be based anywhere and lead nomadic lives as they engage with different aspects of the firm's operations. It is entirely different from the multinational, with very different approaches to marketing. Evidence suggests, though, that, despite serious attempts in many different industries and equally serious investment, there are in fact very few truly global firms.

Marketing in a global firm is unquestionably consistent and centrally produced. It also seems to retain a sense of mystique, stimulating aspiration in its customers. Examples include the thought leadership of McKinsey (the *McKinsey Quarterly*), the 'CEO briefings' of Goldman Sachs and the Christmas campaigns of Coca-Cola. Their work is high-quality, expensive, consistent and impactful.

SUMMARY

Most services firms, whatever their size, will find themselves at some point faced with the possibility of having to work across borders, thanks to a combination of modern technology, with its erosion of time and borders, and the expansion of customers to new markets. So, understanding the issues involved in international work has become imperative for most executives in technology companies. This includes having a clear view of what is the same and what is different in each market in terms of customer expectations, and what impact international work will have managerially and operationally. The creation and marketing of an attractive and valuable service on the world stage is sophisticated and complex work.

Technology companies frequently fall into the offer of a bland, 'global' service that has no differences from its competitors and no emotional appeal to their intended customers. As a result, many of the international services offered by technology firms are little more than commodities, continually struggling to justify their price. As this chapter has shown, these firms must resolve these difficult questions if they are to succeed in profiting from cross-border service business.

Appendix

Marketing Tools

and Techniques

As in other business functions, marketing specialists use a range of concepts, models and tools in their work. These are taught in universities and professional institutes across the world. They are researched, analysed and critiqued by academics; and form a body of knowledge which are the elementary basics of marketing.

This appendix contains, in alphabetical order, descriptions of a number of those tools referred to in the main text of the book. It does not attempt to be an exhaustive review of all the techniques to which marketers can resort, nor does it attempt to provide a detailed description, or critique, of each one. Its purpose is to confine the main body of the book to argument which is directly relevant to marketing the services of technology firms. It describes briefly each tool or concept and its use. Further reading and details on the use of these tools can be found in the publications listed in the reference section.

Service marketers clearly need to be conversant with these tools and have the experience to know when they are applicable. Without that it is unlikely that they have developed the judgement and perspective which can reliably contribute to the creation of value for the firm.

'AIDA' CONCEPT

Application: Advertising Planning

1. The tool

Phillip Kotler cites the source for this venerable tool as *The Psychology of Selling*, published by E.K. Strong in 1925 (Kotler, 2003). Despite its age, it is typical of a number of tools used to design, plan and measure advertising. Although the concepts that they are based on vary, they assume that the buyer 'passes through a cognitive, affective and behavioural stage'; to quote Kotler. In other words, buyers move through a number of states in relation to a proposition (see Figure A.1). This one assumes that they become interested after the proposition has first gained their attention. Properly crafted, it will then cause them to desire it. This, in turn, will cause them to take the action of buying. Advocates of this approach suggest that suppliers break their target market into groups of buyers in these different states and plan their advertising accordingly.

2. Constructing a profile with this tool

In order to gain an empirical understanding of its market, the supplier must first understand the total universe it is addressing and the numbers in it. From this a representative sample can be calculated. Research can then be conducted to understand the numbers in the sample which are in the different states of the model (from attention to action). By multiplying each grouping by the percentage that the sample is of the total market, the number of buyers in

AIDA response model

Figure A.1

each group can be calculated. If this exercise is repeated after a campaign, the effect of the campaign can be estimated.

3. Use of the tool

The concepts behind this tool are useful in themselves. By reminding marketers involved in communications that clients need to move progressively towards a purchase, sophisticated approaches can be planned and communications will be much more precise. It may be, for example, that a campaign is then designed with a number of phases. The first might be to raise awareness whilst others move the potential buyers towards a direct action, like buying. It need not be used on just advertising though. If, say, a high-profile PR launch gains attention for a new idea, and this is followed by large presentations to generate interest and desire, then the final meetings with sales.

The tool can also be used to set communication objectives. If the supplier knows the attitude of its market, then objectives can be set in the light of that knowledge. If the customers are unaware of an offer, for example, then the very first communication objective is to create awareness.

The tool also works quite well in indicating the effect of communications on a large market. If, after a campaign, a number of customers indicate that they have been convinced to engage in the service, then it is likely to have been effective.

ANSOFF'S MATRIX

Application: Strategy Development

1. The tool

The, now, classic representation of Ansoff's matrix is reproduced below in Figure A.2. It suggests that firms distil their strategic options by focusing their thinking through a review of existing markets, new markets, existing products and new products. As such it is a useful simplification to help executives reach consensus during strategy debates.

However, Ansoff's original representation of the concept (see Ansoff, 1957) was more sophisticated and designed to examine diversification options. His work was based on analysis of M&A activity by American businesses in the first half of the 20th century. He suggested that there were two key bases for diversification, which he made the axes of his diagram. They were:

- 'Product lines': referring to both the physical characteristics of the product and its performance characteristics.

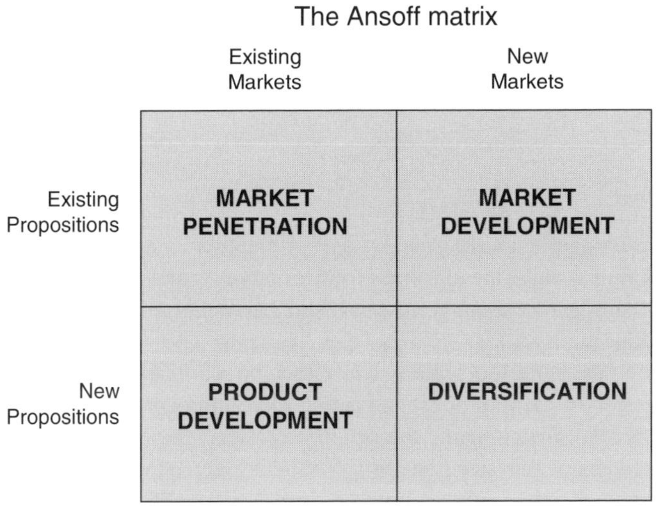

The Ansoff matrix

	Existing Markets	New Markets
Existing Propositions	**MARKET PENETRATION**	**MARKET DEVELOPMENT**
New Propositions	**PRODUCT DEVELOPMENT**	**DIVERSIFICATION**

Figure A.2
Reproduced by permission of H. Igor Ansoff's Estate.

- 'Markets': for the sake of this analysis he referred to markets as 'product missions' rather than buyer segments. By this he meant 'all the different market alternatives' or the various uses for the product and its potential uses.

He proposed his matrix as a way of constructing different 'product/market' strategies; those 'joint statements of a product line and the corresponding set of missions which the products are designed to fulfil'. His original diagram is reproduced below in Figure A.3. (π represents

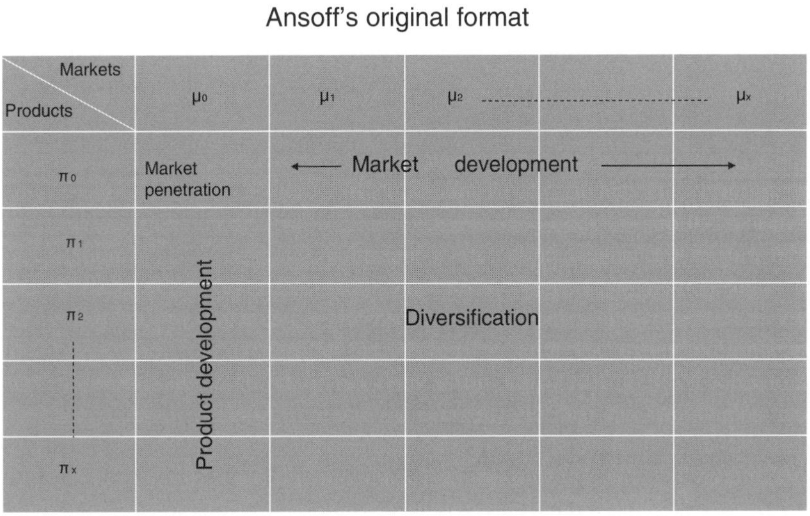

Ansoff's original format

Products \ Markets	μ_0	μ_1	μ_2 -------------------------------- μ_x
π_0	Market penetration	←—— Market	development ——→
π_1			
π_2		Diversification	
π_x			

(Product development labelled vertically on left; Market development labelled horizontally across top rows; Diversification spans lower-right region.)

Figure A.3
Reprinted with permission from *Harvard Business Review*, Ansoff, Sept/Oct 1957.

the product line and μ the 'missions'.) Interestingly, this representation of his work puts less stress on market penetration than the popularly used version and details the many strategic options that arise from thinking broadly about the fusion of product and market possibilities. In other words, it was designed as an innovation tool.

2. Constructing the tool

The first step to constructing the matrix is analysis of product and market opportunities. This may begin with a simple list of all the existing product/market groups in which the company is established. Then, using market reports, research and internal brainstorming, it is possible to identify the other opportunity areas.

Once the analysis is available the options can either be summarised, using judgement, into the simplified version of the tool or crafted into a more thorough analysis by creating a cell for each product/market match, using the original version. The opportunities can either be discussed by the leaders at this stage or prioritised using agreed criteria. Ansoff himself recommended that, due to the risk and cost involved, firms conduct risk analysis of the more likely strategies. The most acceptable programmes should then be developed into full product, marketing and business plans.

3. Use of the tool

The matrix helps leaders think through four different growth strategies, which require different marketing and communications approaches. They are presented below in ascending order of risk:

- Strategy A is 'Market penetration', or increasing market share with existing propositions to current markets.
- Strategy B is 'Market extension', or market development, targeting existing propositions at new markets.
- Strategy C is 'Product development', developing new propositions for existing segments.
- Strategy D is 'Diversification', growing new businesses with new propositions for new markets.

The matrix helps to clarify leaders' thinking and to illustrate the very different strategic approaches needed for each of the four strategies. Ideally, an operational marketing plan should be constructed for each strategic option that is finally approved.

ARR MODEL

Application: Relationship Marketing

1. The model

The 'actors–activities–resources' model was developed, during the early 1980s, by researchers and theorists interested in both business-to-business marketing and network marketing. Yet it appears to be a tool that can also be used in practical marketing within a normal business as well as just pure theoretical research.

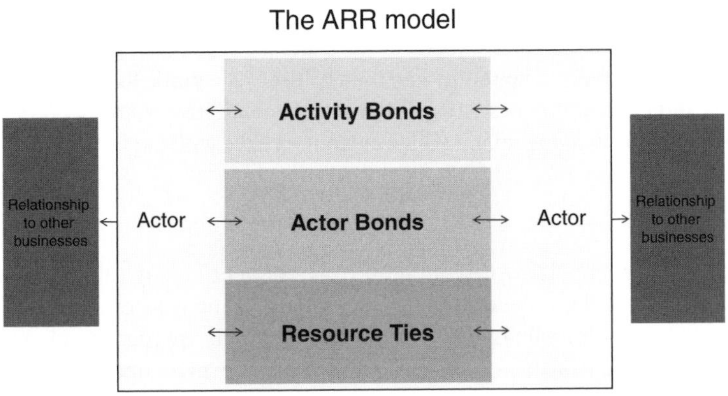

The ARR model

Figure A.4

The model divides business relationships into three layers, see Figure A.4:

(i) 'Actor bonds'. These occur when two business people interact through some professional process. Theorists suggest that there are three components necessary for them to develop. The first is reciprocity during the process, ensuring that both sides give something to the interaction, even if one is a client. The second is commitment and the third is trust. As people interact they form perceptions of each other about: capability, limitations, commitment and trust. If the relationship develops, these perceptions influence the degree and clarity with which the two communicate; and also the degree to which they involve each other in their own professional network. So, for the supplier, the way in which they conduct their work influences the trust their customer develops in them and the degree to which they will be invited further into the organisation, and thus into the possibility of further work.

(ii) 'Activity links' captures the work, or other activity, which is involved in the interaction and business process. These vary with the depth of relationship. They range from simple technical projects through to two firms meshing or adapting their systems and business processes to become more efficient. The latter has created exciting business opportunities in areas such as outsourcing.

(iii) 'Resource ties' are items used by people during business interactions. Resources might include: software, intellectual capital, skilled staff, knowledge, experience and expertise. People who have resources, or control over them, have greater power in professional networks. This power is the basis of the service offer (the customer comes to the supplier because they lack one or more of these resources), but clearly the immediacy or importance of need and the scarcity of resource influences pricing and quality perceptions.

2. Constructing the model

At its very simplest the model can be used as a basis of discussion with internal colleagues to map relationships in a network. Sales people can be asked to complete formats of their customer relationships using the three levels of the model. Actions arising from discussion (e.g. creating more opportunities for non-task-related exchange or making different resources, such as knowledge, available) to strengthen relationships can be put into account plans.

However, the model can be used as a basis of detailed analysis and research. A hypothesis of the professional relationships that exist in a market, and the types of interaction, can be created using the terms of the model. The model can then be used as a guide to designing the research sample and questionnaires. A two-step, qualitative and quantitative research process is likely to reveal powerful insights into the relationships customers have with the firm and its competitors.

3. Use of the model

The thinking behind the model will seem intuitively correct to many sales people. This is its strength, capturing, as it does, the day-to-day experience of many sales staff. It allows a firm, when needed, to use a common process and terminology in its approach to client relationship management. However, it also introduces (perhaps for the first time) a reasonably robust mechanism whereby professionals can analyse and understand in detail what many recognise to be their most important approach to market: relationships with customers.

BLUEPRINTING

Application: Service Design

1. The tool

As services contain a process through which a customer must move, the proactive design of a service through the use of 'organisation and method' (O&M) techniques has been suggested by several writers. Lyn Shostack (Shostack, 1977), a practicing marketer rather than an academic, was first. She pointed out that, whereas it is relatively easy to design a product through engineering specification, there was no 'service engineering' technique to which service designers could turn. Services are therefore often poorly presented to customers.

She suggested that, as a service is a process, it can be 'blueprinted' by using the O&M techniques developed to deal with process improvement. These, she suggested, included:

- Time and motion engineering. Shostack suggested that of eight basic charts used by methods engineers, those which were most applicable were the 'operations process chart', a 'flow process chart' and the 'flow diagram'.
- 'PERT' charting. This is an accepted method of planning detailed projects.
- Systems and software design. It was suggested that many of the software design methodologies could be used to design service processes.

She argued that there needed to be more conceptual work on the application of these process design techniques to service blueprinting because the process needs:

- To show time dimensions in diagrammatic form.
- To identify all the main functions of the service.
- To precisely define tolerances of the model (i.e. the degree of variation from standards).

2. Constructing a blueprint

Shostack offered a 'blueprint for a simple shoeshine service' which is shown in the diagram below, see Figure A.5. She suggested a stepped blueprinting method:

(i) Identify the customer process and map it out.

Blueprinting

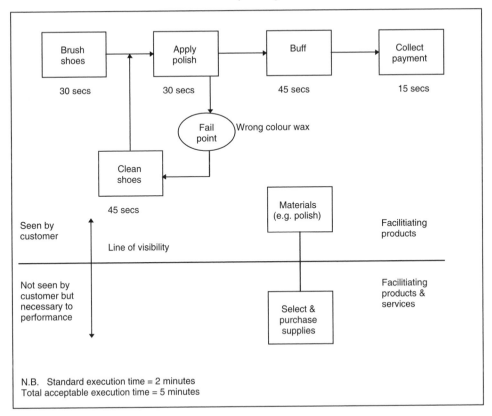

Figure A.5

  (ii) Isolate potential fail points and build in sub-processes to tackle possible errors before they occur.
  (iii) Establish a time frame or a standard execution time for each task.
  (iv) Distinguish between processes which are visible to the client and those which are not. Manage any implications arising from this.
  (v) Analyse profitability.

3. Use of the tool

Originally, Shostack offered the concept of blueprinting alongside molecular modelling and suggested that academics develop and test both for more generic use by business. Later writers on the subject have neglected the former and not substantially developed the use of blueprinting in service design. This may be because the popularity of O&M waned somewhat in the 1980s, although the concept of process mapping has received attention because of emphasis in senior management circles on the concept of 'process re-engineering'.

    Yet the process that clients experience is important to both the delivery of good service and the experience of the client. As explored in this book, it can create opportunities for new services and for new market strategies. Blueprinting is a practical and straightforward method to use in designing these aspects of a service.

'BOSTON MATRIX'

Application: Analysis of Business Portfolios

1. The tool

One of the best known, but poorly understood, portfolio management tools is the 'Boston Consulting Group's growth share matrix', which was developed by the Boston Consulting Group around the concept of the 'experience curve'. The consultancy demonstrated that, over time, companies specialising in an area of expertise became more effective in their market, reducing costs and gaining competitive advantage. A company might be at various points on the experience curve, depending on its maturity and the accumulated investment in its prime area of focus. The 'Boston matrix', shown in Figure A.6 plots relative market share against relative growth, was an attempt to give corporate strategic planners a way of evaluating different business units in different markets.

2. How to construct it

First, the annual growth rate of each business unit, in each market, is calculated. This is plotted on the matrix, depending on whether its growth is high or low. (*Note*: the horizontal axis of the matrix is not positioned at zero on the vertical axis but at 10%.) Second, the relative market share of each unit is calculated and plotted on the matrix. The turnover of each unit is then represented by appropriately sized circles.

The portfolio of business units is then categorised by the matrix into four groups:

(i) The 'question marks' (otherwise called 'problem children' or 'wild cats') have low market share in high-growth markets. A business which has just started operations would be a

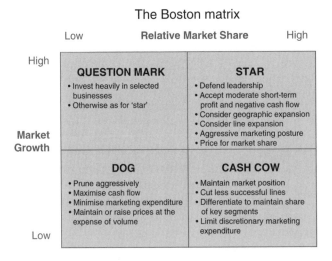

Figure A.6

'question mark' because the ability of the management team to improve on its competence would be unproven. These are businesses with long-term potential which may have been recently launched and are being bought primarily by buyers who are willing to experiment. They need large amounts of cash if they are to be developed to their full potential because the company has to keep adding plant, equipment and personnel to keep up with the fast-growing market.

(ii) The 'rising star' is a company that has established itself in the market and is beginning to thrive. It is a leader, with high share in a high-growth market. It requires significant investment in order to maintain and grow market share. It does not necessarily produce a positive cash flow but stars are usually profitable and can become 'cash cows'.

(iii) The 'cash cows' are companies that are well established and profitable. They have high share in low-growth markets. They are producing profit but were unlikely to achieve much incremental improvement. They are generally cash positive and can be used to fund other initiatives.

(iv) The final group are known as 'dogs'; they have low share in low-growth markets. These are companies that were in decline and worthy of withdrawal. They have weak market share in low-growth markets and tend to be loss-makers, providing small amounts of cash if any.

3. Use of the tool

The matrix can be used to determine strategy for major firms. A multi-business company needs a balanced portfolio of businesses or SBUs (strategic business units) which use cash from the cash cows to invest in other development issues. It is thought that an unbalanced portfolio of businesses can be classified into four areas:

(i) Too many losers causing poor cash flow.
(ii) Too many question marks requiring too much investment.
(iii) Too many profit-producers.
(iv) Too many developing winners.

The matrix is used to develop different business strategies and corporate requirements for each business unit according to its position on the matrix. Objectives, profit targets, investment constraints and even management style are likely to be different according to their position. Strategies set by the leadership are likely to include:

(i) 'Build'. This means increasing the market share using cash, resources, marketing programmes and management attention.
(ii) 'Hold'. This means maintaining the market share and is appropriate for strong cash cows if they are to continue to yield cash.
(iii) 'Harvest'. This means using resources to get as much cash from a business unit as possible regardless of long-term effect. It is appropriate for weakening cash cows whose future is uncertain or dim.
(iv) 'Divest'. This means selling or liquidating the business, and is appropriate for dogs and question marks that are acting as a drag on company resources.

Due to inadequate teaching and lack of understanding, some companies have tried to use this conceptual framework to understand the positioning of their individual products or services rather than business units. Managers can be heard to talk of their offers as either being a 'cash cow' or a 'dog'. This misuse of the concept is dangerous because it muddles two different concepts (the experience curve of a business and the product lifecycle).

More worrying for service firms, though, is the fact that the tool is based on the two axes of growth and market share. It is an assumption of the matrix that growth and share are the two key success criteria of a business. However, that is not always the case in the services industry. Some pursue a strategy of margin maximisation rather than growth. This tool is unlikely to be helpful to such firms.

THE 'MARKETING CONCEPT'

The marketing concept suggests that the buyers and the market, not exclusively the operations of the firm, should be at the centre of management's attention; and that, by putting attention on the market, long-term profits can be achieved. It has been called 'market-orientated' or 'customer-led' but is the antithesis of the internal orientation of many firms. Some are sales-led, some fascinated with technology or product enhancement while others are monopolies, preoccupied with internal rivalry. These approaches can be successful for a time, generating funds for owners, until market conditions change. If new regulation is passed or an aggressive competitor changes market dynamics, firms without a market orientation can fail.

This simple concept is in all marketing textbooks and can sound like wishful thinking by marketing theorists, keen to press for greater influence for their discipline. In practice, the experience of many marketing specialists is almost the opposite; that the function is not appreciated or allowed to contribute as it might. They become absorbed with changing daily priorities, tactical rather than strategic work and the need to engage in internal politics to protect the contribution of their function. But this concept is more than an academic wish, a passing management fad or theoretical fantasy.

Numerous papers (from Professor Benson Shapiro's *Harvard Business Review* article: 'What the hell is market orientated?' to Regis McKenna's 'Marketing is everything') and long-term tracking surveys (like the PIMS database) have shown that market-orientated companies produce long-term shareholder return. Perhaps the clearest example, though, is America's GE. Under Jack Welch this huge conglomerate exceeded expectations in terms of shareholder return and, for over a decade, came top in survey after survey as the most admired company, particularly amongst CEOs. But this success was built over the long term (see M.R. Vaghefi and A.B. Huellmantel's paper: 'Strategic leadership at General Electric'). As far back as 1957, Ralph Cordiner, the then CEO, announced that 'marketing would be both the corporate philosophy and the functional discipline to implement the philosophy'. Up until then all planning had been internally orientated, focused on solving internal problems. This change set the foundation for an externally orientated culture. Evidence suggests that, since then, each CEO has chosen their successor as a means to continuing this philosophy and the consequent enduring legacy has been the foundation of GE's outstanding success.

A marketing-orientated company understands its market and anticipates future trends, exploiting emerging opportunities. It is forward-looking and intimately engaged in understanding the minds and responses of its buyers. Brand management and marketing functions are often lead functions of the business, maintaining a market perspective for all functions.

CONJOINT ANALYSIS

See entry under research.

CONTACT AUDIT

Application: Client Service and Communication

1. The tool

This practical method evolved within the direct marketing industry as it put emphasis on the effect of other interactions of the firm than advertising on the opinion of buyers. The approach of direct marketing is to identify other communications media (such as mail and email) as mechanisms by which the same message can be reinforced. Prior to the evolution of thinking around loyalty, service and integrated communications, it encouraged the thoughtful marketer to think beyond the pushing of messages through broadcast media.

2. Constructing the tool

The marketer should first list all interfaces with the customer. These range from customer service staff, through written materials to attendance at events, see Figure A.7. This, alone, causes leaders to think about the effect of some surprising items. Customers, who visit a scruffy office, are kept waiting by reception staff or who have to negotiate difficult security staff can have their confidence in the firm's services undermined; even if some of these people are clearly employed by contractors.

The tool then causes the firm to think about the frequency with which customers use a particular interface and the reason why. If it is very important or emotionally distressing to them, then the impression it forms is likely to be very influential. If the supplier neglects it, then service issues are likely to arise. Finally, the quality that customers experience and the outcomes they receive are considered. These are best completed in discussion with a sample of them. (Although internal judgement is also effective.)

Contact audit

Type	Frequency	Reasons	Quality	Outcomes
Letters				
Invoices				
Brochures				
Mailings				
Reception				
Website				
Account meetings				
Project management				
Switchboard				
Client service staff				
Security staff				
Advertising				
Hospitality				
Reports				

Figure A.7

3. Use of the tool

This simple tool is a planning mechanism which identifies improvement areas in the service interface. It emphasises all aspects of the customers' interface with the firm and prompts marketers to think beyond technical delivery alone.

DIFFUSION OF INNOVATION

1. The concept

Diffusion of innovation is the socialisation process by which societies learn of new concepts and adjust to them. It is a theory of how, why and at what rate new ideas and technology spread through cultures. Study of the phenomenon was initiated by social scientists like Beal, Bohlen and Rogers after the Second World War. It has been examined in many contexts (from First-World drug use to Third-World water sanitation). Rogers says of it:

> Information about an innovation is often sought from peers, especially information about their subjective evaluations of the innovation. This information exchange about new ideas occurs through a conversation process involving interpersonal networks. The diffusion of innovation is essentially, then, a social process in which subjectively perceived information about a new idea is communicated from person to person. The meaning of an innovation is thus gradually worked out through a process of social construction.

This is, essentially, a communication process which involves word-of-mouth, publicity and subtle forms of marketing. It is from this, not the product lifecycle, that the various groupings of people are made. Some are:

- Innovators: the first to adopt an innovation. They are willing to take risks and interested in scientific or technological ideas (gadgets).
- Early Adopters: the fastest category to adopt an innovation and, generally, a larger group.
- Early Majority: taking up the idea after a degree of time, they tend to be slower in the adoption process.
- Late Majority: adopting the innovation after the average member of the society, these tend to be more sceptical.
- Laggards: the last to adopt an innovation and can take pride in their resistance to new ideas.

2. Constructing the concept

Attitudinal research will reveal the progress of an idea amongst a population. From that the marketers can deduce the state of progress of an innovation in a given market.

3. Using the concept

Marketers need to construct different marketing strategies at different phases in the diffusion of an innovation. The marketing communication programmes to a market dominated by innovators will be different from those of one dominated by the early majority. Marketers should also come to a judgement as to those moments when the innovation can move to the next phase. A market leader that is influential in moving the phase of diffusion for, say, innovators to early majority is likely to reap a growth in sales.

The social scientists who developed this concept recommend the use of opinion leaders, experts and catalysts to help an idea spread. Marketers can adapt these ideas to help the spread of interest in their product or service amongst a market.

DIRECTIONAL POLICY MATRIX

Application: Business Portfolio Tool

1. The tool

Sometimes called the 'GE/McKinsey matrix', this was developed soon after the Boston matrix as a result of inadequacies with the latter. It was typical of several methods of 'multifactor portfolio models' which were developed at the time.

This is a way of categorising businesses against markets and is more flexible than the Boston matrix because it uses criteria created by the management team themselves. As such it is more relevant to the individual strategic position of the company in its marketplace. The grid plots 'market attractiveness' against 'business strength' (see Figure A.8) and allows management to prioritise resources accordingly.

The original GE matrix used the factors of market attractiveness and business position which are listed below in Figure A.9.

GE used these key factors because they believed that, taken together, they had the most influence on return on investment. However, this list should be modified for each company according to its own particular circumstances.

2. Constructing the matrix

There are clear steps in compiling the matrix. They are:

(i) Identify the strategic business units (SBUs).
(ii) Determine the factors contributing to market attractiveness.

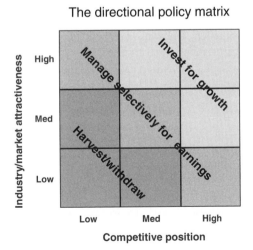

The directional policy matrix

Figure A.8

Factors of market attractiveness and business strength used in the original GE matrix	
Market attractiveness	**Business strength**
Size	Size
Growth rates	Growth rate
Competitive intensity	Market share
Profitability	Profitability
Technology impacts	Margins
Social impacts	Technology position
Environmental impacts	Strengths and weaknesses
Legal impacts	Image
Human impacts	Environmental impact
	Management

Figure A.9

(iii) Determine the factors contributing to business position.
(iv) Rank and rate the market attractiveness and business position features.
 (v) Rank each SBU.
(vi) Plot the SBUs on the matrix.
(vii) Represent the total size of the market and the businesses' share by a pie chart at the appropriate plot on the matrix.

Steps (iv) and (v) involve numerically rating the relative importance of each feature. Multiplying these together and totalling them for each business unit gives a composite score which enables the matrix to be compiled. The total size of each market the firm's businesses operate in and their share of it can be represented by a pie chart centred on each plot. As with the growth/share matrix, the visual presentation enables complex information to be presented in an easily understood form.

3. Use of the tool

Strategy can be deduced from the matrix as follows: where a business unit scores high or medium on business strength or market attractiveness, the firm should maintain or grow investment. Whereas those which score low/low or low/medium should have investment depleted. If possible, cash should be harvested from them. Units scoring high/low or medium/medium should be examined to see if selective investment should be made to increase earnings.

There has been some critical evaluation of the GE matrix. Many people consider the fact that it uses several dimensions to assess business units instead of two, and because it is based on ROI rather than cash flow, it is a substantial improvement on the Boston matrix. However, it is criticised because:

  (i) It offers only broad strategy guidelines with no indication as to precisely what needs to be done to achieve strategy.

 (ii) There is no indication of how to weight the scoring of market attractiveness and business strengths. As such it is highly subjective.

(iii) Evaluation of the scoring is also subjective.

(iv) The technique is more complex than the Boston matrix and requires much more extensive data gathering.

 (v) The approach does not take account of interrelationships between business units.

(vi) It is not supported by empirical research or evidence. For instance, there does not seem to be evidence that market attractiveness and business position are related to ROI.

(vii) It pays little attention to business environment.

It is really powerful, however, as a tool to reach consensus amongst a group of leaders. The definition of business units, the agreement of common criteria and, particularly, the joint scoring exercise stimulate debate which is very valuable.

EXPERIENCE CURVES

Application: New Service Development, Competitive Strategy

1. The tool

This concept, pioneered particularly by the Boston Consulting Group during the early 1960s, suggests that unit costs of a firm fall with experience of operating in an industry and with a company's cumulative volume of production, as shown in Figure A.10. The consultancy invested substantial time over many industries (including service industries) and used 'the

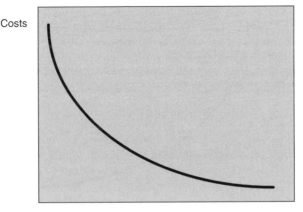

The experience curve

Costs

Time

Figure A.10

scientific method' to validate the concept. Although appearing deceptively simple, and intuitively right, the concept can be used to set exhilarating strategic objectives.

Costs decline due to a combination of: economies of scale, a learning curve and the substitution of technology for labour. The cost decline gives competitive advantage because new competitors can face higher costs if not entering with a major innovative advantage. Some have argued that the advantage is so great that established leaders should drive to gain further advantage through actions such as price cutting.

2. Constructing the tool

Plot the firm's prices or costs against unit volume; projecting back in time as far as is sensible. The resultant curve should reveal the accumulated gains by the firm. In service markets it is also normally possible to obtain estimates of number of employees in competitor firms, so a comparison with competitor gains should be possible.

For the analytically minded, the Boston consulting team recommended using double logarithmic scales because they show percentage gain as a constant distance:

> A straight line ... means, then, that a given percentage change in one factor results in a corresponding percentage change in the other ... reflecting the relationship between experience and costs and experience and prices. (Henderson, 1972)

3. Use of the tool

The tool can be used to identify cost gains and advantages compared to competitors. As a result it can become a benchmark by which the firm can set strategy for business units to improve costs. It can also be used to predict and set prices by giving a directional indication of industry costs and likely competitive responses.

The tool can also be used to plan and sell outsourcing concepts. As illustrated below in Figure A.11, a firm whose business is focused on a particular function is likely to be further down

The experience curve and outsourcing

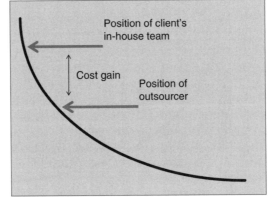

Figure A.11

the experience curve than a client's in-house team. If the in-house operations are passed to the supplier, the client gains advantage of the supplier's 'experience'. When the UK government's audit office first reviewed (in 1990) public sector outsourcing deals initiated by the Thatcher government in the 1980s, gains of up to 20% were found. This sparked the outsourcing trend in much of Europe. Note though that further dramatic gains are unlikely and continuing success depends on the nature of the relationship between the two parties. If the relationship breaks down, the client loses the supplier's experience and begins to build cost back into its business.

FEATURES ANALYSIS

Application: Service Design

Features analysis is a concept used in product design to proactively plan the content of the offer. It is a development of the suggestion by leading marketing thinker Phillip Kotler (Kotler, 2003) that products and services are propositions augmented by intangible marketing concepts such as brand and design.

It suggests that each offer (product or service) comprises three sets of features, see Figure A.12. They are:

- 'The core feature'. This is the hub of the offer and is the prime benefit to buyers. In the case of a briefcase it will be to 'carry documents', in the case of a car it will be 'personal transportation'. Experience shows this to be one of the most difficult aspects of product and service design; it is inordinately difficult to settle on the core proposition.
- 'Augmented features'. These are the physical components of the offer which the product manager chooses to use to represent the core feature. In the case of a briefcase it would include the choice of leather, latches, nature of stitching, internal construction, etc. In the case of a car it would include the engine, the bodywork, the colour and the physical layout of

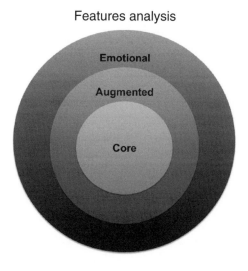

Features analysis

Emotional

Augmented

Core

Figure A.12

the car. This is very much the design and assembly of physical components around identified customer need.

- 'Emotional features'. These are designed to appeal to the buyers' underlying, often unknown and unarticulated, emotional requirements. These are often the most influential aspects of the appeal of the proposition to the buyer and particularly affect perceptions of value. Without them many offers become commodities. Although these are actually offered through the physical (augmented) features, the emotional ring of the planning tool is there to remind designers to proactively plan their presence. They are particularly tied to the firm's brand values. For example, the emotional promise of a briefcase that is labelled 'Gucci' will give a different message than one which is labelled 'Woolworth'.

It is the proactive management of this mix of features that allows managers to design increasingly sophisticated versions of their offer in the light of feedback from markets. This allows evolution of real choice and the ability to create profit through the evolution of differentiated offers. In service businesses, this technique needs to be adjusted to take on board the observation by Lynne Shostack (Shostack, 1977) that propositions from companies are neither all product or all service. Using her goods/services spectrum of offers (see the diagram below, Figure A.13), the subsequent four models can be used with real precision.

Model 1

This represents a proposition where service is primarily an emotional reassurance to a product offer. The core proposition is a product that has been augmented by physical features. However, the supplier has accepted, generally because it is based on new technology, that faults will occur in their product. Service, like warranty, has to be provided as an emotional reassurance to the purchaser of the enduring provision of those benefits, as shown in Figure A.14. Service is therefore an emotional feature of the product. This has been in evidence with

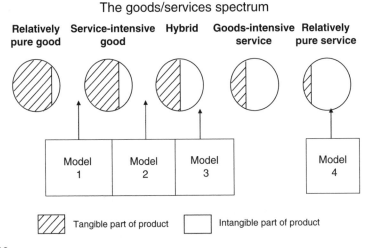

Figure A.13
Reprinted with permission from *The Journal of Marketing*, published by the American Marketing Association, G.L. Shostack, April 1977.

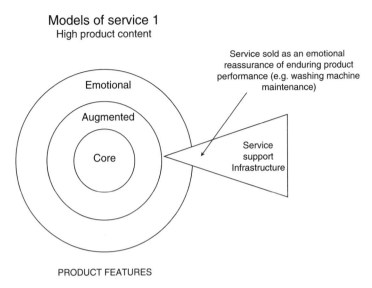

Figure A.14

many product offers over the years, from washing machines and cars through to computers and elevators.

Model 2

This represents an evolution in a market where suppliers begin to build service into the product concept. It occurred, for example, in the computer industry during the latter half of the 20th century. Suppliers began to provide preventative maintenance through a monitored service involving people, procedures and technology. It was sold as part of the product offer so that computers failed less due to self-diagnostic technology and preventative maintenance. (This was an entirely different proposition to the previous maintenance contracts which promised 'Don't worry, if it goes wrong we'll repair it quickly'). In this model, service has become an augmented component of the product offer, see Figure A.15.

Model 3

This, a hybrid, represents a position where people are buying a mix of service and product, as shown in Figure A.16. It is common in industries which offer a high-volume, low-margin product. The fast-food industry, for example, uses service to sell a cheap product. The brand, environmental design, product range, technology support, people behaviour, method of accessing the service and the process through which the service is provided are all integrated into a holistic experience which people buy. This is the core service of fast-food retailers and has evolved over a long period of time.

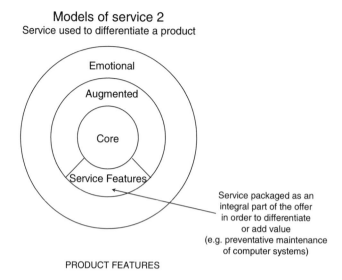

Figure A.15

Model 4

This represents a service offer which applies to many consultancy firms, see Figure A.17. It has almost no physical or product content. Any physical components (e.g. slides or bound reports) are merely an emotional reassurance to the buyer that good quality and high value exist in the offer. The tangible elements are a reassurance of the intangible benefit.

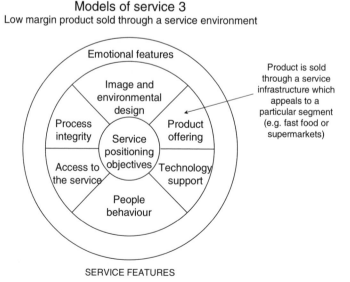

Figure A.16

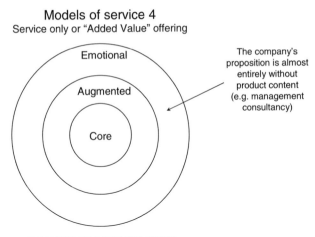

Models of service 4
Service only or "Added Value" offering

Emotional

Augmented

Core

The company's proposition is almost entirely without product content (e.g. management consultancy)

FEATURES OF A SERVICE OFFER

Figure A.17

If a service designer chooses to use features analysis, it is essential that they use the correct design model. It may be, for example, that market conditions have changed and a service which was once associated with a product offer (model 1) can be positioned as an entity which has value in its own right (model 4). In this case, a different features mix must be used.

GAP MODEL

Application: Diagnosis of Client Service Issues, Development of Service Strategy

1. The tool

This model, shown in Figure A.18, was designed and proposed by a group of academics (Parasuraman *et al.*, 1985) who specialise in service marketing studies. It acknowledges that:

- Buyers find service quality more difficult to evaluate than product quality. They have few tangible clues as to quality so must rely on other clues.
- Quality is a comparison between expectation and actual performance. Satisfaction depends on the degree to which the two match or not.
- Quality evaluation by buyers depends on outcomes and processes. Quality can be influenced by technical outcomes and functional or service outcomes. It can also be influenced by physical aspects of the service and by company image as much as by the interaction with client service staff.

The original investigation to substantiate the model was conducted in financial services and product repair services. It has, however, been widely tested and developed since.

The model focuses upon five 'gaps':

- The management perception gap. This is any difference between management's views and those of the customers or the market in general.

The Gap model

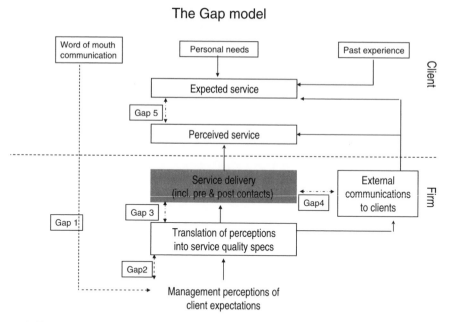

Figure A.18

- The quality specification gap. This exists when quality standards, strategy or plans do not reflect management objectives or views.
- The service delivery gap. This exists if there is a difference between quality strategy or plans and the firm's delivery to customers.
- The market communications gap. This tracks any difference between marketing communications and service delivery.
- The perceived service gap. This exists if there is a difference between the service delivery perceived to be experienced by customers and their expectations. Note the emphasis on perception of experience. What they think they experience may not be reality and might be addressed through marketing campaigns.

2. Constructing the tool

The representation of the tool above is used as a format for analysis, representing as it does major parts of the firm. Research and data collection needs to be undertaken at the point of each gap to compare and contrast opinion or experience.

3. Using the tool

In actual business the tool is best used as a diagnostic for the improvement of service strategy. Its analysis brings into sharp relief the differences between various perceptions and experiences. It allows the leadership to construct very specific improvement programmes in all relevant areas of business.

CULTURAL DIMENSIONS FOR MANAGEMENT AND PLANNING

Application: International Marketing

1. The tool

In the 1980s, Geert Hofstede researched (Hofstede, 1984), developed and published a number of dimensions of cultural difference. He grouped, tested and demonstrated the effect of these dimensions on management practice. They can be used to guide international strategy and planning.

The dimensions were:

(i) Individualism versus collectivism. Some societies are loosely knit, where individuals are supposed to take care of themselves and their immediate families. Work, career, economic provision and progress are centred around the individual. Others are more collective. Individuals can expect their relatives, clan or gang to look after them in exchange for unquestioning loyalty.

(ii) Power distance. This is the extent to which members of a society accept that power in its institutions is distributed unequally. This attitude affects the behaviour of those with and without power. Large power distance cultures accept hierarchical order in which everyone has a place, small power distance cultures strive for equalisation. This issue affects how societies handle inequalities when they occur.

(iii) Uncertainty avoidance. This is the degree to which members of a society feel uncomfortable with risk, ambiguity and uncertainty. Uncertainty avoidance cultures maintain rigid codes of belief and behaviour. They are intolerant towards deviants. Weak uncertainty cultures are the opposite.

(iv) Masculinity versus femininity. In Hofstede's view, masculine cultures prefer achievement, heroism and material success whereas feminine cultures stand for relationships, modesty, caring for the weak and quality of life.

This work shows how different cultures cluster and are similar under different dimensions. The diagram below, Figure A.19, represents just one set of pairings (individualism/collectivism with masculinity/femininity). This clearly shows the clustering of the 'Anglo-Saxon'-influenced cultures of the USA, UK, Australia, New Zealand and Canada. A proposition built on the assumptions of individualistic masculinity (like high-end executive search) is likely to succeed in this group.

2. Constructing the model

Determine the dimensions that have the most profound association with the product service or strategy. Group the firm's existing international operations and any target countries using the clustering on the relevant dimensions. Adjust the programmes to fit key clusters.

3. Use of the model

The model can be used as an aid to almost any international marketing function. It can be used to develop growth and acquisition strategy. It readily reveals compatible cultures that will be low-risk targets. It also shows how communications and product or services need to be adapted to penetrate different cultures.

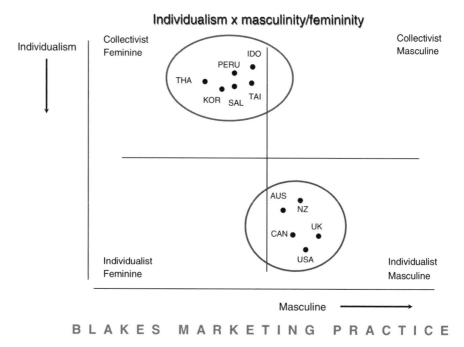

Figure A.19
Reproduced by permission of Geert Hofstede.

INDUSTRY MATURITY

Application: Strategic Insight into Market Development

1. The concept

The industry maturity curve is often mistakenly referred to as the 'product lifecycle concept'. Yet the phenomenon occurs in the sales volume of product groups over time, not individual products. There must be multiple suppliers and multiple buyers in markets that develop over time for it to be observed. (Individual products rarely go through the sales history represented in Figure A.20. Most die soon after launch.)

The concept draws an analogy between biological lifecycles and the sales growth of successful product groups, by suggesting that they are born, introduced to the market, grow in sales, mature (sales growth stops) and then decline (sales fall). In fact, it represents an iterative learning process between the buyers and suppliers in a market.

At 'birth' (the first introduction of the proposition to the world) a new product concept sells poorly. Buyers are unaware of its existence, suspicious of the new idea or experience problems when ordering (with production capacity, effective distribution or product quality). 'Bold' or 'innovative' people buy the new product during this stage as a substitute for an existing product or to meet a newly identified need. Profits may well be low or non-existent because of the high cost of sale.

In phase two, sales growth develops as a consequence of 'word-of-mouth' communication. Early buyers pass on the good experience of the product to others, or re-purchase it. Producers and distributors (whether new to the market or well established) recognise the opportunity and

The phenomenon of industry maturity

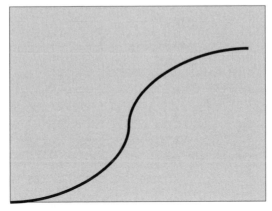

Industry sales

Time

Figure A.20

switch over to produce their own version. The market broadens through policies of product differentiation and market segmentation. Profit margins peak as experience reduces unit costs and promotional expenditure is spread over a larger sales volume.

Maturity occurs because all markets are ultimately finite (in time, volume and geography), and the market becomes saturated. Sales growth becomes more or less flat as sales settle down to a level which reflects the regular volume of new buyers entering the market plus re-purchase rates. Profits decline because of the number of competitive offerings, cost reductions become more difficult and smaller, specialist competitors enter the market.

Decline occurs as buyers switch to new offers which offer advantages or benefits not present in the existing product. Producers therefore initiate a new curve, bringing to an end that of the product group to be displaced. Declining sales are accompanied by falling profit margins as too many competitors fight for the remaining market. Price-cutting is prevalent and marginal competitors move out of the market.

This is a basic description of the industry maturity concept that has been tested, examined, criticised and developed over the past four decades. It was first observed by an economist and then brought to prominence by leading marketing writers. It is also clearly related to the sociological phenomenon: the diffusion of innovations. The history of its development gives clues to its usefulness today, especially in service markets.

2. Constructing the curve

Industry sales figures are often difficult to obtain for service markets. It is often easier to obtain government or industry figures on the number of firms operating in a particular market. By making a judgement about average industry project size and case load, the volume of sales of a type of service can be estimated. If this analysis is backdated, the category's sales curve will be observed. The diagram below, Figure A.21, shows an actual version for the UK executive search market.

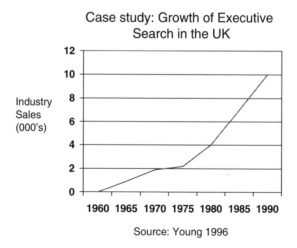

Case study: Growth of Executive
Search in the UK

Source: Young 1996

Figure A.21

3. Use of the concept

As the phenomenon occurs when there are independent variables (i.e. groups of suppliers and buyers), and rarely applies to individual products, it can be used to create marketing strategy. Many have found that the concept can be used to form a judgement about movements in the category's sales growth curve. They can develop marketing strategies appropriate to each stage in the lifecycle.

The concept is also useful in indicating the maturity of a market (i.e. the relationship of the group of customers to the group of suppliers, and the level of maturity of the understanding of the concept). An individual product being offered into that marketplace can then be adjusted in the light and understanding of that relationship. The likely strategies in each phase are summarised in Table A.1 below.

Table A.1   Strategic implications of different phases of industry maturity

Characteristics	Introduction	Growth	Maturity	Decline
Sales	Low	Fast growth	Slow growth	Decline
Profits	Negligible	Peak levels	Declining	Low or zero
Cash flow	Negative	Moderate	High	Low
Customers	Innovative	Mass market	Mass market	Laggards
Competitors	Few	Growing	Many rivals	Declining
Responses				
Strategic focus	Expand market	Penetration	Defend share	Efficiency
Marketing spend	High	Declining	Falling	Low
Marketing emphasis	Product awareness	Brand preference	Brand loyalty	Selective
Distribution	Patchy	Intensive	Intensity	Selective
Price	High	Lower	Lowest	Rising
Product	Basic	Improved	Differentiated	Rationalised

Source: Doyle (1976).

Service firms can, therefore, use this concept to understand the position of their firm and service in the light of a total market's evolution. Business strategy for their business should be developed in the light of it.

MARKET AUDIT

1. The tool

The market audit is an attempt to conduct dispassionate, objective and logical analysis of a market. The aim is to think through a number of aspects of a market. As each step is completed, 'insights' from that analysis are distilled into a summary of market perspective. This, in turn, can be used as a basis for the development of marketing strategy.

2. Constructing the tool

The analysis is conducted in the following steps. Not all need to be started sequentially but each yields insights that supplement others and may cause some to be re-evaluated. Analysis of the macro market and a lack of good data might, for instance, prompt the need for detailed customer or competitor research, unforeseen at the start of the process.

Step 1. Analysis of external forces affecting the market

This helps the firm to understand the macro-economic forces shaping markets and which are creating or destroying opportunities within them. They include:

- Understanding the raw forces affecting the market. Changes in economics, social demography and technology affect the prosperity of the market and there is little the firm can do to influence them.
- Understanding 'moderating forces'. Politics, law and industry-specific regulation moderate the impact of raw forces on the market. These can be influenced by the firm.

By trawling through published data on these issues and drawing them into a perspective on change, surprisingly powerful insights can be found. They could, for instance, highlight an issue on which firms might need to lobby regulators.

These external forces are often remembered with the mnemonic 'PESTEL'.

Step 2. Understand market structure

- Decide the market's definition.
- Plot the maturity of the market.
- Determine purchasing power. What is the balance of power between suppliers and buyers?
- Examine the competition. Who are they and what are their strategies?
- Analyse substitutes. Can buyers get the benefits of the service in any other way?
- Segment the market.

Step 3. Detailed buyer analysis and research

This involves analysis of research into the needs and aspirations of existing and intended buyers. It is wise to start by collecting all published research and previously conducted research projects in order to identify gaps in knowledge. The team may then come to the conclusion that a specific research study is needed to fill important gaps in knowledge. If so, this is likely to be the most costly and longest aspect of the audit, yet the most valuable. Reassess segmentation, the service offer, pricing and communication approaches in the light of this research.

Step 4. Internal analysis

This is about understanding the internal position of the firm within its market by detailed analysis of both the firm's own competencies and the profile of its clients. It includes the source of business, revenue and income trends and the potential for growth. This analysis can yield surprising insights. Some firms, for example, believe that their buyers are chief executives, when analysis shows them to be lower-level specialists. Others have been surprised to find that larger corporate accounts are less profitable than mid-market, smaller customers. Such insights can yield real benefit for the firm's approach to its market.

3. Use of the tool

The market audit tends to be used in large organisations as a basis for logical, linear market planning. Done properly, it can yield profound market insights. It is, though, based on an economic view of a market, so there is a danger that behavioural insights will be missed.

MARKET MATURITY

See entry under industry maturity.

MAVENS

1. The concept

The idea of market mavens was popularised by Malcolm Gladwell (Gladwell, 2000). They are intense gatherers of information and so are often the first to pick up on new or nascent trends. Often found amongst innovators, they tend to not only enjoy gadgets and technology but also to try to sell it to others in their personal social network. They will be motivated by the belief that a particular innovation will help their friends and family.

2. How to construct it

Attitudinal research and test marketing programmes will identify mavens in any intended market.

3. How to use it

Create a specific marketing programme to test ideas on mavens and encourage them to educate others. This might take the form of a club or social group. Ideas can be tested and word-of-mouth started by pre-launching new ideas to these 'super users'.

MARKETING MIX

Application: Planning

1. The tool

This concept focuses on the aspects of marketing which need to be coordinated in order to influence the buyers. Although several academics have previously pointed out that marketing was a 'mix' of activities aimed at a buyer, it was Jerome McCarthy (McCarthy, 1960) who reduced them to the, now famous, 'four P's' of marketing, see Figure A.22. They are:

- The product; or the offer to clients.
- The price at which the product is offered.
- The promotion of the product to the target buyers.
- The placing of the product in the market through sales and distribution channels.

  Classic marketing training emphasises that all these elements need to be planned in order to achieve success. However, there are two other ingredients. The first is a clear knowledge of the target market. Suppliers need to know, in detail, the attributes and benefits that the buyers will value. The second is the 'mix' of components that will most appeal to the buyers. These need to be planned and balanced carefully.

  In reality, few marketers have direct-line responsibility for all the components of the mix. They therefore need to influence these other areas in order to achieve their objectives and to create value for their employers. Experience suggests that they will fail to have impact and

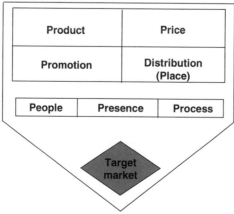

Figure A.22

their work may as well not be attempted if they are restricted to just short-term tactical aspects of one or two aspects of the mix.

Service marketing academics later demonstrated that the mix for a service business is different. Firstly, the offer is not a tangible product but a proposition which is likely to be a mix of intangibles and tangibles. This changes marketing dramatically. However, it is generally accepted that there are three further aspects of the marketing mix for services, three extra 'P's'. They are:

- The people who deliver the service because the buyer often cannot separate them from the value they buy.
- The physical evidence, or tangible aspects of the offer designed to help deliver perceived value to the buyer.
- The process through which the buyer moves while using or buying the service.

Again, all aspects of the mix need to be designed to match the aspirations of the intended buyers.

There is, of course, complexity behind this concept. Each 'P' has many detailed aspects. This book has, for example, complete chapters dedicated to the 'P' of promotion and that of service proposition.

2. Constructing the tool

In any planning situation simply list the elements of the mix and ensure that they have been considered for the particular target market; try to understand how each set of variables might influence their purchase intent.

3. Use of the tool

The marketing mix is most often used in management dialogue or communications. Those involved can use it as an informal checklist to ensure that all aspects of a proposition have been properly considered. However, it can also be used in detailed marketing planning. Once strategy has been decided, a full campaign which comprises all elements of the mix should be created for each target market.

MOLECULAR MODELLING

Application: Service Design

1. The tool

This is a method of planning the detailed components of product or service offers. It was suggested as a technique, based on actual experience, by Lyn Shostack (Shostack, 1977).

The technique allows planners to create a picture of the total proposition, whether service or product-dominant. It reflects the fact that propositions might have different degrees of physical or service components without diminishing the importance of either. It also allows marketers to adjust their technique according to the degree of service content in the total offer.

The method breaks down the offer into 'tangible' and 'intangible' elements. Tangible elements are represented by a firm circle whereas intangible elements are represented by dotted lines. The outer rings represent various aspects of marketing such as price, distribution

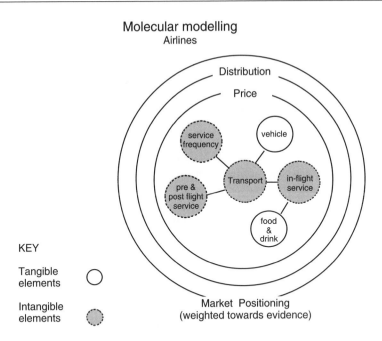

Figure **A.23**

and market positioning. Lines interconnecting the various elements show the interrelationship of process in delivering the service. The classic representation of the technique (for airlines) is reproduced in the diagram above, see Figure A.23.

The benefit of this technique is that it acknowledges that there are tangible and intangible elements to the offer. It allows the service designer to vary components to integrate service components into a product.

2. Constructing a molecular model

  (i) Identify the 'nucleus' of the proposition. (In the case of cars it is personal transportation.)
 (ii) Identify physical and intangible elements.
(iii) Link the elements.
(iv) Ring the total entity and define it by a set value.
 (v) Circumscribe it by its distribution method, so that its relationship to the market is clarified.
(vi) Describe its brand positioning or 'face'.

3. Use of the tool

Molecular modelling has suffered from a lack of communication rather than being an impractical or irrelevant method. Managers responsible for the creation of new services might start to use the technique with a service which is well known in their company. By breaking it down into its components they may identify new elements of the offer which need to be designed and also ways of adjusting the offer to make it more relevant to new markets or to new segments. Having experimented with the technique, and tested it in anger, they are likely to find it a practical method for the detailed design of new service offers.

PERCEPTUAL MAPS

Application: Positioning, Competitive Strategy

1. The tool

Perceptual maps are used by brand managers in many different businesses to set strategy and to gain insight. There are two axes which, in most companies' work, are a derivative of critical issues or success factors in the market. They are normally issues which are uppermost in the minds of most buyers. By understanding these and ranking the buyers' views of all the significant offers, marketers can deduce and set direction for their own offer.

The resultant diagram, Figure A.24 (adapted from Lambin, 2000), is a 'perceptual' map of buyer views because it concentrates on their perception.

The map represents buyers' views of the issues that are important to them and the way suppliers respond to them. These views may not be technically correct or factually accurate because they represent perception. It may be that there are offers which provide excellent technical performance (one version of quality) but are not perceived to be the leading offer or the best-quality offer by the majority of the market. (This insight in itself shows a supplier that they need to change perception.)

The Figure A.24 is a generic model which uses two axes: quality and price. Together they form buyers' views of value (i.e. value f price $\times$ quality). The market leader normally takes centre stage. It has the offer by which all are judged. It sets the price/quality expectations and has the power to change the whole market. The premium position (normally a heritage brand) is taken by a features-rich offer at a high price, whilst the least-cost supplier strips out all that is possible to achieve low price. Various niche providers set themselves against the market leader and survive by providing a different offer.

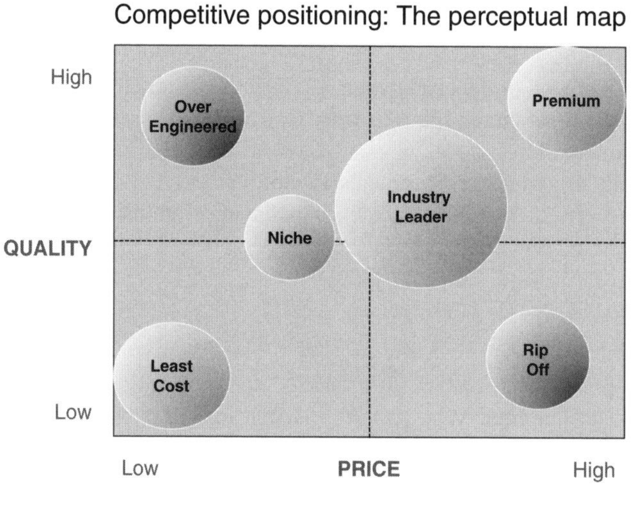

Competitive positioning: The perceptual map

Figure A.24

There are two unsustainable positions. The 'rip off' has low quality at high price. It normally exists because of some distortion or recent trauma in the market. However, it cannot keep this position in the long term because buyers will compare offers and desert it for other suppliers with an offer nearer to their values. The 'over-engineered' supplier might be a naïve new entrant or a recently privatised or de-monopolised supplier. Again, it must move in the long term.

2. Constructing the model

The tool is best constructed using detailed conjoint research amongst representatives of all buyers in the market. Contrary to popular belief, they will not all want the cheapest offer. They will want a mix of features and price which represents value to them. Some seek a features-rich offer with a high price and some a basic offer at a low price.

The likely output of this research, reflecting buyers' needs and requirements, is likely to be near to the diagram below, see Figure A.25. The buyers' ideal purchases will be scattered around a line (the line on which suppliers can achieve long-term position). In a market which is not distorted by monopoly or regulation, most buyers cluster around the middle, a position that the market leader can dominate.

The research will also reveal buyers' perceptions of where all suppliers are positioned.

3. Use of the tool

The prime use of the tool is to work out where the target buyers' values lie and to move the firm's offer towards them. 'Positioning' the firm in this way is a guide for NPD and NSD policy. It also leads communication and marketing programmes. The tool can be used to create both corporate and competitive strategy. The leadership can use it to reach consensus on the position that they want the firm to take in the market. It is therefore a very powerful input to strategy debate amongst leadership teams.

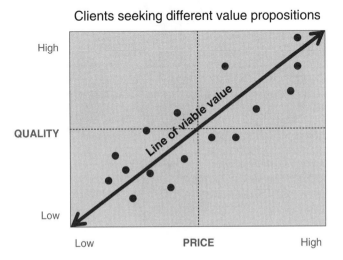

Figure A.25

The tool can also be used as an internal communication aid. Firstly, it is a very simple and clear summary of the market and the firm's ambitions which can be used effectively in internal meetings. Secondly, it can be used as a diagnostic tool to develop internal communications programmes. By using it as a basis for internal research of employee opinion, it can be contrasted against customers' views. Training and communications programmes can then be designed to address any gap in view.

PESTEL

See entry under market audit.

PORTER'S COMPETITIVE FORCES

Application: Competitive Analysis, Market Analysis, Strategy Development

1. The tool

Amongst Michael Porter's impressive and prodigious work on competitive strategy he offered a powerful conceptual framework (Porter, 1979) which works well as part of the market analysis and strategy development process. His 'five forces' of competition are a useful checklist for marketers to work through when analysing a market, see Figure A.26. They are:

● The power of buyers. Buyers can influence a market by forcing down prices, by demanding higher service and quality or by playing competitors off against each other. Porter suggested that there are a number of circumstances when a buyer group is powerful, including: if they

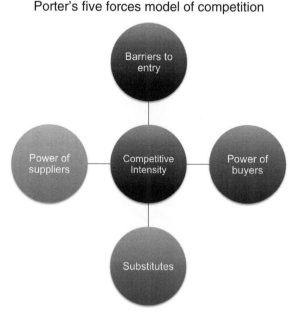

Figure A.26
Reprinted with permission from *Harvard Business Review*, Porter, March/April 1979.

are concentrated, if buying commodities, or components, if driven to get price cuts or if the purchase is unimportant to them.

- The power of suppliers. Suppliers can execute power by raising prices or reducing the quality of the offer. They can squeeze profitability out of the industry. They are powerful if: dominated by a few, have unique offers, are not obliged to compete, threaten forward integration or are not part of an important industry to the buyers.
- The threat of new entrants. These bring new capacity, the desire for market share and resources. The seriousness of the threat depends on barriers to entry which have six sources (economies of scale, product differentiation, capital requirements, cost advantages, access to distribution and government policy).
- The threat of substitute offers. These affect the profit of an industry by placing a ceiling on what it can charge through offering an alternative price/feature option.

2. Constructing the tool

The concept can simply be used as a checklist to prompt planners to cover relevant issues during market analysis and strategy development. However, it is most powerful when good analysis is put behind the thinking so that judgements can be made with the benefit of real data. Industry reports and original research can be summarised into the model and used as criteria by which to develop competitive responses or critical success factors.

3. Use of the tool

The tool can be used to guide debate and is also effective as a communications device. Its clarity summarises graphically and quickly the competitive landscape and can be used as part of the rationale for competitive programmes. It is best used, though, as a background planning tool in the market planning process.

RESEARCH

Application: Client, Competitor or Market Insight

1. The tool

Field research is familiar not only to marketers but also to many business leaders. They may have seen the results of research presented at internal meetings or read research reports. Unfortunately, familiarity can breed contempt, making the processes, the techniques and the outcomes seem deceptively simple. As a result, there are many unconvincing or poor research reports resulting from poor specification or poor use of the research industry.

Yet, undertaken properly, field research yields insights into human needs, customer views (which can be different from needs), competitor performance and market trends. It can reveal the different elements of a service which customers value and how they combine with different price points to form packages that they will buy. Moreover, it can reveal how these vary between different customer groups, creating opportunities through variation of offer in different market segments. It can also save money by stopping new service ideas or marketing initiatives which the market will reject. Yet to do all this it has to be properly specified and managed. It needs a brief and a managerial process if it is going to produce results. This needs to ensure that the sample frame, the approach, the technique and the questionnaires are appropriate.

2. Constructing research projects

There are two main types of research. The first is the qualitative or 'in-depth' approach. This involves spending time with a relatively small number of people and seeking deep answers to questions. It gives colour to views and can reveal underlying feelings and motivations that can be enormously valuable. Quantitative research, on the other hand, involves a wider number of contacts, normally to investigate trends. Both have their strengths and weaknesses.

A typical process includes:

1. Agree objectives and research needs.
2. Write brief for agencies. Briefs tend to include:
   - Research objectives.
   - Summary description of the market.
   - Description of the research problem and desired output.
   - Description of existing knowledge and previous research.
   - Budget constraints.
   - Time scales.
   - Report requirements.
   - Constraints (e.g. interviews must be arranged via client relationship managers).
3. Shortlist potential agencies.
4. Contact and invite to pitch.
5. Create selection criteria. These might include:
   - Technical skills.
   - Previous experience.
   - Interpretation of the brief.
   - Proposed approach.
   - Team fit (will the firm's people be able to work with them?)
6. Hold presentations by agencies to selection team.
7. Choose and confirm agency.
8. Negotiate contract.

Methods used to collect data vary enormously. They range from face-to-face interviews and observed discussion groups to telephone, postal or Internet surveys. All are used in the service industry.

There are, however, several different research techniques.

(i) Conjoint research. This uses questions (either in face-to-face meetings or via mail or telephone) designed to trade off different pairs of values or ideas. Interviewees are forced to choose. It mimics the thought processes of customers when considering purchase and yields the type of detailed output illustrated in the diagram on next page, see Figure A.27. It can provide powerful insight into new service needs and adjustments needed to customer service.

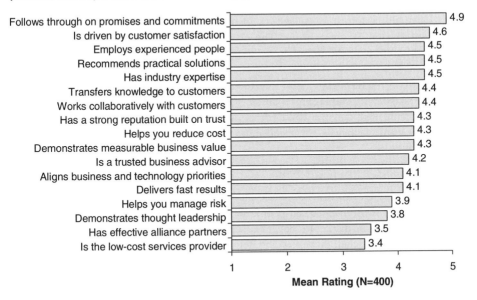

Importance Ratings of IT professional Services Firms' Attributes

When you and your company are selecting a professional services consultant or solution provider, how important is it that this vendor ___?

Attribute	Rating
Follows through on promises and commitments	4.9
Is driven by customer satisfaction	4.6
Employs experienced people	4.5
Recommends practical solutions	4.5
Has industry expertise	4.5
Transfers knowledge to customers	4.4
Works collaboratively with customers	4.4
Has a strong reputation built on trust	4.3
Helps you reduce cost	4.3
Demonstrates measurable business value	4.3
Is a trusted business advisor	4.2
Aligns business and technology priorities	4.1
Delivers fast results	4.1
Helps you manage risk	3.9
Demonstrates thought leadership	3.8
Has effective alliance partners	3.5
Is the low-cost services provider	3.4

Mean Rating (N=400)

Figure A.27

(ii) Observational research. As the name suggests, this involves a researcher observing the behaviour of people. It can give real insight into behaviours which reflect customer views of service quality and future service needs.

(iii) Explorative research. This is normally used in developing a new proposition or identifying a new customer segment, this technique follows issues until a trend suggests that they are likely to be substantive. The work can be iterative, checking back and adjusting the idea as interviewees respond.

(iv) Concept testing. This involves testing ideas for new propositions or approaches with customers before launch. They are shown an idea or marketing programme and asked to comment on it in a structured way.

3. Use of the tool

Field research ought to carry an arrogance warning. Managers and leaders in all sectors of industry can be very dismissive of it. Many have been heard to remark that there is little that field research can tell them about their buyers. Yet they are almost always wrong. In fact, some very senior business leaders have been chastened by the direct comments they have heard their buyers make when sitting behind two-way mirrors watching focus groups.

Research does, however, need to be used properly. It needs to fill a gap in knowledge. It is sensible to first conduct an exhaustive 'desk review' to see if an Internet search can point to libraries, professional societies or academic institutions that have already conducted research or provided commentary on the subject in question. In fact, many people in large companies which lack a structured research library find, after a brief search, that their own company has conducted research near to the subject in question on previous occasions.

Once this preliminary work has been completed, the gap in knowledge ought to be clearly defined in a brief to a specialist research agency. This should specify the purpose of the research. It might be to test a new service idea, test segmentation dimensions, understand a new concept or identify client needs. The exercise will be confused if there are too many objectives. In particular, the information yield that is expected must be made clear to the supplier.

Once results come through, careful interpretation is needed. It is important to understand, not only the statistically valid representation of results but also their meaning. Human beings often do not know what they want and, sometimes, why they behave in a particular way. They will say they want cheapness yet spend outrageous sums on a consultancy project from a branded supplier.

There are numerous examples of mistakes due to poor interpretation. In the late 1980s, for example, one American telecoms supplier commissioned a leading consultancy firm to judge the ultimate size of the worldwide mobile phone market. The consultancy, which was not a specialist research firm, interviewed people in various parts of the world and estimated the size of the mobile phone market to 'never exceed a million handsets'. They did not understand people's social need to chat, move fast and get a grip on personal communications.

Research therefore needs careful commissioning, good execution and enlightened interpretation. Yet the expense and effort is more than worthwhile. It can lead to profitable new insights that build strong future revenue streams and it can obviate mistakes.

SEGMENTATION

1. The concept

The segmentation of markets into groups of buyers which can be easily reached by suppliers is a powerful concept which has improved the profit of many businesses. It suggests that buyers can be grouped around common needs. Then, by customising the firm's offer to meet these common needs, suppliers can both gain competitive advantage and save costs because they are only addressing a portion of the market.

There have been several methods of segmentation developed and publicised over the past few decades. They include:

Demographics and socioeconomics. The grouping of people according to physical characteristics (age, sex) or circumstances (income, occupation or education). This is commonly used in developed nations. In the USA and Europe, for example, there is currently much emphasis on the design of products for ageing populations.

Life stage. This is a more precise form of demographics. It groups buyers according to the phase they have reached in their life, such as 'married', 'home building' or 'retired'. They might become 'freedom seekers', 'dropouts' or 'traditionalists' according to their phase of life.

Psychographics. The grouping of people according to various personal characteristics such as personality or social class. In the 1980s, for instance, the British National Readership Survey categorised the population as 'A' (higher marginal), 'B' (middle class), 'C' (lower middle class) and 'D' (working class). However, this is now breaking down.

Geographic/location. Grouping people according to their country of birth or area of residence. This can focus on the region, population density and climate. It can involve county, town or even street.

Behavioural or attitudinal. Grouping according to a particular behaviour which may affect product usage or price sensitivity, or values and attitudes. A good example was created by the marketing agency McCann-Erickson in the latter half of the 20th century. It identified: 'avant guardians' (concerned with change and well-being), 'pontificators' (who have strongly held traditional views) and 'self-exploiters' (who have high self-esteem).

'Tribal'. A specific example of behavioural segmentation which groups customers according to the social groups or cultures with which they identify. For example, in the 1990s, one of Europe's premier television companies started to commission programmes for tribes in society (such as young, independent women) based on how they communicate and live.

Benefits sought. The grouping of people according to the advantages they are seeking from the product or service. For instance, as early as 1968, Russell Haley (Haley, 1968) published segmentation for the toothpaste market based on this approach. Customers were in the 'sensory' segment (seeking flavour or product appearance) or the 'sociable' segment (seeking brightness of teeth) or the 'worriers' (seeking decay prevention).

Lifestyle. Grouping customers by a common approach to life. One famous example of this type of segmentation was developed by Young & Rubican in the 1980s. It was this advertising agency which developed, among others, the famous, but now defunct term, 'YUPPIE' (Young Urban Professional). Incidentally, this also illustrates an important point about customer segmentation: it dates easily. Whereas people revelled in being a Yuppie in the early 1980s, it is now considered out-of-date and unattractive.

Context. Proposed by Dr Paul Fifield in the early 1990s (Fifield, 1992), this method groups customers according to the context in which they use a product or service. It focuses attention on things that bring people together, exploiting shared interests. For instance, one cursory glance at people on a fishing bank will show that they have little in common other than the sport itself.

Business-to-business segmentation types include:

- **Industry sector.** Grouping businesses according to the industry in which they specialise. These sectors are often formally set by government economists as a means of defining and recording activities in different areas of the economy.
- **Organisation style.** Grouping businesses according to the culture or prevailing climate of the company. They may be centralised or de-centralised or 'innovative' versus 'conservative'. The Myers Briggs organisational types ('fraternal', 'collegial', 'bureaucratic' or 'entrepreneurial') have, for example, been used as a basis for segmentation.
- **Organisation size.** Grouping businesses according to the number of employees, assets or revenue.
- **Company lifecycle.** Companies, like product groups, have a 'lifecycle' through which they evolve. They go from birth to death at different rates, struggling to get through 'inflection points' to increase revenue and margin. They have similar characteristics (e.g. management style) in each phase and this has been used as a basis for segmentation.

- **Industry maturity.** Industries also move through different phases. For instance, in developed economies, their agricultural or manufacturing industries are at a different phase of evolution to, say, biotechnology. The phase of growth affects the behaviour of suppliers in it and has also been used as a basis for segmentation.
- **Context.** As with consumers, grouping businesses according to the context in which they use the product or service.
- **Needs/benefits.** This is based on underlying needs or benefits sought by the company from its suppliers.

Clear tests have been developed to check whether a particular segmentation is appropriate for a particular company in a particular market. They include:

- **Homogeneity.** To what extent will the members of the segment act in the same way?
- **Measurability.** How big and valuable is the segment?
- **Accessibility.** Is it possible to reach the segment with marketing or sales programmes?
- **Profitability.** Is the segment substantial enough for the supplier to profit from?
- **Attractiveness or relevance.** Is the segment something customers will want to identify with?

The last point is particularly powerful. If the target group can be expanded by people who aspire to belong, demand for the proposition will be increased.

2. Constructing the tool

There appear to be few developed and well-accepted approaches to segmentation. The following process is an amalgamation of several suggested by researchers.

Step 1 – Review all known segmentation methods. A group of experienced leaders should be drawn together to discuss segmentation as a subject and the types previously created, their benefits and drawbacks.

Step 2 – Create a hypothesis. In discussion, the team will create an idea of how they think their market might segment. If they have existing customers in the market, they will need to think about how they behave. They may have to customise or examine market research or industry reports to get to the heart of this. In particular, they will need to discuss different attitudes or behaviours that they have observed. Eventually they will reach an idea of which previous segmentation type best fits their view of buyers in the market and how it might be adapted to their market. In doing this they are creating a hypothesis which can be tested.

Step 3 – Create segmentation dimensions. Segmentation dimensions are the ways in which the buyers will behave towards the firm and its services. As far as possible, they should be values, beliefs or cultural biases (whether in consumers or organisations) because they determine behaviour. If the segmentation is effective, each group will exhibit these in different ways.

Through discussion, the team should create a set of segmentation dimensions by which these differences will manifest themselves. These can then be scaled using sensible scoring of the extremes. For example, if a supplier was segmenting a business-to-business market on the basis of 'organisation style', they might hypothesise that there are centralised and de-centralised organisations. This would manifest itself in different business practices, one of which would be purchasing style and it would become a 'segmentation dimension'. For a centralised organisation, buying would be controlled by a central purchasing department, whereas in a de-centralised organisation it would be devolved to business units.

Step 4 – First test: use data to test the segment dimensions. By examining any existing customers and scoring their behaviour against segmentation dimensions, it is possible to conduct a fast, inexpensive test on the validity of the dimensions and where different clusters appear. If no clustering appears in this first test, then new dimensions, or maybe a new hypothesis, need to be created.

Step 5 – Second test: research. Any clustering should be confirmed with direct research. This is best conducted in two phases: first, a qualitative phase, testing the dimension in depth with a few potential customers; second, a large quantitative project using a trade-off technique such as conjoint research. Through this method different clusters of potential buyer groups will become evident. Again, if clustering does not appear the team should revisit its initial hypothesis.

Step 6 – Third test: test marketing programmes. Research itself is probably not sufficient to confirm such an important subject as customer segmentation. Potential segments can also be confirmed in a more practical way, imitating as far as possible the rough and tumble of the real marketplace. A number of test marketing programmes could be designed in order to ensure that the buyers identify with the proposed groups that they are in and respond to propositions specifically designed for them.

Step 7 – Create a full investment and market entry plan. Segmentation has implications for: the market proposition for each group, customer service methods for each group, the ideal method of marketing communication, sales strategy, IT systems, operational processes and pricing. Taken seriously, it affects every aspect of the way the firm approaches its intended market. Each aspect needs to be carefully thought through, costed and built into the market entry plan.

Step 8 – Gain approval for the investment. This needs to be treated like any other hard-headed investment strategy. The pros and cons, benefits and return on investment need to be assessed and drawn together into the market entry plan, which should be submitted to the appropriate leadership team for formal approval.

Step 9 – Implementation. As part of implementation of the plan, everyone in the new service organisation will eventually need to be familiar with the new segments and how they should be handled.

3. How to use the tool

Customer segmentation is one of the fundamental building blocks of marketing. It is the basis of: effective marketing communication, innovative new services and appealing brands. The human insight from deep knowledge of a group of human beings is the source of real competitive advantage. Segmentation, to some extent, is marketing.

SWOT ANALYSIS

Application: Strategy Development

1. The tool

Probably the best known of the strategy tools, this matrix helps structure discussion by summarising a firm's strategic position into: strengths, weaknesses, opportunities and threats, see Figure A.28.

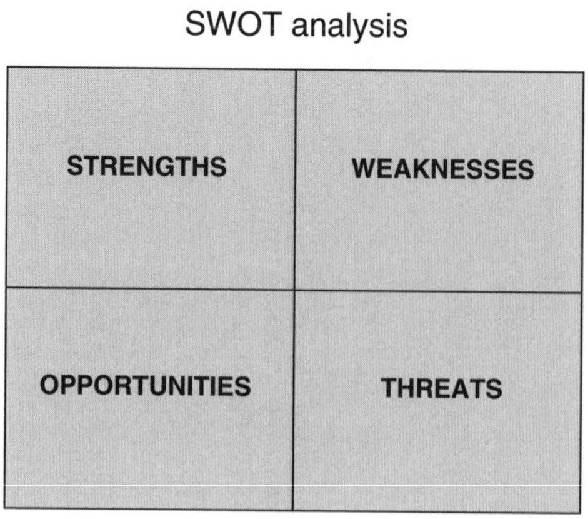

SWOT analysis

Figure A.28

2. Constructing the matrix

A SWOT analysis can easily be created during discussion amongst a management team. Valuable insight and debate can emerge from its succinct summary of the firm.

It can, however, be constructed using detailed analysis. The market analysis techniques outlined in Chapter 3 can be summarised into it. Competitive analysis will reveal 'threats', for example, and client research will give insight into 'opportunities'. On the other hand, an environmental analysis, which reviews 'PEST' factors, can contribute to both of these.

Those wishing to take a thorough, analytical approach can use the 'TOWS' method. This suggests that each item in the matrix is numbered. Then each threat is compared against each weakness and each opportunity against each strength, in a systematic search for strategic options. As the strengths and weaknesses arise from debate about the firm's competencies, the strategist is, in fact, checking these against market developments through this process.

3. Use of the tool

The tool is best used in the strategy development process to summarise analysis. It allows senior people to focus debate and decision-making.

References

Aaker, D. (1996). *Building Strong Brands*. The Free Press.
Alderson, W. (1957). *Marketing Behavior and Executive Action*. Homewood.
Ambler, T. (2003). *Marketing and the Bottom Line*. FT/Prentice-Hall, 2nd edn.
Anderson, E. and Onyemah, V. (2006). 'How right should customers be?', *Harvard Business Review*, **Jul/Aug**.
Ansoff, I. (1957). 'Strategies for diversification', *Harvard Business Review*, **Sept/Oct**.
Auguste, B.G., Harmon, E.P. and Pandit, V. (2006). 'The right service strategies for product companies', *The McKinsey Quarterly*, No. 1.
Bartels, R. (1988). *The History of Marketing Thought*. Publishing Horizons.
Bateson, J.E.G. (1985). 'Perceived control and the service encounter'. In *The Service Encounter*, Czepiel, J.A., Solomon, M.R. and Surprenant, C.F. (eds). Lexington Books.
Bateson, J.E.G. and Hoffman, D.K. (1999). *Managing Services Marketing*. Thomson Learning.
Berry, L. (1988). 'In services what's in a name'. *Harvard Business Review*, **Sept/Oct**.
Berry, L. (1995). *On Great Service*. The Free Press.
Berry, L. and Parasuraman, A. (1991). *Marketing Services: Competing through quality*. The Free Press.
Branson, R. (2005). *Losing my Virginity*. Virgin Books.
Bulmore, J. (2001). *Posh Spice & Persil*. Published lecture to the British Brands Group.
Bundschuh, R.G. and Dezvane, T.M. (2003). 'How to make after sales services pay off', *The McKinsey Quarterly*, No. 4.
Colletti, J.A. and Fiss, M.S. (2006). 'The ultimately accountable job', *Harvard Business Review*, **Jul/Aug**.
Cowell, D. (1984). *The Marketing of Services*. Heinemann.
Darby, M.R. and Karni, E. (1973). 'Free competition and the optimal amount of fraud'. *Journal of Law and Economics*, **16**(Apr).
Davidow, W.H. and Uttal, B. (1989). 'Service companies: focus or falter', *Harvard Business Review*, **Jul/Aug**.
De Brentani, U. (1991). 'Success factors in developing new business services', *European Journal of Marketing*, **25**(2).
Doyle, P. (1976). 'The realities of the product life cycle', *Quarterly Review of Marketing*, **Summer**.
Edgerton, D. (2007). *The Shock of the Old*. Profile Books.
Evans, M.R., Ribeiro, C.D. and Salmon, R.L. (2003). 'Hazards of healthy living: bottled water and salad vegetables as risk factors for *Campylobacter* infection', *The Journal of Emerging Infectious Diseases*, **9**(10).
Fifield, P. (1992). *Marketing Strategy: How to Prepare it – How to Implement it*. Butterworth-Heinemann.
File, F.K., Mack, J.L. and Prince, R.A. (1995). 'The effect of interactive marketing on commercial satisfaction in international finance markets', *Journal of Business & Industrial Marketing* **10**(2).
Ford, D. (1992). *Understanding business marketing and purchasing*. Thomson Learning
Friedel, R. (2007). *A Culture of Improvement; technology and the Western millennium*. MIT Press.
Gerstner, L.V. (2002). *Who Says Elephants Can't Dance? Inside IBM's Historic Turnaround*. Harper Business.
Gladwell, M. (2000). *The Tipping Point*. Little, Brown.

Grönroos, C. (2003). *Service Management and Marketing*. John Wiley & Sons Ltd.

Gruglis, I. and Wilkinson, A. (2001). *British Airways: culture and structure*. Published research paper, Loughborough University.

Håkansson, H. and Johanson, J. (1992). 'A model of industrial networks: A new view of reality'. In *Understanding Business Marketing and Purchasing. An Interaction Approach* (ed). D. Ford. Thomson Learning.

Haley, R.I. (1968). 'Benefit segmentation: a decision-orientated tool', *Journal of Marketing*, **July**.

Halliday, S.V. (2004). 'How "placed trust" works in a service encounter', *Journal of Service Marketing*, **8**(1).

Hamel, G. and Prahalad, C.K. (1989). 'Strategic intent', *Harvard Business Review*, **May/Jun**.

Hamel, G. and Prahalad, C.K. (1990). 'The core competence of the organisation', *Harvard Business Review*, **68**.

Hammer, M. and Champy, J. (1993). *Reengineering the Corporation*. HarperCollins.

Harris, T.A. (1973). *The Book of Choice* (later called *I'm OK, You're OK*). Jonathan Cape.

Hart, W.L., Heskett, J.L. and Sasser, W.E. (1990). 'The profitable art of service recovery', *Harvard Business Review*, **Jul/Aug**.

Harte, H.G. and Dale, B.G. (1995). 'Improving quality in professional service organizations: a review of the key issues', *Managing Service Quality*, **5**(3).

Henderson, B.D. (1972). *Perspectives on Experience*. Boston Consulting Group Publication.

Heskett, J.L., Sasser, W.E. and Schlesinger, L.A. (1997). *The Service Profit Chain*. The Free Press.

HM Treasury. (1992). *Her Majesty's Treasury Guide to the UK Privatisation Programme*. HM Treasury.

Hobsbawm, E. (1999). *Industry and Empire*. Penguin.

Hofstede, G. (1984). 'Cultural dimensions in management and planning', *Asia Pacific Journal of Marketing*, **Jan**.

Humble, J. (1989). 'Satisfying the customer', *Harvard Business Review*, **Jul/Aug**.

Joachimsthaler, E. and Aaker, D.A. (1997). 'Building brands without mass media', *Harvard Business Review*, **Jan/Feb**.

Javalgi, R.G., Martin, L.M. and Todd, P.R. (2004). 'The export of e-services in the age of technology transformations', *Journal of Services Marketing*, **18**.

Johnston, R. and Clark, G. (2001). *Service Operations Management*. Prentice-Hall.

Johnston, R. and Clark, G. (2005). *Service Operations Management*, 2nd edn. Prentice-Hall.

Keynes, M. (1920). *The Economic Consequences of the Peace*.

Klein, N. (2001). *No Logo*. Flamingo.

Kotler, P. (2003). *Marketing Management*. Prentice-Hall.

Kotler, P. and Armstrong, G. (2003). *Principles of Marketing Management*, 10th edn. Prentice-Hall.

Kotler, P., Rackham, N. and Krishnaswamy, S. (2006). 'Ending the war between sales and marketing', *Harvard Business Review*, **Jul/Aug**.

Lambin, J. (2000). *Market-driven Management*. Macmillan.

Levitt, T. (1966). 'Innovative imitation'. *Harvard Business Reviews*. **Sept/Oct**.

Levitt, T. (1966). 'Market myopia', *Harvard Business Review*.

Levitt, T. (1972). 'Production line approach to service', *Harvard Business Review*, **Sept/Oct**.

Levitt, T. (1976). 'The industrialization of service', *Harvard Business Review*, **Sept/Oct**.

Levitt, T. (1983). 'The globalization of markets', *Harvard Business Review*, **May/June**.

Maister, D.H., Green, C.H. and Galford, R.M. (2000). *The Trusted Advisor*. The Free Press.

McCarthy, E.J. (1960). *Basic Marketing*. Richard D. Irwin Inc.

McDonald, M. (2002). *Marketing Plans*. Butterworth-Heinemann.

McKenna, R. (1991). 'Marketing is everything', *Harvard Business Review*, **Jan/Feb**.

Mintzberg, H., Ahlstrand, B. and Lambel, J. (2005). *Strategy bites back*. Prentice Hall.

Monnoyer, E. and Spang, S. (2005). 'Manufacturing lessons for service industries: An interview with AXA's Claude Brunet', *The McKinsey Quarterly*, May.

Moore, G.A. (1991). *Crossing the Chasm*. Harper Business.

Palmer, A. (2005). *Principles of Services Marketing*. McGraw-Hill.

Parasuraman, S., Zeithaml, V.A. and Berry, L. (1985). 'A conceptual model of service quality and its implications for future research', *Journal of Marketing*, **49**(Fall).

Parasuraman, S., Zeithaml, V.A. and Berry, L. (1991). 'Understanding customer expectations of service', *Sloan Management Review*, **Spring**.

Parker, D. (2009). *The Official History of Privatisation*. Routledge.

Pendergrast, M. (1994). 'For God, country and Coca-cola'. Phoenix.

Peters, T.J. and Waterman, R.H. (1982). *In Search of Excellence: Lessons from America's Best Run Companies*. Harper & Row.

Piercy, N. (2001). *Market Led Strategic Change: Transforming the Process of Going to Market*. Butterworth-Heinemann.

Porter, M. (1979). 'How competitive forces shape strategy', *Harvard Business Review*, **Mar/Apr**.

Rackham, N. (1995). *SPIN Selling*. Gower.

Reicheld, F.F. (2003). 'The only number you need to grow', *Harvard Business Review*, **Dec**.

Reicheld, F.F. (2006). *The Loyalty Effect*. Harvard Business School Press.

Reicheld, F.F. and Sasser, W.E. (1990). 'Zero defections: quality comes to services', *Harvard Business Review*, **Sept/Oct**.

Ringland, G. (1997). *Scenario Planning: Managing for the Future*. John Wiley & Sons Ltd.

Rodgers, B. (1986). *The IBM Way*. Harper & Row.

Rogers, E.M. (2003). *Diffusion of Innovations*. The Free Press.

Ruskin Brown, I. (2005). *Marketing your Service Business*. Thorogood.

Sampson, H. (1875). *A History of Advertising*. Chatto and Windus.

Shamir, B. (1980). 'Service and servility: role conflict in subordinate service roles', *Human Resources*, **33**(10).

Shapiro, B. (1987). 'What the hell is market orientated?', *Harvard Business Review*, **Aug**.

Shostack, G.L. (1977). 'Breaking free from product marketing', *Journal of Marketing*, **Apr**.

Shostack, G.L. (1982). 'How to design a service', *European Journal of Marketing*, **16**(1).

Sieff, M. (1986). *Don't Ask the Price: The Memoirs of the President of Marks and Spencer*. Weidenfeld and Nicolson.

Silk, A.J. (1996). 'Brand valuation, a simple example', *Harvard Business School Note*, N9.

Stone, M. and Young, L. (1994). *Competitive Customer Care*. Croner.

Trailer, B. and Dickie, J. (2006). 'Understanding what your sales manager is up against', *Harvard Business Review*, **Jul/Aug**.

UN (2008). *UNCTAD Handbook of Statistics*, United Nations Conference on Trade and Development.

Üstüner, T. and Godes, D. (2006). 'Better sales networks', *Harvard Business Review*, **Jul/Aug**.

Vaghefi, M.R. and Huellmantel, A.B. (1998). 'Strategic leadership at General Electric', *Long-Range Planning*, 31.

Vandermerwe, S. (1988). 'Scandinavian airlines system', *Harvard Case Study*, N8.

Zaltman, G. (2003). *How Customers Think*. Harvard Business Press.

Zeithaml, V. and Bitner, M.J. (2003). *Services Marketing*. McGraw-Hill.

Zoltners, A.A., Sinha, P. and Lorimer, S.A. (2006). 'Match your sales force structure to your business life cycle', *Harvard Business Review*, **Jul/Aug**.

Index